191

Advances in Polymer Science

Advances in Polymer Science

Recently Published and Forthcoming Volumes

Surface-Initiated Polymerization II
Volume Editor: Jordan, R.
Vol. 198, 2006

Surface-Initiated Polymerization I
Volume Editor: Jordan, R.
Vol. 197, 2006

Conformation-Dependent Design of Sequences in Copolymers II
Volume Editor: Khokhlov, A. R.
Vol. 196, 2006

Conformation-Dependent Design of Sequences in Copolymers I
Volume Editor: Khokhlov, A. R.
Vol. 195, 2006

Enzyme-Catalyzed Synthesis of Polymers
Volume Editors: Kobayashi, S., Ritter, H., Kaplan, D.
Vol. 194, 2006

Polymer Therapeutics II
Polymers as Drugs, Conjugates and Gene Delivery Systems
Volume Editors: Satchi-Fainaro, R., Duncan, R.
Vol. 193, 2006

Polymer Therapeutics I
Polymers as Drugs, Conjugates and Gene Delivery Systems
Volume Editors: Satchi-Fainaro, R., Duncan, R.
Vol. 192, 2006

Interphases and Mesophases in Polymer Crystallization III
Volume Editor: Allegra, G.
Vol. 191, 2005

Block Copolymers II
Volume Editor: Abetz, V.
Vol. 190, 2005

Block Copolymers I
Volume Editor: Abetz, V.
Vol. 189, 2005

Intrinsic Molecular Mobility and Toughness of Polymers II
Volume Editor: Kausch, H.-H.
Vol. 188, 2005

Intrinsic Molecular Mobility and Toughness of Polymers I
Volume Editor: Kausch, H.-H.
Vol. 187, 2005

Polysaccharides I
Structure, Characterization and Use
Volume Editor: Heinze, T.
Vol. 186, 2005

Advanced Computer Simulation Approaches for Soft Matter Sciences II
Volume Editors: Holm, C., Kremer, K.
Vol. 185, 2005

Crosslinking in Materials Science
Vol. 184, 2005

Phase Behavior of Polymer Blends
Volume Editor: Freed, K.
Vol. 183, 2005

Polymer Analysis/Polymer Theory
Vol. 182, 2005

Interphases and Mesophases in Polymer Crystallization II
Volume Editor: Allegra, G.
Vol. 181, 2005

Interphases and Mesophases in Polymer Crystallization I
Volume Editor: Allegra, G.
Vol. 180, 2005

Interphases and Mesophases in Polymer Crystallization III

Volume Editor: Giuseppe Allegra

With contributions by

G. Allegra · D. Frenkel · M. Hikosaka · W. Hu
M. Imai · K. Kaji · T. Kanaya · T. Konishi · G. Matsuba
S. V. Meille · M. Muthukumar · K. Nishida · K. Okada
K. Watanabe · S. Yamazaki · T. Yamamoto

Springer

The series *Advances in Polymer Science* presents critical reviews of the present and future trends in polymer and biopolymer science including chemistry, physical chemistry, physics and material science. It is adressed to all scientists at universities and in industry who wish to keep abreast of advances in the topics covered.
As a rule, contributions are specially commissioned. The editors and publishers will, however, always be pleased to receive suggestions and supplementary information. Papers are accepted for *Advances in Polymer Science* in English.
In references *Advances in Polymer Science* is abbreviated *Adv Polym Sci* and is cited as a journal.

Springer WWW home page: http://www.springeronline.com
Visit the APS content at http://www.springerlink.com/

ISSN 0065-3195
ISBN-10 3-642-42422-8 Springer Berlin Heidelberg New York
ISBN-13 978-3-642-42422-9 Springer Berlin Heidelberg New York
DOI 10.1007/11537656

Springer is a part of Springer Science+Business Media

springeronline.com

Softcover re-print of the Hardcover 1st edition 2005

Cover design: *Design & Production* GmbH, Heidelberg
Typesetting and Production: LE-TeX Jelonek, Schmidt & Vöckler GbR, Leipzig

Printed on acid-free paper 02/3141 YL – 5 4 3 2 1 0

Volume Editor

Prof. Giuseppe Allegra
Dipartimento di Chimica
Materiali e Ingegneria Chimica "Giulio Natta"
Politecnico di Milano
Via L. Mancinelli, 7
20131 Milano, Italy
giuseppe.allegra@polimi.it

Editorial Board

Advances in Polymer Science
Also Available Electronically

For all customers who have a standing order to Advances in Polymer Science, we offer the electronic version via SpringerLink free of charge. Please contact your librarian who can receive a password or free access to the full articles by registering at:

springerlink.com

If you do not have a subscription, you can still view the tables of contents of the volumes and the abstract of each article by going to the SpringerLink Homepage, clicking on "Browse by Online Libraries", then "Chemical Sciences", and finally choose Advances in Polymer Science.

You will find information about the

- Editorial Board
- Aims and Scope
- Instructions for Authors
- Sample Contribution

at springeronline.com using the search function.

Preface

In polymer crystallization the challenge is to identify and clarify the transformations by which chain molecules pass from a disordered, molten state to the ordered supra-molecular organization known as the semi-crystalline state. The subject is highly relevant in terms of both basic science and technology; it is indeed clear that many modern applications require complete control of the structure and the morphology of polymers from macroscopic dimensions down to below the nanoscale. As a simple example, making the crystallites in a polymer fiber equally oriented and reducing the number of chain folds (or hairpins) therein, usually turn out to be very favorable requisites for mechanical performance.

Is the onset of polymer crystallization, at least in some instances, preceded or accompanied by partial ordering of the system, possibly with influence on the kinetics and the equilibrium at both the molecular and the supra-molecular level? We may look at this issue, addressed in the present collection of contributions to *Advances in Polymer Science*, from very different sides. Modern microscopic techniques enable us to explore localized morphological aspects down to the observation of individual molecules, whereas X-ray, neutron and electron scattering provide molecular structure information down to the atomistic level. Experimental techniques allow us to explore kinetic aspects and are paralleled nowadays by fast-expanding molecular simulation approaches, increasingly able to give clues to the many open problems relating to structure development and morphology. Besides, the statistical-mechanical viewpoint may help to make sense out of the many experimental results and related simulations.

While in volumes 180 and 181 of this series several basic aspects of morphology, inter-phase structure and disorder were addressed, in the present volume, molecular interactions, modeling, phase transformation and crystallization kinetics are considered (see the subject index including keywords from volumes 180 and 181 at the end of the book). Needless to say, in spite of substantial success over 60 years or more we are still far from having a complete and unambiguous picture of polymer crystallization. We firmly believe that a fruitful approach to such a complex problem requires one to give way to many different and sometimes conflicting viewpoints, as we have attempted to do in these volumes. We do hope that they are not only a time-capsule left for future scientists, but that they also contain the seeds of a coherent view that will eventually develop.

I would like to renew my gratitude to Valdo Meille for his very creative, intelligent and active co-operation.

Milan, October 2005 *Giuseppe Allegra*

Contents

Contents of Volume 180

Interphases and Mesophases in Polymer Crystallization I

Volume Editor: Giuseppe Allegra
ISBN: 3-540-25345-9

Contents of Volume 181

Interphases and Mesophases in Polymer Crystallization II

Volume Editor: Giuseppe Allegra
ISBN: 3-540-25344-0

Adv Polym Sci (2005) 191: 1–35
DOI 10.1007/12_011

Published online: 5 October 2005

Polymer Crystallization Driven by Anisotropic Interactions

Wenbing Hu[1] (✉) · Daan Frenkel[2]

[1]Department of Polymer Science and Engineering, State Key Lab of Coordination Chemistry, College of Chemistry and Chemical Engineering, Nanjing University, 210093 Nanjing, P.R. China
wbhu@nju.edu.cn

[2]FOM Institute for Atomic and Molecular Physics, Kruislaan 407, 1098 SJ Amsterdam, The Netherlands
frenkel@amolf.nl

Abstract In this review, we consider a variety of aspects of polymer crystallization using a very simple lattice model. This model has three ingredients that give it the necessary flexibility to account for many features of polymer crystallization that have been observed experimentally. These ingredients are (1) a difference in attraction between neighboring (nonbonded) components, (2) attraction between parallel bonds, and (3) temperature-dependent flexibility due to the energy cost associated with kinks in the polymer chain. We consider this model using both dynamic Monte Carlo simulations and a simple mean-field theory. In particular, we focus on the interplay of polymer crystallization and liquid–liquid demixing in polymer solutions. In addition, we study the factors that are responsible for the characteristic crystal morphologies observed in a variety of homopolymer and statistical-copolymer crystals. Finally, we consider how the freezing of polymers in the bulk can be related to the crystallization of a single polymer chain.

Keywords Crystallization · Lattice statistics · Melting · Monte Carlo simulations · Phase diagram

1
Introduction

The building blocks of liquid-crystalline polymers are anisometric, and many of them form liquid-crystalline mesophases, even in monomeric form. Monomers that have this property are called mesogens. The molecular driving force to form a nematic phase can be due to anisotropic steric repulsions between the anisometric hard cores of the mesogens. This mechanism was proposed by Onsager [1]. It provides a successful description of many lyotropic disorder–order phase transitions. Alternatively, nematic ordering can be induced by the anisotropy of the polarizability of the mesogens, making the parallel orientation of mesogens energetically favorable. This mechanism for the isotropic–nematic transition was proposed by Maier and Saupe [2, 3]. It provides a useful description of thermotropic disorder–order phase transitions. In many cases of practical interest, both interactions play a role and should be taken into account in a description of the isotropic–nematic transition [4–11].

The building blocks of nonmesogenic polymers are also nonspherical; however, their degree of nonsphericity may be insufficient to induce nematic ordering. As already pointed out by Flory [12], the rigidity of a polymer chain – and thereby the anisometry of the Kuhn segments—tends to increase with decreasing temperature. Flory argued (on basis of the Onsager model) that, at sufficiently low temperatures, the anisometry of the Kuhn segments becomes so large that the isotropic (disordered) state is no longer stable and spontaneous ordering—in this case crystallization—must occur [12]. Note that this freezing mechanism is rather different from the one considered in simple liquids: there it is assumed that freezing occurs simply because the molecules can pack more densely in the solid state than in the liquid. The density change on freezing of simple liquids is typically much less than that observed in the orientational ordering of hard rods. Moreover, most lattice models cannot be used to describe a freezing transition driven by packing alone. However, this does not imply that a lattice model cannot properly describe polymer crystallization other than as an isotropic–nematic transition driven by anisotropic excluded-volume effects. In fact, it is possible to describe polymer freezing by taking into account the enhanced attraction between bonds with parallel orientation.

A lattice model that takes such attractions between parallel bonds into account provides a reasonable prediction of polymer melting points [13] and of their interplay with liquid–liquid demixing in polymer solutions [14]. The same factors that favor freezing do affect to a greater or lesser extent the formation of mesophases; hence, there is a close relation between polymer crystallization and the formation of mesophases, which are frequently observed before polymer crystallization (see other papers in this issue).

In this review, we focus on the effect of anisotropic interactions, in particular parallel attractions, and demonstrate that the inclusion of such interactions in a model leads to a great richness in possible polymer phase behavior. From a practical point of view, the model that we describe has the advantage that it is computationally very cheap—although this advantage comes at the price of sacrificing the greater realism of an off-lattice model.

In what follows, we use simple mean-field theories to predict polymer phase diagrams and then use numerical simulations to study the kinetics of polymer crystallization behaviors and the morphologies of the resulting polymer crystals. More specifically, in the molecular driving forces for the crystallization of statistical copolymers, the distinction of comonomer sequences from monomer sequences can be represented by the absence (presence) of parallel attractions. We also devote considerable attention to the study of the free-energy landscape of single-chain homopolymer crystallites. For readers interested in the computational techniques that we used, we provide a detailed description in the "Appendix."

2 Lattice Model for Polymer Crystallization

2.1 Flory's Treatment for Semiflexible Polymers

The structure of a simple mixture is dominated by the repulsive forces between the molecules [15]. Any model of a liquid mixture and, *a fortiori* of a polymer solution, should therefore take proper account of the configurational entropy of the mixture [16–18]. In the standard lattice model of a polymer solution, it is assumed that polymers "live" on a regular lattice of n sites with coordination number q. If there are n_2 polymer chains, each occupying r consecutive sites, then the remaining n_1 single sites are occupied by the solvent. The total volume of the incompressible solution is $n = n_1 + rn_2$. In the case $r = 1$, the combinatorial contribution of two kinds of molecules to the partition function is

$$Z_{\text{comb}} = \frac{n!}{n_1!n_2!} \approx \left(\frac{n}{n_1}\right)^{n_1} \left(\frac{n}{n_2}\right)^{n_2} . \tag{1}$$

This expression accounts for the configurational entropy of an ideal binary mixture with identical molecular sizes, but not for that of a polymer solution, since polymer chains are large and flexible. For that case, more contributions arise from the chain conformational entropy, first considered by Meyer [19] and then derived by Huggins [20] and Flory [21]. In analogy with a nonreversing random walk on a lattice, the conformational contribution of polymer chains to the partition function is given by

$$Z_{\mathrm{conf}} = \frac{\left(\frac{q}{2}\right)^{n_2} z_{\mathrm{c}}^{(r-2)n_2}}{a^{(r-1)n_2}}, \tag{2}$$

where the first factor 1/2 is the symmetry factor of chain ends. This factor accounts for the fact that the calculation can start from either of two chain ends. In Eq. 2, q is the number of possible ways to put the second chain unit along the chain, z_c $(= q - 1)$ is the number of possible ways to place each subsequent chain unit of the rest, and a is a correction term for each step of random walk due to the presence of other chains. Flory showed that if one assumes random mixing (i.e., ignores all local structural correlations), $a = \mathrm{e}$. Huggins used a somewhat more sophisticated procedure to estimate the probability of finding two consecutive vacant sites and obtained the estimate $a = (1 - 2/q)^{-(q/2-1)}$ [22]. To account for semiflexibility, Flory introduced a potential energy penalty E_c for every "kink" in the lattice polymer. The presence of this kink energy changes z_c, the intramolecular part of the partition function, to $z_c = 1 + (q - 2)\exp[-E_c/(k_B T)]$, where k_B is Boltzmann's constant and T the temperature [12]. For the fully disordered state at very high temperatures, the so-called "disorder parameter" d, defined as the mean fraction of consecutive bonds that are not collinear, should be

$$d = \frac{(q-2)\exp\left(-\frac{E_c}{k_B T}\right)}{1 + (q-2)\exp\left(-\frac{E_c}{k_B T}\right)}. \tag{3}$$

As the temperature is decreased, the chains become increasingly rigid: z_c then approaches 1 if we assume that there is only one fully ordered crystalline structure and Z_{conf} for the liquid becomes smaller than 1. This means that, at this level of approximation, the disordered state becomes less favorable than the crystalline ground state. A first-order disorder–order phase transition is expected to occur under these conditions. Flory interpreted this phase transition as the spontaneous crystallization of bulk semiflexible polymers [12]. However, since the intermolecular anisotropic repulsion essential in the Onsager model is not considered in the calculation, only the short-range intramolecular interaction is responsible for this phase transition.

The calculation of Z_{conf} makes use of the random mixing approximation for the fully disordered state. Several authors [23–27] have reported improved estimates of Z_{conf} that take into account the effect of local ordering at low temperatures; however, the resulting improvement in the prediction of the melting point is not very large [22]. Another approach in the calculation of configurational entropy of semiflexible lattice chains was suggested by Di-Marzio [28] and was expanded by Ronca [29], and this has been found useful in the study of orientational relaxation of stretched polymer liquids [30–32].

A number of Monte Carlo simulations have verified the spontaneous disorder–order phase transition of semiflexible polymers in 3D lattice models [33–36]. In molecular dynamics simulations, even the metastable chain-folding in the supercooled melt has been observed [37]. However, the ordering transition studied in these simulations was the one from the isotropic to the nematic phase, rather than the actual crystallization transition [38]. At high densities, cooling results in the formation of a glassy disordered state rather than a crystal [39].

2.2 Implications of Parallel Attractions in Polymer Systems

In Monte Carlo simulations, it has been found that introducing a parallel attraction between the polymer bonds, in addition to the bending-energy penalty, could significantly enhance the first-order nature of the isotropic–nematic phase transition at high concentrations [40, 41]. In fact, the inclusion of attraction between parallel bonds has been found to be useful in many studies of nonmesogenic polymers. Such attractions between parallel bonds can mimic the short-ranged orientational order in polyethylene melts that was observed in molecular dynamics simulations [42], in agreement with experimental observations on *n*-alkane liquids [43]. The anisotropic interactions have been considered in the study of orientational relaxation of stretched polymer melts [30–32] and of local order in polymer networks [44, 45].

An early study on the role of parallel attraction in polymer crystallization was made by Bleha [46], who considered the enthalpic effect of parallel packing on the melting point of polymers. In addition, Mansfield [47] took parallel interactions into account in his Monte Carlo calculation of the chain-folding probability at the interphase zone between lamellar crystals and amorphous liquid. Monte Carlo simulations by Yoon [48] showed that parallel attraction can lead to the formation of ordered domains and a density-functional theory study of melt crystallization by McCoy et al. [49] revealed the existence of an effective "chain straightening force" originating from attractive potentials [50]. In Monte Carlo simulations of AB-copolymer crystallization, parallel attractions were used to distinguish the crystallizable sequences from the noncrystallizable sequences [50–52]. Parallel attractions were also applied in the Monte Carlo study of polymer crystallization from dilute solutions on 2D [53] and 3D lattices [54], as well as from the homopolymer melt in 2D [55] and 3D [56] lattices. In earlier work, we showed that the incorporation of attraction between adjacent, unconnected bonds allows us to reproduce the sectorization of chain-folding in a single lamellar crystallite [57] and the shish-kebab morphology of polymer crystallites induced by a single pre-aligned chain [58]. More details of some simulation results are discussed in Sects. 3 and 4.

2.3 Mean-Field Treatment of Parallel Attractions

We now consider a lattice model for a polymer solution that has both isotropic and anisotropic interactions. A mean-field expression for the free energy of the system can be obtained by approximating the local concentration of polymer chain units by its average value. We consider a solution of polymers consisting of r units on a cubic lattice. The volume fraction occupied by the polymers is denoted by ϕ. Two energetic interaction parameters play a role. One is the "mixing energy" B. It is a measure for the energetic cost (relative to the unmixed situation) for having a solvent particle and a polymer chain unit on adjacent lattice sites: $B = E_{us} - (E_{ss} + E_{uu})/2$, where E_{ab} represents pair interactions of the chain units (u) and the solvent particle (s). The second interaction energy E_p denotes the energy cost to break up a pair of adjacent, parallel polymer bonds. The mixing interactions act between sites and are isotropic, while the parallel attractions act between bonds and are anisotropic.

In the fully disordered state, the probability to find a bond at a given bond site is simply given by the ratio of the total number of bonds $[n_2(r-1)]$ to the total number of bond positions $(nq/2)$. The probability that a given bond has a specific parallel neighbor is therefore given by $2n_2(r-1)/(nq)$. Every bond has $q-2$ neighbors, since two consecutive neighbors along the chain should be subtracted from the coordination number. Unless a neighboring site is occupied by a parallel bond, its energy cost equals E_p. The average potential energy cost due to nonparallel packing is therefore $\ln(z_p) = -1/2(q-2)[1-2n_2(r-1)/(nq)]E_p/(k_BT)$, where the factor 1/2 eliminates double counting of pair interactions. At the mean-field level, the potential energy due to nonparallel packing reduces the partition function by a factor of $z_p^{n_2(r-1)}$. Similarly, most chain units can have $q-2$ neighbors occupied by solvent. The probability of finding a solvent molecule on a specific neighboring site is n_1/n. It then follows that the total mixing potential energy per chain unit is $\ln(z_m) = -(q-2)n_1B/(nk_BT)$. The corresponding contribution to the partition function is $z_m^{n_2r}$.

Combining all contributions to the partition function of the disordered state of a lattice polymer solution, we obtain

$$\begin{aligned} Z &= Z_{comb} Z_{conf}\, z_p^{n_2(r-1)} z_m^{n_2r} \\ &= \left(\frac{n}{n_1}\right)^{n_1} \left(\frac{n}{n_2}\right)^{n_2} \left(\frac{q}{2}\right)^{n_2} z_c^{n_2(r-2)} e^{-n_2(r-1)} z_p^{n_2(r-1)} z_m^{n_2r}, \end{aligned} \tag{4}$$

where $z_c = 1 + (q-2)\exp\left(-\frac{E_c}{k_B T}\right)$, $z_p = \exp\left\{-\frac{q-2}{2}\left[1-\frac{2n_2(r-1)}{qn}\right]\frac{E_p}{k_B T}\right\}$, and $z_m = \exp\left[-\frac{(q-2)n_1}{n}\frac{B}{k_B T}\right]$.

The mean-field expression for the free-energy density of the polymer solution is therefore [13, 14]

$$
\begin{aligned}
f(\phi) = {} & (1-\phi)\ln(1-\phi) + \frac{\phi}{r}\ln\phi - \phi\ln\left(\frac{qr}{2}\right) \\
& - \phi\left[-\left(1-\frac{2}{r}\right)\ln z_c + 1 - \frac{1}{r} + (q-2)B + \frac{q-2}{2}\left(1-\frac{1}{r}\right)E_p\right] \\
& - \phi^2\left[(q-2)B + \frac{q-2}{q}\left(1-\frac{1}{r}\right)^2 E_p\right].
\end{aligned}
\tag{5}
$$

In the perfectly ordered crystalline ground state, all polymer bonds are parallel and no solvent–polymer contacts are present. If we ignore disorder (vacancies, kinks) in the polymer crystal at finite temperatures, the free-energy density of the crystalline state is zero.

2.4 Predictions of the Polymer Melting Point

Inspection of the mean-field free-energy density given in the previous paragraph allows us to see the relationship between the (microscopic) molecular parameters of the lattice-polymer model and its (macroscopic) phase diagram. Let us first focus on the equilibrium melting point, i.e., the temperature at which the crystalline phase and the isotropic liquid phase are in thermodynamic equilibrium. We first consider the effect of the energy parameters in the model and of the polymer chain length on the melting point of bulk homopolymers. Polymer solutions and mixtures will be discussed in the next section.

At coexistence, the chemical potentials of given species must be equal. In a plot of $f(\phi)$ versus the polymer concentration ϕ, this equality leads to the familiar common-tangent condition: at coexistence, the tangents to the free-energy densities of the solid and liquid phases must coincide. In the lattice model that we use, the partition function for the fully ordered ground state is given by $Z = 1$ and hence its free-energy density is zero. At finite temperatures, the presence of defects will change the free-energy density of the solid. We ignore this effect. In addition, the lattice model ignores the effect of the vibrational degrees of freedom of the polymers.

In a pure homopolymer system, the free-energy density only depends on E_c (the quantity that determines the chain rigidity) and E_p (the quantity that determines the tendency of backbone chains to form parallel, close-packed structures). Let us first consider the relative stability of the pure polymer melt and the polymer solid in the limit of infinitely long chains. In that case, we find that the free energies of the liquid and solid are equal when

$$1 + (q-2)\exp\left(-\frac{E_c}{k_B T_m}\right) = \exp\left[1 + \frac{(q-2)^2}{2q}\frac{E_p}{k_B T_m}\right]. \tag{6}$$

If $E_c < k_B T$ at melting, we can ignore the first term on the left-hand side and we obtain

$$T_m = \frac{E_c + \frac{(q-2)^2}{2q} E_p}{k_B[\ln(q-2) - 1]}. \tag{7}$$

Equation 7 shows that both an increase in chain rigidity and an increase in the interaction between parallel chains will lead to an increase in the melting point, in agreement with experiments [59–62]. For example, semirigid chains that contain aromatic groups in the chain backbone usually have high melting points. Similarly, aliphatic polyamides that have strong interchain interactions, due to hydrogen bonding, tend to have higher melting points than aliphatic polyesters. In addition, strong interchain interactions are only possible in the absence of steric obstructions. For example, polypropylene has smaller side branches than poly(1-butene) and, *a fortiori*, than poly(1-pentene). And indeed, polypropylene has a higher melting point (460.7 K) compared with poly(1-butene) (411.2 K) and poly(1-pentene) (403.2 K) [63]. Bunn [64] has observed a linear dependence of T_m on the cohesive energy density of the same series of homologues [64]. This observation is understandable because both E_c and E_p contribute to the cohesive energy density of solid polymers in a linear way, and in addition the compounds in the same homologous series should have similar E_c and E_p values.

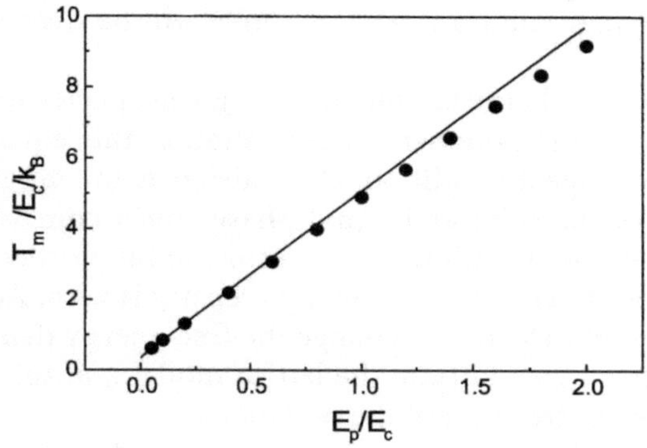

Fig. 1 Melting temperatures of polymers ($k_B T_m/E_c$) with variable E_p/E_c values. The *line* is calculated from Eq. 10 and the *circles* are the simulation results obtained from the onset of crystallization on the cooling curves of disorder parameters, in a short-chain ($r = 32$) system (occupation density is 0.9375 in a 32-sized cubic box) with a template substrate (Hu and Frenkel, unpublished results)

The quality of the mean-field approximation can be tested in simulations of the same lattice model [13]. Ideally, direct free-energy calculations of the liquid and solid phases would allow us to locate the point where the two phases coexist. However, in the present studies we followed a less accurate, but simpler approach: we observed the onset of freezing in a simulation where the system was slowly cooled. To diminish the effect of supercooling at the freezing point, we introduced a terraced substrate into the system to act as a crystallization seed [14]. We verified that this seed had little effect on the phase coexistence temperature. For details, see Sect. A.3. At freezing, we have

$$\mu^{c} = \mu^{s}, \tag{8}$$

where μ^c and μ^s are the chemical potentials of the polymers in the crystal and solution, respectively. We ignore disorder in the polymer crystal, so $\mu^c = 0$. As the free-energy expression of the polymer solution is approximated by

$$\begin{aligned}\frac{\Delta F}{k_B T} = \frac{F^s}{k_B T} &= n_1 \ln n_1 + n_2 \ln n_2 - n_1 \ln n - n_2 \ln n - n_2 \ln \frac{q}{2} \\ &+ n_2(r-1) - n_2(r-2)\ln z_c \\ &+ n_2(r-1)\frac{q-2}{2}\left[1 - \frac{2(r-1)n_2}{qn}\right]\frac{E_p}{k_B T} \\ &+ \frac{n_1 n_2 r}{n}\frac{(q-2)B}{k_B T},\end{aligned} \tag{9}$$

the condition that the chemical potentials in the solid and the liquid are equal yields

$$\begin{aligned}&(1-r)\frac{n_2 r}{n} + \ln\frac{qn}{2n_2} + (r-2)\ln\left[1 + (q-2)\exp\left(-\frac{E_c}{k_B T_m}\right)\right] \\ &= \frac{(r-1)(q-2)}{2}\left[1 - \frac{2(r-1)n_2(n+n_1)}{qn^2}\right]\frac{E_p}{k_B T_m} + \frac{r n_1^2}{n^2}\frac{(q-2)B}{k_B T_m}.\end{aligned} \tag{10}$$

The melting point T_m is computed by solving this equation iteratively. It is often convenient to use $E_c/(k_B T_m)$ as our unit of (inverse) temperature. The phase diagram of the polymer solution then depends on the molecular parameters r, q, B/E_c, and E_p/E_c, the composition parameters n_1 and n_2, and on the temperature parameter $E_c/(k_B T_m)$.

Figure 1 shows a comparison of the simulation data with the corresponding theoretical predictions. The figure shows that, over a range of E_p/E_c values, the theoretical predictions are in good agreement with the simulation results. Note that the curve in Fig. 1 is close to the straight line expected on the basis of Eq. 7.

In addition to variations in E_p/E_c, we can change the polymer chain length. In particular for small chain lengths, the melting point can be quite sensitive to this parameter [65, 66]. Flory and Vrij [67] analyzed this effect by treating polymer melting as a virtual two-step process: the first step involves the melting of infinitely long chains and the second step corresponds to the cutting of an infinitely long polymer into chains of finite length. The second step leads to an additional free-energy change Δf_e upon melting, as shown in the equilibrium condition

$$\Delta f_m = r\Delta f_u + \Delta f_e - k_B T_m \ln r = 0\,, \tag{11}$$

where Δf_u is the free-energy change of each chain unit and Δf_e is the additional free-energy change associated with the breakup of the infinite chain. If we assume that the terms in Eq. 7 correspond to the terms in $T_m = \Delta h_u/\Delta s_u$ for each chain unit, we can arrive at the approximate expression

$$\begin{aligned}\Delta f_u &= \Delta h_u - T_m \Delta s_u \\ &= E_c + \frac{(q-2)^2}{2q} E_p - k_B T_m [\ln(q-2) - 1]\,.\end{aligned} \tag{12}$$

The fusion free energy of both chain ends can be calculated from the equilibrium condition $Z = 1$ in Eq. 4 by setting the chain length $r = 2$ in the melt phase. The additional contribution is thus given by

$$\Delta f_e = \frac{(q-2)(q-1)}{2q} E_p - k_B T_m (\ln q - 1) - 2\Delta f_u\,. \tag{13}$$

Figure 2 shows that, for all but the shortest chains, the Flory–Vrij analysis predicts a slightly higher melting temperature than the present mean-field model. Both approximations are give values higher than the simulation results, but the overall agreement is reasonable.

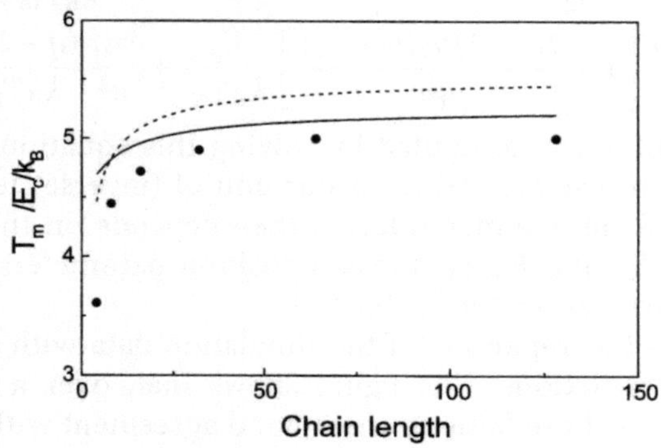

Fig. 2 Melting temperatures of polymers ($k_B T_m/E_c$) with variable chain lengths. The *solid line* is calculated from Eq. 10, the *dashed line* is calculated from Flory–Vrij analysis (Eq. 11), and the *circles* are the simulation results in the optimized approach. In simulations, the occupation density is 0.9375, and the linear size of the cubic box is set to 32 for short chains and 64 for long chains (Hu and Frenkel, unpublished results)

3
Interplay of Polymer Crystallization and Liquid–Liquid Demixing

3.1
Thermodynamic Interplay in Polymer Solutions

When B is positive, the polymer solution can exhibit liquid–liquid demixing. By adjusting the relative values of B and E_p, we can tune the phase diagrams of polymer solutions and study the interplay of freezing, on the one hand, and demixing on the other [14].

To calculate the liquid–liquid coexistence curves, we cast Eq. 5 in the standard Flory–Huggins form where all terms linear in ϕ are subtracted. The free-energy expression then becomes

$$
\begin{aligned}
\frac{\Delta F_{\text{mix}}}{k_B T} &= n_1 \ln n_1 + n_2 \ln n_2 - n_1 \ln n - n_2 \ln n + n_2 \ln r \\
&\quad + n_2(r-1)\left(\frac{q}{2}-1\right)\frac{2(r-1)}{qr}\left(1-\frac{n_2 r}{n}\right)\frac{E_p}{k_B T} + \frac{n_1 n_2 r}{n}\frac{(q-2)B}{k_B T} \\
&= n_1 \ln \frac{n_1}{n} + n_2 \ln \frac{n_2 r}{n} \\
&\quad + \frac{n_1 n_2 r}{n}\left[\left(1-\frac{2}{q}\right)\left(1-\frac{1}{r}\right)^2 \frac{E_p}{k_B T} + \frac{(q-2)B}{k_B T}\right].
\end{aligned}
\tag{14}
$$

This allows us to introduce an effective χ parameter, through

$$
\frac{\Delta F_{\text{mix}}}{k_B T} = n_1 \ln \frac{n_1}{n} + n_2 \ln \frac{n_2 r}{n} + \frac{n_1 n_2 r}{n}\chi_{\text{eff}}, \tag{15}
$$

where $\chi_{\text{eff}} = (q-2)B/(k_B T) + (1-2/q)(1-1/r)^2 E_p/(k_B T)$. As usual, the liquid–liquid coexistence curves can be calculated from the chemical-potential equivalence of two mixtures.

Figure 3a shows the mean-field predictions for the polymer phase diagram for a range of values for E_p/E_c and B/E_c. The corresponding simulation results are shown in Fig. 3b. As can be seen from the figure, the mean-field theory captures the essential features of the polymer phase diagram and provides even fair quantitative agreement with the numerical results. A qualitative flaw of the mean-field model is that it fails to reproduce the crossing of the melting curves at $\phi = 0.73$. It is likely that this discrepancy is due to the neglect of the concentration dependence of χ_{eff}. Improved estimates for χ_{eff} at high densities can be obtained from series expansions based on the lattice-cluster theory [68, 69].

Figure 3 illustrates the thermodynamic interplay of polymer crystallization and liquid–liquid demixing in polymer solutions. The liquid–liquid binodal curve is primarily determined by the B value. With the increase of E_p values, the liquid–liquid binodal curves shift slightly upward. On the other hand, the liquid–solid coexistence curves are primarily controlled by the E_p value. With the decrease of B values from positive to negative, the melting point depression due to dilution will be enhanced. Proper choices of energy parameters can lead to an intersection of two phase-coexistence curves: the phase diagram then exhibits a (monotectic) triple point. Experimental studies of the intersection of freezing and demixing curves in solutions and blends have been reported[70–74].

Flory proposed a semiempirical expression to predict the concentration dependence of the melting curve of long-chain polymers mixed with small solvent molecules [75]:

$$\frac{1}{T_m} - \frac{1}{T_m^0} = \frac{k_B}{\Delta h_u}[1 - \phi - \chi(1 - \phi)^2] , \tag{16}$$

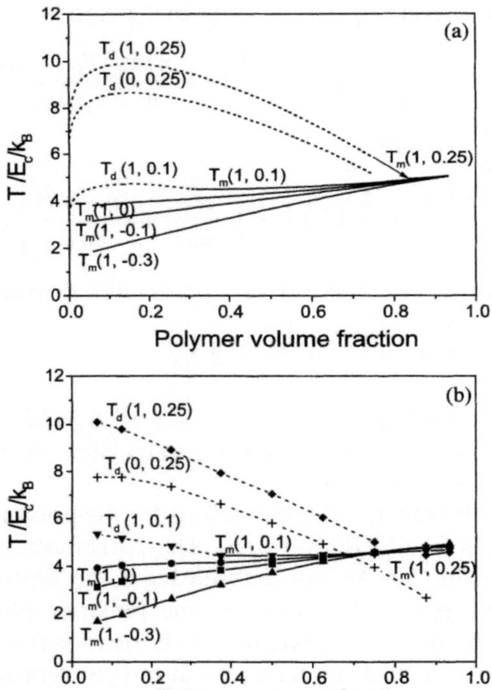

Fig. 3 Liquid–liquid demixing curves (*dashed lines* denoted by T_d) and liquid–solid transition curves (*solid lines* denoted by T_m) of polymer solutions with variable energy parameter sets [denoted by $T(E_p/E_c, B/E_c)$]. The solution system is made of 32-mers in a 32-sized cubic box. **a** Theoretical curves; **b** simulation results in the optimized approach [14]

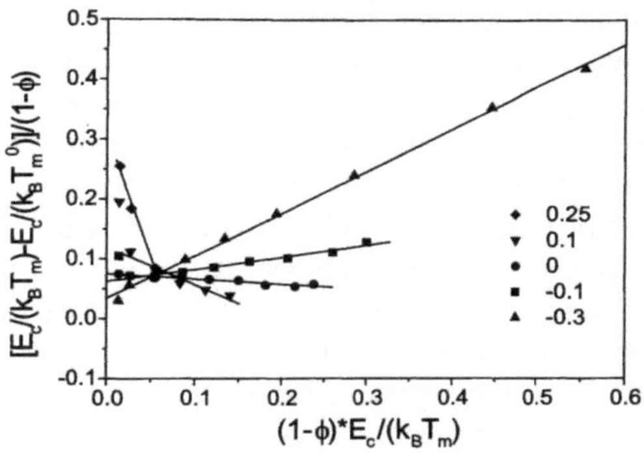

Fig. 4 Rescaled data from Fig. 3b to show the linear relationship predicted by Eq. 16. The bulk equilibrium melting temperature $E_c/k_B T_m^0$ is chosen to be approximately 0.2. The *lines* are the results of linear regression, and the *symbols* are for the variable values of B/E_c [14]

where T_m^0 is the melting point of pure polymers and Δh_u is the heat of fusion per chain unit. This semiempirical equation accounts well for numerous experimental data [76]. In Fig. 4, we have transformed the simulation results of Fig. 3b in such a way that, according to Eq. 16, a linear plot should result. However, the values Δh_u and χ that follow from a fit to the numerical data differ from the expressions that follow from Eq. 16 [14].

3.2 Kinetic Interplay in Polymer Solutions

In the dynamic Monte Carlo simulations described earlier, we used a crystalline template to suppress supercooling (Sect. A.3). If this template is not present, there will be a kinetic interplay between polymer crystallization and liquid–liquid demixing during simulations of a cooling run. In this context, it is of particular interest to know how the crystallization process is affected by the vicinity of a region in the phase diagram where liquid–liquid demixing can occur.

Simulations [77] and theoretical analysis [78, 79] indicate that the rate of homogeneous crystal nucleation may be significantly increased in the one-phase region near a metastable liquid–liquid critical point. In simple polymer solutions and melts, crystallization often occurs after the system has entered the region where fluid–fluid spinodal decomposition takes place. The density modulations that occur during this spinodal decomposition are "frozen in" during subsequent crystallization and affect the morphology of the resulting crystalline phase [80]. This phenomenon is, in fact, of considerable practical importance for the control of sol-gel transitions, in particular in the context of membrane formation [81–83].

To study the interplay between demixing and crystallization, we considered three different model systems, all at a polymer volume fraction of 15% [84]. The interaction parameters of the models were chosen such that all systems had the same (equilibrium) melting temperatures but different demixing temperatures at the chosen concentration. One case (C1) had its critical demixing temperature very close to the melting curve, the second case (C2) had its critical point at a temperature where primary crystal nucleation would occur in the absence of liquid–liquid demixing, and for the third case (C3), the critical demixing temperature was located far below the melting curve. In case C1, we expect to observe demixing prior to crystallization, in case C2 crystallization and liquid–liquid demixing may be strongly coupled, and in case C3, crystal nucleation should proceed without any effect of liquid–liquid demixing. The theoretical (i.e., mean-field) phase diagrams for these three cases are shown in Fig. 5. Figure 6 shows the simulation results for slow cooling runs of homogeneous polymer solutions corresponding to models C1, C2, and C3.

As can be seen from Fig. 6, liquid–liquid demixing clearly precedes crystallization in case C1. Moreover, crystallization in this case occurs at a higher temperature than in cases C2 and C3. Apparently, the crystallization takes place in the dense disordered phase (which has a higher melting temperature than the more dilute solution; Fig. 5). In case C2, the crystallization temperature is close to the expected critical point of liquid–liquid demixing, but higher than in case C3. This suggests that even pre-critical density fluctuations enhance the rate of crystal nucleation.

The different pathways for crystallization have consequences for the resulting crystal morphology. This can be seen in Fig. 7, where we compare the

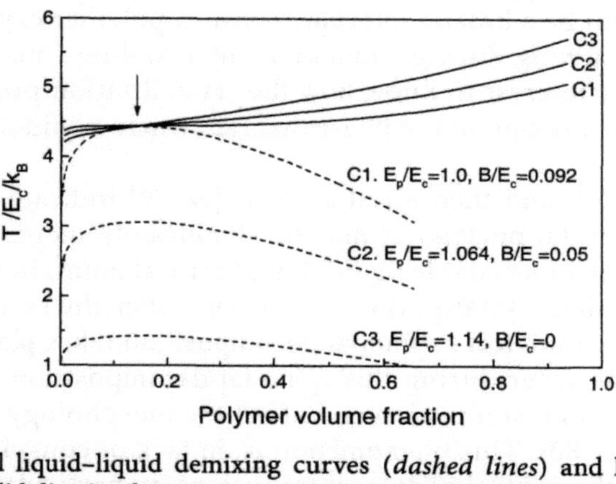

Fig. 5 Theoretical liquid–liquid demixing curves (*dashed lines*) and liquid–solid transition curves (*solid lines*) of 32-mers in a 64-sized cubic box. Three sets of energy parameters are denoted by *C1*, *C2*, and *C3*, respectively. The *arrow* indicates the cooling trajectory of the simulations [84]

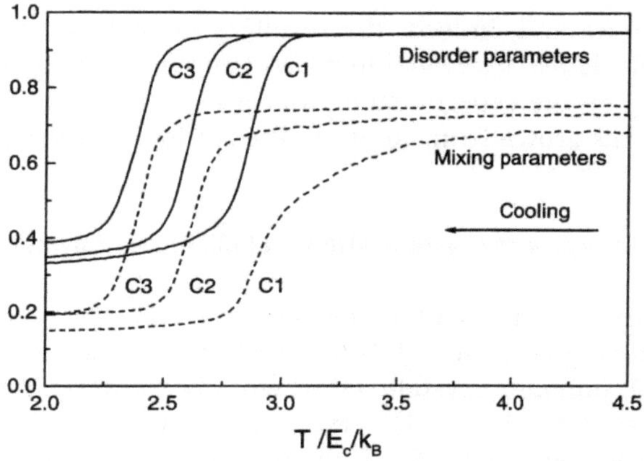

Fig. 6 Simulational cooling curves of disorder parameters (*solid lines*) and mixing parameters (*dashed lines*) for 32-mers with different sets of energy parameters in a 64-sized cubic box (the concentration is fixed at 0.150). The mixing parameter is defined as the mean fraction of neighboring sites occupied by the solvent for each chain unit [84]

crystal morphologies that result if systems C1, C2, and C3 are all quenched to the same temperature ($T = 2.857E_c/k_B$). The figure shows that, for system C1, small crystallites are homogeneously distributed throughout the simulation box. This is the result of liquid–liquid demixing under conditions of a deep spinodal quench (short-wavelength instability), followed by freezing of the high-density domains. In case C2, larger crystallites are formed. This is the result of liquid–liquid demixing under conditions of a shallow spinodal quench (long-wavelength instability), again followed by the freezing of the high-density domains. In case C3, liquid-liquid demixing cannot occur at the

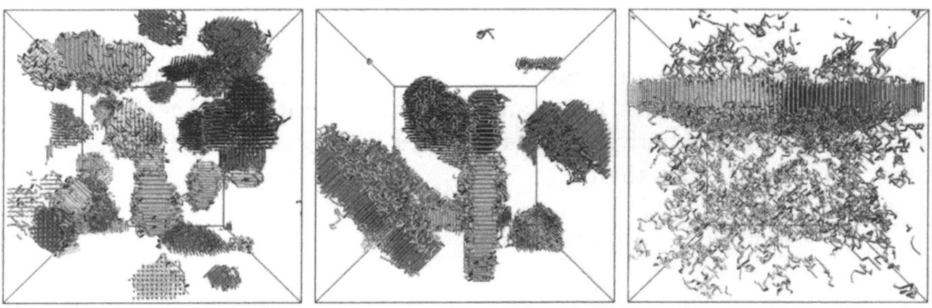

C1 C2 C3

Fig. 7 Snapshots of the simulation systems for C1, C2, and C3 after an isothermal crystallization following the quenching from the infinite temperature to a temperature of 2.857 E_c/k_B [84]

chosen temperature and, indeed, the resulting crystal morphology is quite different: a single, lamellar crystallite grows from the homogeneous solution. For more details on the experimentally observed differences in morphology of polymer crystals grown from solution, the reader is referred to Ref. [85].

3.3 Thermodynamic Interplay of Crystallization and Mixing in Polymer Blends

In polymers mixed with small molecules, there is a very strong entropy penalty for demixing (see Eq. 15). If, in our lattice model, we put B equal to zero and $E_\mathrm{p} \neq 0$, polymer crystallization will always pre-empt liquid–liquid demixing (see case C3 in Fig. 5). However, in polymers mixed with long-chain polymers, the entropy penalty for demixing becomes so small that a difference in E_p for the two polymer species may lead to liquid–liquid demixing before crystallization [86]. Recently, the monomeric geometrical assymmetry between two species has been found to raise a positive entropic contribution to the mixing free energy [87]. This was achieved by the lattice-cluster theory with calculations beyond the random mixing approximation [88]. Here, this assymmetry may also be absent for seeing demixing. If we view E_p as a measure for the crystallizability of a polymer, then one could argue that, in such polymer blends, the liquid–liquid demixing is driven solely by the difference in crystallizability of two components [86].

To see this, consider the mixing free energy expression for a polymer blend with symmetrical chain lengths and with only one crystallizable component (i.e., $E_\mathrm{p} = 0$ for one component and $E_\mathrm{p} \neq 0$ for the other). In that case the (mean-field) partition function for the liquid mixture is

$$Z = \left(\frac{n}{n_1}\right)^{n_1} \left(\frac{n}{n_2}\right)^{n_2} \left(\frac{q}{2}\right)^{n_1+n_2} z_\mathrm{c}^{(n_1+n_2)(r-2)} e^{-(n_1+n_2)(r-1)} z_\mathrm{p}^{n_2(r-1)} z_\mathrm{m}^{n_2 r}, \qquad (17)$$

where z_c and z_p are defined as in Eq. 4, $z_\mathrm{m} = \exp\left[-\frac{n_1 r}{n}\frac{(q-2)B}{k_\mathrm{B}T}\right]$, n_1 and n_2 denote the number of noncrystallizable and crystallizable polymer chains, all containing r units, $n = n_1 r + n_2 r$, and B is the net potential-energy exchange for a site–site contact between units of types 1 and 2. The mixing free energy of this polymer blend is then

$$\frac{\Delta f_\mathrm{mix}}{k_\mathrm{B}T} = \frac{\phi_1}{r}\ln\phi_1 + \frac{\phi_2}{r}\ln\phi_2 + \phi_1\phi_2\left[(q-2)\frac{B}{k_\mathrm{B}T} + \left(1-\frac{2}{q}\right)\left(1-\frac{1}{r}\right)^2 \frac{E_\mathrm{p}}{k_\mathrm{B}T}\right], \qquad (18)$$

where ϕ_1 and ϕ_2 are the volume fractions of noncrystallizable and crystallizable polymer chains, respectively. When $r \gg 1$, the mixing entropy is very small, and hence a small contribution of E_p may already make the mixed state unstable. Then, liquid–liquid demixing pre-empts polymer crystallization on

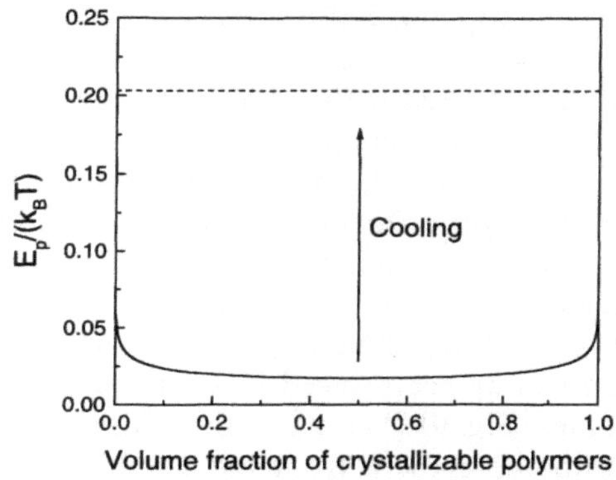

Fig. 8 Theoretical liquid–liquid demixing curve (*solid line*) and the bulk melting temperature (*dashed line*) of a flexible-polymer blend with one component crystallizable and with athermal mixing. The chain lengths are uniform and are 128 units, the linear size of the cubic box is 64, and the occupation density is 0.9375 [86]

cooling (Fig. 8). Such liquid–liquid demixing has been verified by simulation of a cooling process passing through the critical point of the symmetrical-polymer blend [86].

One practical example of demixing that might be attributed to a difference in crystallizability is the incompatibility in blends of polymers with different stereochemical compositions. The stereochemical isomers contain both chemical and geometrical similarities, but differ in the tendency of close packing. In this case, both the mixing energy B and the additional mixing entropy due to structural asymmetry between two kinds of monomers are small. However, the stereochemical differences between two polymers will result in a difference in the value of E_p. Under this consideration, most experimental observations on the compatibility of polymer blends with different stereochemical compositions [89–99] are tractable. For more details, we refer the reader to Ref. [86].

4 Some Applications of Parallel Attractions in Molecular Simulations

4.1 Characteristic Morphologies of Polymer Crystallites

One of the most remarkable features of polymer crystallization is that such chain molecules can form lamellar crystals that contain heavily folded polymer chains. In experiments, the structural analysis of these lamellar crystals became possible when polyethylene single crystals were first prepared from a solution [100–102]. It was found that the orientation of the polymer chains was perpendicular to the top and bottom faces of the lamellar crystal. However, as the crystal thickness is typically much smaller than the polymer length, this can only be realized if the chains fold back at the top and bottom surfaces. The fold ends prefer to align in parallel to the crystal growth front. However, there is no correlation between the positions of fold ends in successive crystalline layers. During growth, a single crystal can develop several facets. Each facet corresponds to a sector in which the fold ends are preferably parallel to this facet. This leads to the sectorization of chain folding in the single crystal of polymers [103–106]. Interestingly, this sectorization phenomenon can be reproduced in simulations of the simple lattice model described before (Fig. 9) [57]. As in the experiments, we find that the folds are aligned with the growth front but exhibit little correlation from one crystalline layer to the next. The simulations provide molecular-level detail on how a single chain can be incorporated into the growth front. Multiple steps have been found and can be attributed to a limited size of the growth front. A detailed observation of a single chain attaching to the smooth growth front has been reported by Muthukumar's group [107]. There is experimental evidence for sectorization on the surface of thin films of bulk polymers [108, 109]. However, the available simulations have, thus far, not reproduced the sectorization of lamellae grown in the melt away from any surface [56].

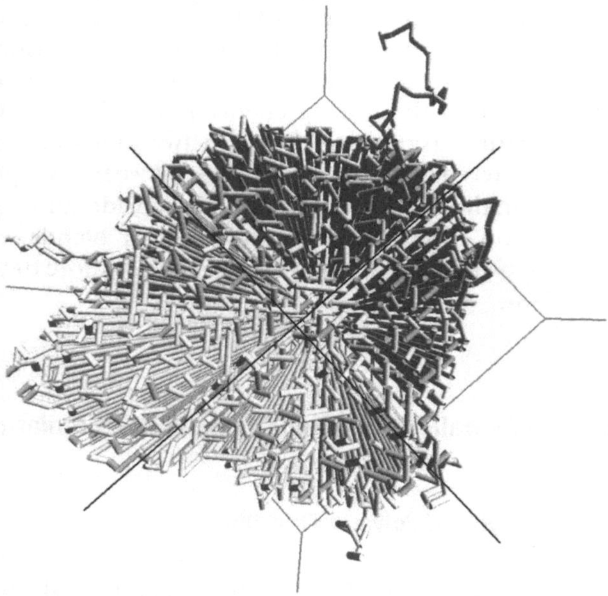

Fig. 9 Snapshot of a single crystal of lattice polymers viewed from the chain direction. The bonds are drawn as *solid cylinders*. The viewing angle is large for better observation of folds. The chain length is 512 units and the thickness of the crystallite is about 12 units. The dissolved chains are not shown for clarity [57]

Polymer crystallization is usually initiated by nucleation. The rate of primary nucleation depends exponentially on the free-energy barrier for the formation of a critical crystal nucleus [110]. If we assume that a polymer crystallite is a cylinder with a thickness l and a radius R, then the free-energy cost associated with the formation of such a crystallite in the liquid phase can be expressed as

$$\Delta F_c(\nu) = -\pi R^2 l|\Delta\mu| + 2\pi R^2\sigma_e + 2\pi Rl\sigma_l\,, \qquad (19)$$

where $\nu = \pi R^2 l$ is the total number of chain units in the crystallite, $\Delta\mu$ is the free-energy difference per chain unit between solid and liquid, and σ_e and σ_l are the surface free energies for the fold surface and the lateral surface, respectively. On the right-hand side of Eq. 19, we can see that there are competing terms: the first one is the thermodynamic driving force for crystallization, and the remaining two terms are the surface free-energy penalties. Accordingly, there are two basic ways to accelerate the polymer nucleation rate, i.e., enhance the driving force or decrease the barrier. An interesting way to enhance the driving force is to decrease the polymer conformational entropy in the liquid through pre-aligning or stretching of the chains under an extensional or shear flow. Under those conditions, one often observes the formation of a stack of lamellae around a central fiber. The resulting morphology has been given the name "shish kebab" [105, 111–114]. The central fiber can be a substrate for the nucleation of lamella growth as in a heterogeneous nucleation. As can be seen in Fig. 10, in simulations even a single pre-aligned chain can facilitate the nucleation of lamellar crystallites, leading to the shish-kebab structure [58]. In this case, the remaining chains in the liquid were not pre-aligned; hence the central fiber acted as a template for the nucleation of lamellar crystals.

The critical size of crystallites can be calculated from the condition $\partial\Delta F(\nu)/\partial\nu = 0$. Beyond the critical size, the thermodynamic condition for crystal growth is $\partial\Delta F(\nu)/\partial\nu \le 0$. Since at the later stage of crystal growth, $R \gg l$, the last term on the right-hand side of Eq. 19 can be omitted. The thermodynamic growth condition, therefore, gives $l \ge 2\sigma_e/\Delta\mu \equiv l_{\min}$. This means that there is a minimum thickness of lamellae for the lateral crystal growth. The linear relationship between l^{-1} and T_m (or T_c) has been observed by small-angle X-ray scattering measurements in many polymer systems [115, 116].

In the classical Lauritzen–Hoffman theory for the mechanism of polymer crystal growth [106], it is assumed that the observed lamellar thickness corresponds to those crystallites that happen to have the largest growth velocity. However, this picture is hard to reconcile with the experimental observation that the thickness of polyethylene single crystals can be modulated by varying the temperature at which they are grown [117, 118]. In fact, simulations by Doye et al. [119, 120] suggest that the observed lamellar thickness does

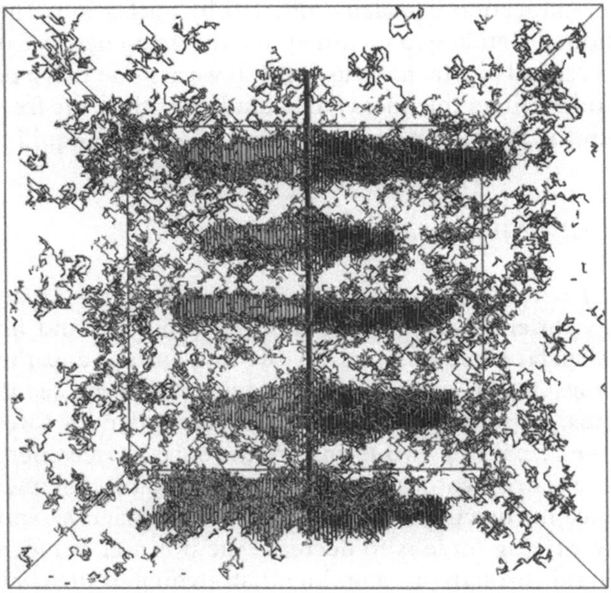

Fig. 10 Snapshot of a shish-kebab crystallite induced by a pre-aligned single chain (drawn much thicker than other chains for better visibility) in a solution. The chain length is 32 units and the thickness of crystallites is about 7 units. The bonds are drawn in *solid cylinders* [58]

not correspond to a maximal growth velocity, but rather to a condition of dynamic stability during growth.

The tip of a growing lamellar crystal has a thickness close to $l_{\min}$. However, behind the tip, the crystallite tends to thicken, as this increases the thermodynamic stability of the crystallite. Whether or not such crystallite thickening occurs depends on the ability of the polymer chains in the crystallites to undergo sliding diffusion [121–123]. High *c*-slip mobility such as what is observed in the hexagonal phase of polyethylene, can even lead to the formation of extended chain crystallites, while low *c*-slip mobility such as is observed in the orthorhombic phase of polyethylene prohibits the “stretching out” of folded-chain crystallites. When the polymer chains are sufficiently short, one can observe that the thickening proceeds in steps: from one state of integral folding to the next [106]. As our dynamic Monte Carlo simulations allow for sliding diffusion, we can study the phenomenon of crystal thickening by simulation. In order to do so, we should take account of the fact that long stems experience more friction during sliding diffusion than short stems. It is possible to account for this length-dependent friction in a way that satisfies detailed balance [57]. Figure 11 shows the thickening from a twice-folded to a once-folded layer for 32-mers.

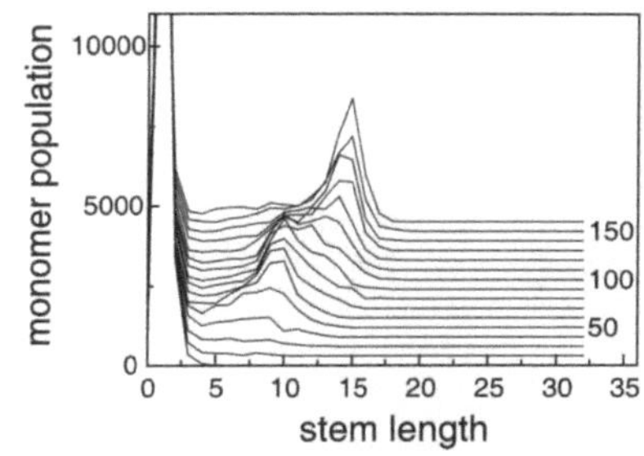

Fig. 11 Monomer distributions of 32-mers with $E_f/E_c = 0.1$ at $E_c/k_B/T = 0.174$ vs. variable crystalline-stem lengths changing with time during isothermal crystallization at a specific temperature. The evolution time is denoted by the *numbers* (times 1000 Monte Carlo cycles) near the curves. The curves are shifted vertically with an interval of 300 for clarity. We can see that with time the peak shifts from one third to half of the chain length [56]

4.2 Crystallization and Melting of Statistical Copolymers

Irregularities in the structure of the polymer backbone will make it difficult for a polymer to be incorporated in a regularly packed crystal structure. This phenomenon is particularly pronounced in systems of random copolymers that consist of a mixture of crystallizable monomers and noncrystallizable comonomers. A very simple way to represent the difference in crystallizability is to assume that crystallizable monomers have a parallel bond–bond interaction energy E_p, while no such interaction is present in pairs of bonds involving the comonomers [52]. In addition, we account for the difference in the size of monomers and comonomers by assuming that the comonomers cannot diffuse through a crystalline region of monomers. With these ingredients, we can perform dynamic Monte Carlo simulations to study how the statistical nature of the copolymers affects the crystal morphology. Three kinds of statistical sequences were then generated, namely, homogeneous (randomly sequenced) copolymers, homogeneous (slightly alternating) copolymers, and heterogeneous (a product in batch reaction with a significant compositional shift) copolymers. On cooling from the melt and then on reheating, crystallization and melting of bulk statistical copolymers were monitored through the absolute crystallinity, which was defined as the fraction of monomer bonds having more than five parallel neighbors of the same type. We find that the phase transition temperatures depend not only on the comonomer content but also on the sequence distribution. Figure 12 shows that on cooling a copolymer system, almost all monomers eventually end up in crystalline domains, irrespective of the composition of the copolymer. Upon reheating a partially crystallized system, we first observe crystallization, rather than melting (cold crystallization). The crystallites that form first upon cooling tend to contain predominantly long monomer sequences. Hence there is

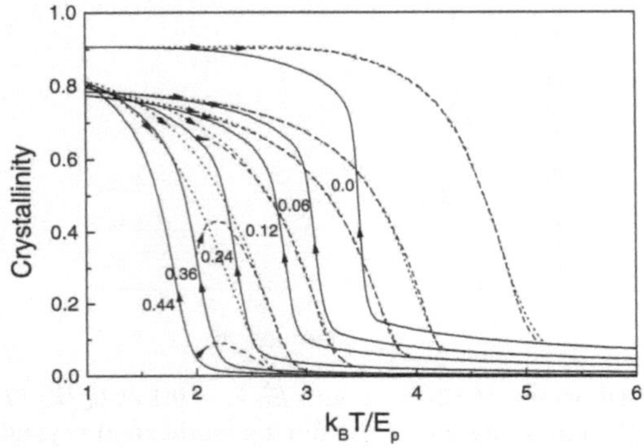

Fig. 12 Cooling (*solid lines*) and heating (*dotted* and *dashed lines*) crystallinity curves of random copolymers with variable comonomer mole fractions as denoted near the curves. The *dashed lines* start from the reduced temperature of 2 and meet the *dotted curves* at high temperatures [52]

a sequence-length segregation during crystallization (Fig. 13) [124]. As the comonomer content of the polymer is increased, the morphology of the crystallites changes from lamellar to granular (Fig. 14). A more detailed analysis can be found in Ref. [52]. Furthermore, there exists a liquid–liquid demixing in the heterogeneous copolymers, but not in the homogeneous copolymers, before the crystallization occurs on cooling (Fig. 15). Since the heterogeneous copolymers are some kinds of polymer blend, the principle of this prior demixing was actually discussed in Sect. 3.3.

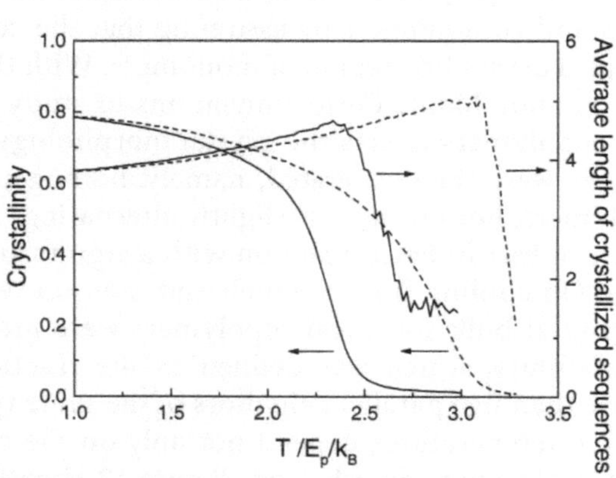

Fig. 13 Cooling (*solid line*) and heating (*dashed lines*) curves of crystallinity and averaged length of crystallized sequences for slightly alternating copolymers with a comonomer mole fraction 0.24. The crystallized sequences are defined as the monomer sequences more than half of whose bonds are in crystalline states [124]

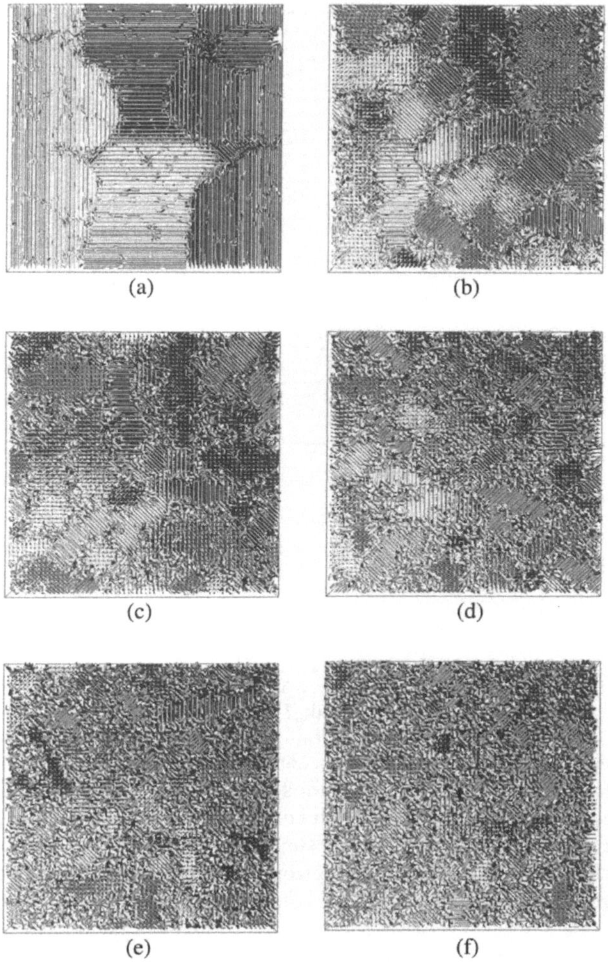

Fig. 14 Snapshots of random copolymers with variable comonomer mole fractions at the reduced temperature of 1 in the cooling process of Fig. 12. **a–f** Comonomer contents of 0, 0.06, 0.12, 0.24, 0.36, and 0.44, respectively. Polymer bonds are drawn in *cylinders* and the bonds containing comonomers are shown in *double thickness* [52]

4.3 Free-Energy Barrier for Melting and Crystallization of a Single-Homopolymer Model

Thus far, we have been discussing the crystallization of a multichain system. However, under suitable conditions, crystallization can even occur in a single-chain system. Using a combination of biased sampling, multihistogram techniques, and parallel tempering [125], we can directly compute the

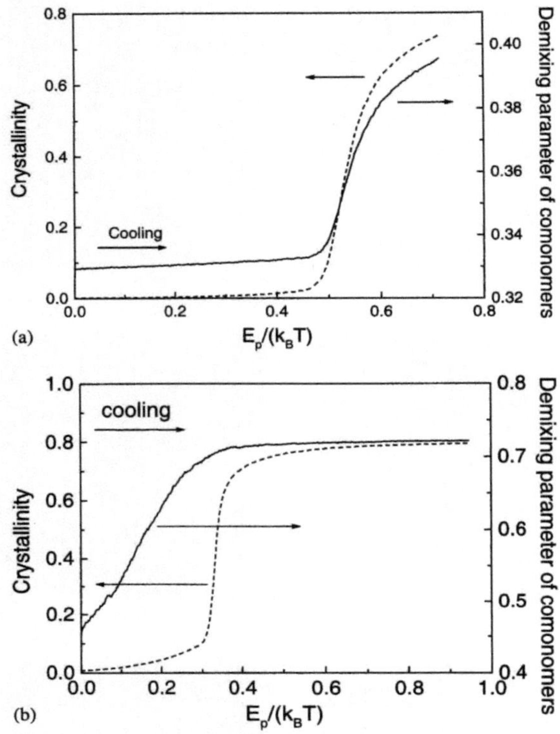

Fig. 15 Cooling curves of crystallinity (*solid line*) and demixing parameter of comonomers (*dashed line*). The latter is defined as the mean fraction of neighboring sites occupied by other comonomers around each comonomer. The cooling program is a stepwise increase of $E_p/(k_B T)$ from zero with a step length of 0.002 and a step period of 300 Monte Carlo cycles. **a** The slightly alternating copolymer with a comonomer mole fraction 0.36; **b** the heterogeneous copolymer with a comonomer mole fraction of 0.36 [52]

free-energy barrier (if any) that separates the crystalline state of the single chain from the disordered "coil" state [126]. Technical details can be found in Sect. A.4. The simulations showed that, at coexistence, there can be a quite high free-energy barrier between the crystalline and molten states of a single chain. The height of this free-energy barrier depends on chain length [127]. As can be seen from Fig. 16, the chain-length dependence can be described by a simple nucleation-like model that takes into account the bulk and surface contributions to the free energy change of single-chain melting:

$$\Delta F_{\mathrm{melt}} = n_{\mathrm{m}} \Delta f_{\mathrm{melt}} + \sigma (N - n_{\mathrm{m}})^{2/3}, \tag{20}$$

where n_{m} is the number of molten units, Δf_{melt} is the free-energy change of each chain unit on melting of bulk polymers, σ represents surface free energy of the crystallite, and N is the chain length.

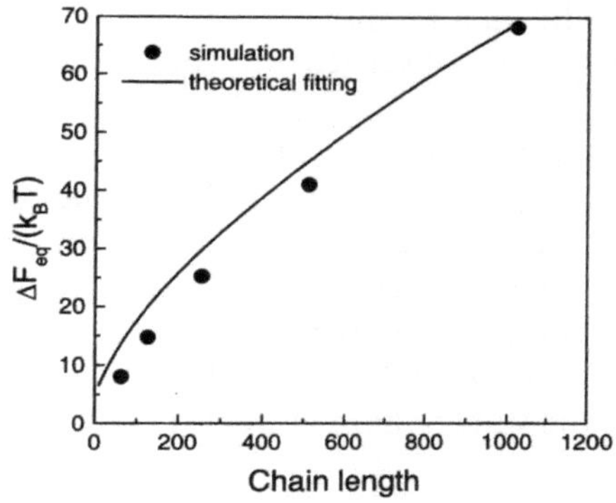

Fig. 16 Height of the equilibrium free-energy barrier for melting and crystallization vs. the chain length in single-chain systems. The *circles* are the simulation results, and the *solid line* is calculated from Eq. 20 with fitting parameter $\sigma = 15E_p$ [127]

Equation 20 predicts a free-energy barrier for primary crystal nucleation (i.e., the free-energy difference between the top of the barrier and the initial coil state) as

$$\Delta F_c = \frac{4\sigma^3}{27\Delta f_{melt}^2} . \tag{21}$$

Interestingly, this barrier does not depend on chain length. This result coincides with experimental observations on the primary nucleation rate of bulk polymers [128–130]. For secondary nucleation of crystallization on a smooth growth front, a similar free-energy expression can be obtained for 2D nucleation:

$$\Delta F_{melt,2D} = n_m \Delta f_{melt,2D} + \sigma_{2D}(N - n_m)^{1/2} , \tag{22}$$

where $\Delta f_{melt,2D}$ and σ_{2D} have slightly different values from Eq. 20. The free-energy barrier for secondary nucleation is still independent of chain length, and is given by

$$\Delta F_{c,2D} = \frac{\sigma^2}{4\Delta f_{melt,2D}} . \tag{23}$$

This result also happens to be compatible with the experimental observations on the crystal growth rate of bulk polymers [131, 132]. In addition, both Eqs. 21 and 23 give a reasonable temperature dependence of the free-energy barriers for primary nucleation and secondary nucleation, compared with the the cases for bulk polymers. It is therefore tempting to speculate that the rate of crystallization of bulk polymers is determined by intramolecular nucleation, similar to the macromolecular nucleation mechanism suggested by Wunderlich [133]. The model suggests that both primary nucleation and secondary nucleation of long-chain polymers are dominated by an intramolecular process. Incidentally, such an intramolecular nucleation model provides a natural explanation for the observed molecular-size fractionation during polymer crystal growth.

The free-energy barrier for melting and crystallization of single chains is also dependent on the quality of the solvent. The same concepts that apply to the interplay of polymer crystallization and liquid–liquid demixing in polymer solutions are also relevant in the freezing of a single-chain system: a large positive B will drive a coil–globule collapse transition [134], while a large E_p drives crystallization [126]. The mean-field theory developed for a polymer solution is still meaningful for the single-chain system. Figure 17 shows that if B/E_p is

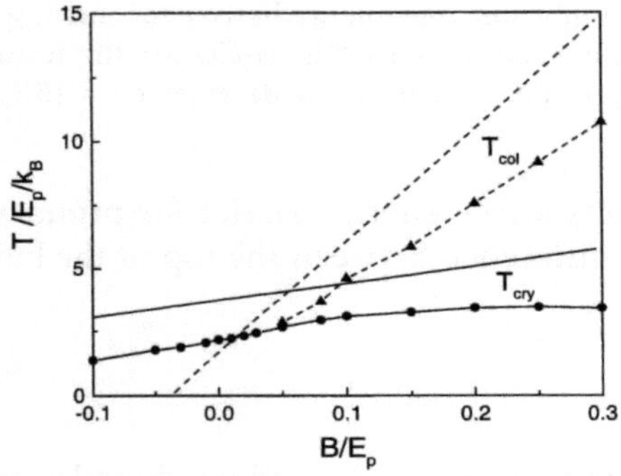

Fig. 17 B/E_p dependence of the critical temperatures of liquid–liquid demixing (*dashed line*) and the equilibrium melting temperatures of polymer crystals (*solid line*) for 512-mers at the critical concentrations, predicted by the mean-field lattice theory of polymer solutions. The *triangles* denote T_{col} and the *circles* denote T_{cry}; both are obtained from the onset of phase transitions in the simulations of the dynamic cooling processes of a single 512-mer. The segments are drawn as a guide for the eye (Hu and Frenkel, unpublished results)

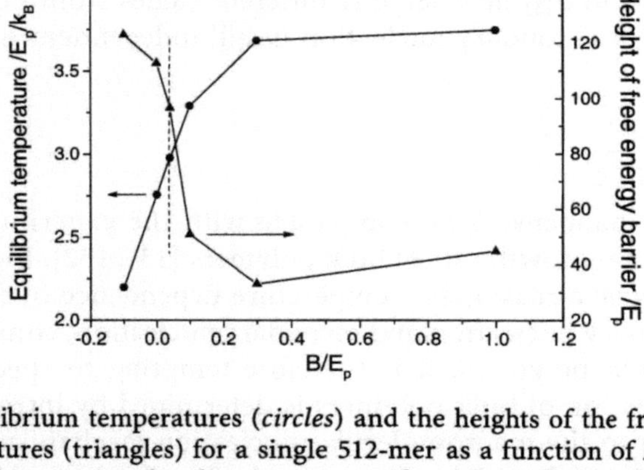

Fig. 18 The equilibrium temperatures (*circles*) and the heights of the free-energy barrier at these temperatures (triangles) for a single 512-mer as a function of B/E_p. The *dashed line* shows the demarcation for the occurrence of a prior collapse transition (Hu and Frenkel, unpublished results)

small (or even negative) there will be a direct transition from the coil state to the single-chain crystal. In contrast, for large positive values of B/E_p a coil–globule transition will precede the transition to the crystalline state (Hu and Frenkel, unpublished results). The free-energy barrier for crystallization decreases with increasing B/E_p, but levels off once there has been a prior coil–globule collapse transition (Fig. 18). These results suggest that single-chain crystallization is easiest if the polymer has undergone a prior coil-to-globule transition, yet the temperature is not so low that the globule has effectively vitrified.

Appendix

A Dynamic Monte Carlo Simulations of Lattice Polymers

A.1 Microrelaxation Model

In the dynamic Monte Carlo simulations that we describe, polymers "live" on a lattice [135]. They can move either by local jumps or by "sliding" moves that involve a longer stretch of the polymer. The ability to perform such sliding moves greatly increases the rate at which the polymers can sample configuration space. Moreover, it mimics the real dynamics of polymers in dense media. For this reason, the present "microrelaxation model" allows us to gain some insight into the dynamics by which an initial nonequilibrium state of the polymer system relaxes. The first microrelaxation model for lattice polymers was suggested by Verdier and Stockmayer [136], who allowed the change of local chain conformation through end-bond twisting, kink jumping, and crankshaft rotation. The sliding diffusion model was developed to simulate chain diffusion on a lattice [137]. However, all of these models tend to be rather inefficient in changing the orientational distribution of bond vectors, as new bond orientations are predominantly generated at the chain ends. These models are therefore not very efficient in achieving conformational relaxation. In contrast, the kink-generation model allows new bond vectors to be generated in the middle of the chains [138]. This kink-generation model was later developed into the well-known bond-fluctuation model [139, 140]. Actually, combining both kink generation and sliding diffusion together provides higher efficiency for chain relaxation. Such a hybrid approach was first suggested by Lu and Yang [141]. In this hybrid model, sliding diffusion that extends to the end of the chain is allowed during kink generation. In the algorithm that we use, we assume that sliding diffusion takes place between two defects: it is terminated by smoothening out the nearest kink along the chain (Fig. 19) [134]. In this respect, the algorithm is a numerical implementation

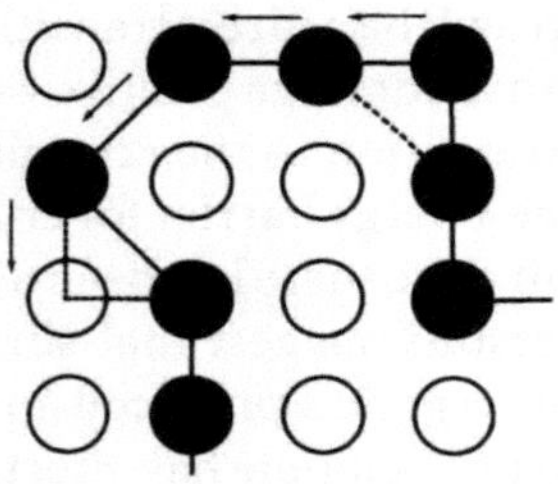

Fig. 19 Our hybrid microrelaxation model. The *solid circles* are occupied by a polymer chain. The *dashed lines* show the new bond positions produced by a move consisting of kink generation and partial sliding diffusion along the chain. The *arrows* indicate the directions of monomer jumping [134]

of the kink-defect diffusion mechanism proposed by de Gennes [142]. Since the polymer bonds are allowed to stay either along the lattice axis or along the body and face diagonals, the coordination number of such a cubic lattice includes all the neighbors along these directions, namely, $6 + 8 + 12 = 26$. In all simulations, we used periodic boundary conditions. In order to map the length and time scales of the lattice model onto those of real polymer systems, one can study the behavior of the radius of gyration (or the end-to-end distance) of the polymer chain (for static properties) and the polymer diffusion coefficient (for dynamics) [143]. Figure 20 shows the mean-square end-to-end distance of lattice polymers vs. the chain length r for polymer solutions over a wide range of concentrations. In very dilute solutions, the polymer size (in three dimensions, in an athermal solvent) is expected to scale as $\langle h^2 \rangle \sim r^{1.2}$, while in the melt, it scales as $\langle h^2 \rangle \sim r$ [144]. Figure 20

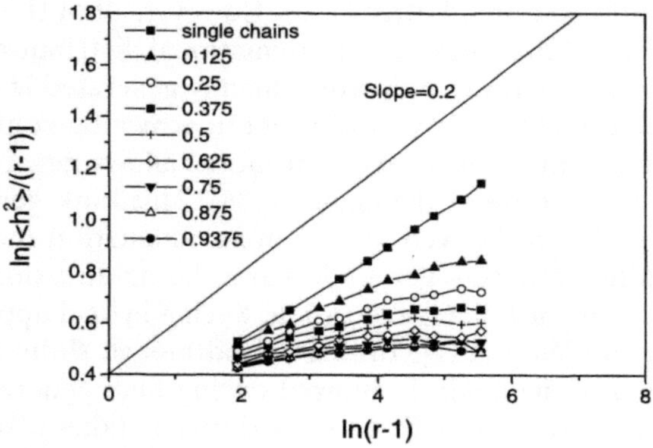

Fig. 20 Mean-square end-to-end distance of chains vs. chain length in a 32 (or above)-sized cubic lattice. The data are those of the polymer volume fractions (Hu and Frenkel, unpublished results)

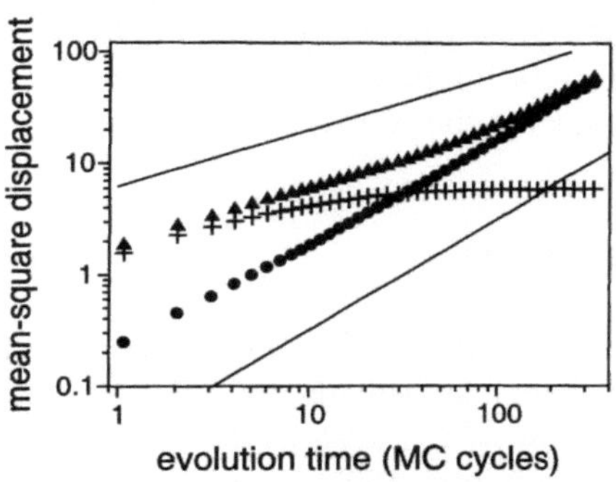

Fig. 21 Mean-square displacement vs. evolution time for 16-mers with an occupation density of 0.9375 in a 32-sized cubic lattice. The *triangles* are for four middle chain units, the *circles* are for the mass center, and the *crosses* are for the chain units relative to the center of mass. The *lines* with slopes of 1.0 and 0.5 indicate the scaling expected according to the Rouse model of polymer chains [56]

shows the slopes of $\ln(\langle h^2\rangle/r)$ vs. $\ln(r)$ changing from 0.2 to zero with the increase of polymer concentrations. Figure 21 shows the time dependence of the mean-square displacements of individual chain units, of the chain units relative to the mass center, and of the center of chain mass. For the relatively short chains studied, we expect to observe Rouse dynamics [145], as is indeed the case.

A.2 Sampling Strategy

In our simulations, we use the Metropolis method to accept or reject trail moves [146]. Moves are rejected if they cause hard-core overlaps or bond crossing; otherwise, they are accepted with a probability equal to $\min\{1, \exp[-\Delta E/(k_B T)]\}$, where $\Delta E/(k_B T) = (bB + pE_p + cE_c)/(k_B T) = (bB/E_c + pE_p/E_c + c)E_c/(k_B T)$. The meaning of the quantities B, E_p, and E_c is described in the main text, b denotes the change in the number of polymer–solvent contacts, p is the change in the number of nonparallel pairs of neighboring bonds, and c accounts for the change in the number of noncollinear connections between consecutive bonds along the chain. Within the same model, we can add a frictional energy penalty E_f for the local sliding diffusion of chains in the crystalline region. Note that this penalty is present in forward and reverse moves; therefore, it does not affect the detailed balance condition. Rather, it acts like a kinetic pre-factor that slows down the sliding diffusion of long polymer stems in the crystallites, compared with that of short ones [56]. The potential energy barrier can be expressed in the reduced parameters. $k_B T/E_c$ is often used as the reduced temperature, B/E_c is a measure for the solvent quality, and E_p/E_c reflects the flexibility of the chains. For fully flexible chains, $E_c = 0$. Then, $k_B T/E_p$ is used as the reduced temperature. Usually, we fix $E_p/E_c = 1$ for the semiflexible chains.

A.3
Temperature Scanning Program

To perform temperature scans, we increase/decrease the value of $E_c/(k_B T)$ in steps of size 0.002. Every step takes 300–500 Monte Carlo cycles [14]. In monitoring the properties of the system during heating or cooling, we discard the results of the first 100–400 Monte Carlo cycles because the system is far from equilibrium at the early stages of every step. Statistics on thermal or structural properties of the system are then collected during the remainder of the step. We use the "disorder parameter" described in the main text to monitor the progress of crystallization. To monitor the phase separation, we follow the behavior of the mixing parameter, defined as the mean fraction of solvent sites around a chain unit. To facilitate comparison with experiment, we use the same (rather ad hoc) methods to detect the onset of phase transitions: it is defined as the crossing point of two lines extrapolated from the transition region and from the one-phase region on either side of the phase transition, respectively. To observe primary nucleation on cooling, a large supercooling is usually required. This can delay the onset of crystallization well beyond the equilibrium freezing point, especially in the case of dilute solutions. To suppress such overshooting effects we introduce one molecular layer of solid substrate consisting of fully extended chains. This substrate serves as a template for primary nucleation of both crystallization and liquid–liquid demixing. Figure 22 compares two cooling curves: one was obtained

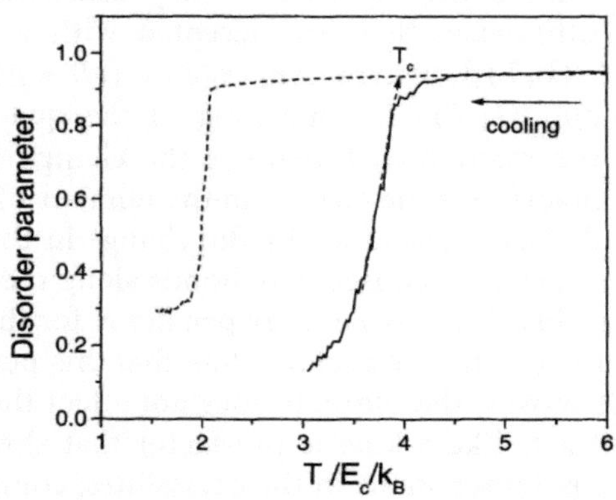

Fig. 22 Cooling curves of the disorder parameter for 32-mers in 32-sized cubic lattice with a conventional cooling program (*dashed line*) and an optimized cooling program (*solid line*). Polymers have a volume fraction of 0.0625 with $B/E_c = 0$ and $E_p/E_c = 1$ [14]

by cooling a homogeneous phase; the other employed the substrate to suppress overshooting. Clearly, the presence of the template significantly raises the onset temperature of crystallization in dilute solutions.

A.4
Biased Sampling and Multihistogram Parallel Tempering

The formation of crystal nuclei in a moderately supersaturated solution is a rare event. In order to probe the frequency of such fluctuations, we used umbrella sampling [148]. In particular, we bias the formation of crystallites by increasing their Boltzmann weight. In fact, during a single simulation, we favor the formation of crystallites with crystallinity x_1 in a window around x_0 by lowering their potential energy with $W = k(x_1 - x_0)^2$, where k determines the width of the window. To recover the free energy of the clusters in the unbiased system, we have to correct for the bias [149, 150]. In practice, about 15 overlapping windows were employed to calculate the free-energy barrier separating the crystalline and disordered states of a single-chain system [126]. The multiple histograms in the simulations of these windows are then merged to form a single, smooth curve. Parallel tempering was used to enhance equilibration between the different windows. An example is shown in Fig. 23.

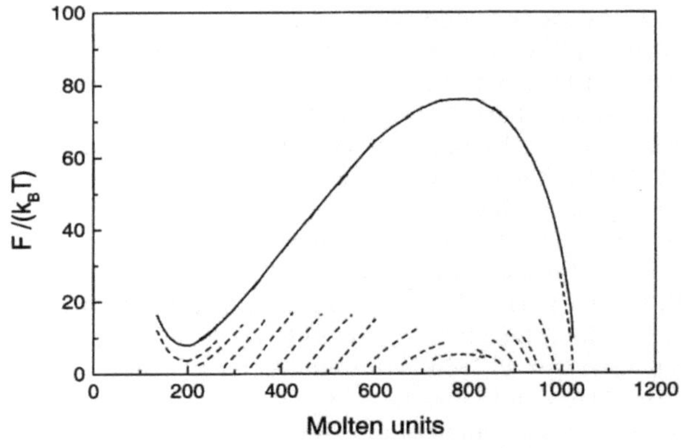

Fig. 23 Parallel tempering of the free-energy curves in the overlapping windows as a function of the number of molten units for a single 1024-mer at a temperature of $2.967E_p/k_B$. The y-axis is not for the absolute value of the free energy but for the relative distribution of the free energy (Hu and Frenkel, unpublished results)

Acknowledgements This work is part of the research program of the Stichting voor Fundamenteel Onderzoek der Materie (FOM), which is financially supported by the Nederlandse organisatie voor Wetenschappelijk Onderzoek (NWO). We thank S. van Albada for critically reading the manuscript. W. H. thanks the funding of National Natural Science Foundation of China (Grant No. 20474027).

References

1. Onsager L (1949) Ann NY Acad Sci 51:627
2. Maier W, Saupe AY (1959) Z Naturforsch 14a:882
3. Maier W, Saupe AY (1959) Z Naturforsch 15a:287
4. Jaehnig F (1979) J Chem Phys 70:3279
5. Ronca G, Yoon DY (1982) J Chem Phys 76:3295
6. Ronca G, Yoon DY (1984) J Chem Phys 80:925
7. ten Bosch A, Maissa P, Sixon P (1983) J Chem Phys 79:3462
8. ten Bosch A, Maissa P, Sixon P (1983) J Phys Lett (Paris) 44:L105
9. Khokhlov AR, Semenov AN (1985) J Stat Phys 38:161
10. Gupta AM, Edwards SF (1993) J Chem Phys 98:1588
11. Lekkerkerker HNW, Vroege GJ (1993) Philos Trans R Soc Lond Ser A 344:419
12. Flory PJ (1956) Proc R Soc Lond Ser A 234:60
13. Hu WB (2000) J Chem Phys 113:3901
14. Hu WB, Frenkel D, Mathot VBF (2003) J Chem Phys 118:10343
15. Rowlinson JS (1970) Faraday Disc Chem Soc 49:30
16. Guggenheim EA (1952) Mixtures. Clarendon, Oxford
17. Flory PJ (1953) Principles of polymer chemistry. Cornell University Press, Ithaca, NY p 495
18. Prigogine I (1957) The molecular theory of solution. North-Holland, Amsterdam
19. Meyer KH (1939) Z Phys Chem B 44:383
20. Huggins ML (1942) Ann NY Acad 43:1
21. Flory PJ (1942) J Chem Phys 10:51
22. Flory PJ (1982) Proc Natl Acad Sci USA 79:4510
23. Nagle JF (1974) Proc R Soc Lond Ser A 337:569
24. Gordon M, Kapadia P, Malakis A (1976) J Phys A 9:751
25. Gujrati PD (1980) J Phys A 13:L437
26. Gujrati PD, Goldstein MJ (1981) J Chem Phys 74:2596
27. Nagle JF, Gujrati PD, Goldstein MJ (1984) J Phys Chem 88:4599
28. DiMarzio EA (1961) J Chem Phys 35:658
29. Ronca G (1983) J Chem Phys 79:6326
30. Jarry JP, Monnerie L (1979) Macromolecules 12:316
31. Doi M, Pearson D, Kornfield J, Fuller G (1989) Macromolecules 22:1488
32. Watanabe H, Kotaka T, Tirrell M (1991) Macromolecules 24:201
33. Baumgaertner A, Yoon DY (1983) J Chem Phys 79:521
34. Yoon DY, Baumgaertner A (1984) Macromolecules 17:2864
35. Boyd RH (1986) Macromolecules 19:1128
36. Mansfield ML (1994) Macromolecules 27:4699
37. Meyer H, Mueller-Plathe F (2002) Macromolecules 35:1241
38. Weber H, Paul W, Binder K (1999) Phys Rev E 59:2168
39. Weber H, Paul W, Kob W, Binder K (1997) Phys Rev Lett 78:2136
40. Baumgaertner A (1984) J Chem Phys 81:484; (1986) J Chem Phys 84:1905
41. Kolinsky A, Skolnick J, Yaris R (1986) Macromolecules 19:2560
42. Weber TA, Helfand E (1979) J Chem Phys 71:4760
43. Fischer EW, Strobl GR, Dettenmaier M, Stamm M, Steidle N (1979) Faraday Disc Chem Soc 68:26
44. Walasek J (1990) J Polym Sci Part B Polym Phys 28:1075
45. Walasek J (1990) J Polym Sci Part B Polym Phys 28:2473
46. Bleha T (1985) Polymer 26:1638

47. Mansfield (1983) Macromolecules 16:914
48. Yoon DY (1989) Polym Prepr Am Chem Soc Div Polym Chem 30:63
49. McCoy JD, Honnell KG, Schweizer KS, Curro JG (1991) J Chem Phys 95:9348
50. van Ruiten J, van Dieren F, Mathot VBF (1993) In: Dosiere M (ed) Crystallization of polymers. Kluwer, Dordrecht, p 481
51. Hu WB, Yu TY (1995) Chem J Chin Univ 16:1140
52. Hu WB, Frenkel D, Mathot VBF (2003) Macromolecules 36:2165
53. Toma L, Toma S, Subirana JA (1998) Macromolecules 31:2328
54. Chen CM, Higgs PG (1998) J Chem Phys 108:4305
55. Hu WB, Yu TY, Bu HS (1994) Chem J Chin Univ 15:1099
56. Hu WB (2001) J Chem Phys 115:4395
57. Hu WB, Frenkel D, Mathot VBF (2003) Macromolecules 36:549
58. Hu WB, Frenkel D, Mathot VBF (2002) Macromolecules 35:7172
59. Lenz RW (1967) Organic chemistry of synthetic high polymers. Interscience, New York, p 91
60. Wunderlich B (1973) Macromolecular physics, vol 1. Crystal structure, morphology, defects. Academic, New York, p 68
61. Tadokoro H (1979) Structure of crystalline polymers. Wiley, New York, p 15
62. Sperling LH (1992) Introduction to physical polymer science, 2nd edn. Wiley, New York, p 261
63. Pan Y, Cao MY, Wunderlich B (1989) In: Brandrup J, Immergut EH (eds) Polymer handbook, 3rd edn. Wiley, New York, p VI/376
64. Bunn CW (1955) J Polym Sci 16:323
65. Wunderlich B (1980) Macromolecular physics, vol 3. Crystal melting. Academic, New York, p 27
66. Mandelkern L (2002) Crystallization of polymers,vol 1, 2nd edn. Equilibrium concepts. Cambridge University Press, Cambridge, p 42
67. Flory PJ, Vrij A (1963) J Am Chem Soc 85:3548
68. Bawendi MG, Freed KF (1988) J Chem Phys 88:2741
69. Buta D, Freed KF, Szleifer I (2000) J Chem Phys 112:6040
70. Richards RB (1946) Trans Faraday Soc 42:10
71. Flory PJ, Mandelkern L, Hall HK (1951) J Am Chem Soc 73:2532
72. He XW, Herz J, Guenet JM (1987) Macromolecules 20:2003
73. Aerts L, Berghmans H, Koningsveld R (1993) Makromol Chem 194:2697
74. Wang H, Shimizu K, Kim H, Hobbie EK, Wang ZG, Han CC (2002) J Chem Phys 116:7311
75. Flory PJ (1949) J Chem Phys 17:223
76. Mandelkern L (2002) Crystallization of polymers,vol 1, 2nd edn. Equilibrium concepts. Cambridge University Press, Cambridge, p 70
77. ten Wolde PR, Frenkel D (1997) Science 277:1975
78. Talanquer V, Oxtoby DW (1998) J Chem Phys 109:233
79. Sear RP (2001) Phys Rev E 63:066105
80. Inaba N, Sato K, Suzuki S, Hashimoto T (1986) Macromolecules 19:1690
81. Lee HK, Myerson AS, Levon K (1992) Macromolecules 25:4002
82. Guenet JM (1996) Thermochim Acta 284:67
83. Berghmans H, de Cooman R, de Rudder J, Koningsveld R (1998) Polymer 39:4621
84. Hu WB, Frenkel D (2004) Macromolecules 37:4336
85. Schaaf P, Lotz B, Wittmann JC (1987) Polymer 28:193
86. Hu WB, Mathot VBF (2003) J Chem Phys 119:10953
87. Dudowicz J, Freed KF, Douglas JF (2002) Phys Rev Lett 88:095503

88. Dudowicz J, Freed KF (1991) Macromolecules 24:5076; 24:5112
89. Clancy TC, Putz M, Weinhold JD, Curro JG, Mattice WL (2000) Macromolecules 33:9452
90. Xu GQ, Clancy TC, Mattice WL (2000) Macromolecules 35:3309
91. Maier RD, Thomann R, Kressler J, Muelhaupt R, Rudolf B (1997) J Polym Sci B Polym Phys 35:1135
92. Haliloglu T, Mattice WL (1999) J Chem Phys 111:4327
93. Thomann R, Kressler J, Setz S, Wang C, Muelhaupt R (1996) Polymer 37:2627
94. Silvestri R, Sgarzi P (1998) Polymer 39:5871
95. Wang ZG, Phillips RA, Hsiao BS (2000) J Polym Sci B Polym Phys 38:2580
96. Lohse D (1986) J Polym Eng Sci 26:1500
97. Yeh GSY, Lambert SL (1972) J Polym Sci A-2 10:1183
98. Ermer H, Thomann R, Kressler J, Brenn R, Wunsch J (1997) Macromol Chem Phys 198:3639
99. Woo EM, Lee ML, Sun YS (2000) Polymer 42:883
100. Keller A (1957) Philos Mag 2:1171
101. Fischer EW (1957) Z Naturforsch 12a:753
102. Till PH Jr (1957) J Polym Sci 24:301
103. Lindenmeyer PH (1963) J Polym Sci C 1:5
104. Geil PH (1963) Polymer single crystals. Wiley-Interscience, New York
105. Wunderlich B (1973) Macromolecular physics, vol 1. Crystal structure, morphology, defects. Academic, New York, pp 178–379
106. Bassett DC (1981) Principles of polymer morphology. Cambridge University Press, London
107. Welch P, Muthukumar M (2001) Phys Rev Lett 87:218302
108. Lotz B (1994) Philos Trans R Soc Lond Ser A 348:19
109. Bu Z, Cheng SZD, Putthanarat S, Eby RK, Reneker DH, Lotz B, Magonov S, Hsieh ET, Johnson TW, Geerts RG, Plackal SJ, Hawley GR, Welch MB (2000) Macromolecules 33:6861
110. Kelton KF (1991) In: Ehrenreich H, Turnbull D (eds) Crystal nucleation in liquids and glasses. Academic, Boston, p 75
111. Pennings AJ (1977) J Polym Sci Polym Symp 59:55
112. McHugh AJ (1982) Polym Eng Sci 22:15
113. Barham PJ, Keller A (1985) J Mater Sci 20:2281
114. Keller A, Odell JA (1985) Colloid Polym Sci 263:181
115. Al-Hussein M, Strobl G (2002) Macromolecules 35:1672
116. Heck B, Strobl G, Grasruck M (2003) Eur Phys J E 11:117
117. Bassett DC, Keller A (1961) Philos Mag 6:1053
118. Dosiere M, Colet M, Point JJ (1986) In: Sedlacek B (ed) Morphology of polymers. De Gruyter, Berlin, p 171
119. Doye JPK, Frenkel D (1998) Phys Rev Lett 81:2160
120. Doye JPK, Frenkel D (1999) J Chem Phys 110:7073
121. Hikosaka M (1987) Polymer 28:1257
122. Hikosaka M (1990) Polymer 32:458
123. Hikosaka M, Rastogi S, Keller A, Kawabata H (1992) J Macromol Sci Phys B31:87
124. Hu WB, Mathot VBF (2004) Macromolecules 37:673
125. Frenkel D, Smit B (2002) Understanding molecular simulations, 2nd edn. Academic, San Diego, p 167
126. Hu WB, Frenkel D, Mathot VBF (2003) J Chem Phys 118:3455
127. Hu WB, Frenkel D, Mathot VBF (2003) Macromolecules 36:8178

128. Nishi M, Hikosaka M, Ghosh SK, Toda A, Yamada K (1999) Polym J 31:749
129. Ghosh SK, Hikosaka M, Toda A (2001) Colloid Polym Sci 279:382
130. Umemoto S, Hayashi R, Kawano R, Kikutani T, Okui N (2003) J Macromol Sci B 42:421
131. Umemoto S, Okui N (2002) Polymer 43:1423
132. Umemoto S, Kobayashi N, Okui N (2002) J Macromol Sci B 41:923
133. Mehta A, Wunderlich B (1974) J Polym Sci Polym Phys Ed 12:255
134. Hu WB (1998) J Chem Phys 109:3686
135. Kremer K, Binder K (1988) Comput Phys Rep 7:259
136. Verdier PH, Stockmayer WH (1962) J Chem Phys 36:227
137. Wall FT, Mandel F (1975) J Chem Phys 63:4592
138. Larson RG, Scriven LE, Davis HT (1985) J Chem Phys 83:2411
139. Carmesin I, Kremer K (1988) Macromolecules 21:2819
140. Deutsch HP, Binder K (1991) J Chem Phys 94:2294
141. Lu JM, Yang YL (1993) Sci Chin A 36:357
142. de Gennes PG (1971) J Chem Phys 55:571
143. Binder K (1995) In: Binder K (ed) Monte Carlo and molecular dynamics simulations in polymer science. Oxford University Press, New York, p 22
144. de Gennes PG (1979) Scaling concepts in polymer physics. Cornell University Press, Ithaca, NY, p 29
145. Kremer K, Grest GS (1990) J Chem Phys 92:5057
146. Metropolis N, Rosenbluth AW, Rosenbluth MN, Teller AH, Teller E (1953) J Chem Phys 21:1087
147. Binder K (1979) In: Binder K (ed) Monte Carlo methods in statistical physics. Springer, Berlin Heidelberg New York, p 1
148. Torrie GM, Valleau JP (1974) Chem Phys Lett 28:578
149. ten Wolde PR (1998) PhD thesis, Amsterdam University, p 72
150. Auer S, Frenkel D (2001) Nature 409:1020

Adv Polym Sci (2005) 191: 37–85
DOI 10.1007/12_012

Published online: 5 October 2005

Molecular Dynamics Modeling of the Crystal-Melt Interfaces and the Growth of Chain Folded Lamellae

Takashi Yamamoto

Department of Physics, Biology, and Informatics, Yamaguchi University,
753-8512 Yamaguchi, Japan
yamamoto@mms.sci.yamaguchi-u.ac.jp

Abstract The molecular mechanism of polymer crystallization is one of the most difficult problems and has defied innumerable efforts to understand the process over the last fifty years in spite of its great importance both from the academic and the industrial point of view. We have been studying this historical problem by use of the molecular dynamics simulation method. In this chapter of the book, we review our recent work on the crystal growth of polymers with special focus on polymer behavior at the crystal surface, either at crystal-vapor or crystal-melt interfaces. Our starting molecular model is a bead-spring chain, or a wormlike chain, made of methylene-like united atoms; the zigzag structure of polymethylene is here neglected in order to accelerate crystallization. We proceed with stepwise revisions of the model toward the realistic modeling of polymer crystallization from the dense melt. We start our discussion with the crystallization of polymers on a two-dimensional surface, which is a model of the chain strongly adsorbed on the growth surface. Then we treat the three-dimensional process of crystallization of a single chain from a vapor phase: the adsorption to and the ordering on the growth substrate. Lastly, polymer crystallization from the dense melt is investigated. We also report on fiber formation from a highly oriented amorphous state. Various important issues concerning the molecular mechanism of polymer crystallization are discussed in the light of findings from our direct molecular simulations.

Keywords Chain folding · Computer modeling · Crystal growth · Crystal-melt interfaces · Molecular dynamics · Polymer crystallization

1 Introduction

Polymer crystallization controls the structural formation processes of polymeric materials and thereby dominates the properties of the final polymer products. A deep understanding of the molecular mechanisms of polymer crystallization is indispensable if we want to design favorable microscopic and macroscopic structures of crystalline polymers [1]. Polymer crystallization is also a great academic challenge. The constituent atoms are linked by strong chemical bonds and are therefore forced to move quite cooperatively, while they interact with atoms far distant along the chain through weak van der Waals forces. The very specific nature of the polymers must be the origin of their peculiar way of ordering into chain-folded crystals [2, 3], but how do the macromolecules with huge internal degrees of freedom perform the remarkable feat of chain folded crystallization?

Polymer crystallization is usually divided into two separate processes: primary nucleation and crystal growth [1]. The primary nucleation typically occurs in three-dimensional (3D) homogeneous disordered phases such as the melt or solution. The elementary process involved is a molecular transformation from a random-coil to a compact chain-folded crystallite induced by the changes in ambient temperature, pH, etc. Many uncertainties (the presence of various contaminations) and experimental difficulties have long hindered quantitative investigation of the primary nucleation. However, there are many works in the literature on the early events of crystallization by var-

detailed inspection of the molecular processes is expected to give decisive answers to many long-standing arguments of the last half century. The growth of the polymer crystal is, however, a very slow process. An experimental estimation has shown that the velocity of the step propagation along the growth surface is of the order of μm/min [36]. The steps advance in a few ns, the possible duration of the MD simulation, is estimated to be only a very small fraction of a nm; straightforward MD simulations of polymer crystallization by use of very realistic models are thus beyond execution. We must devise a simple and effective molecular model.

We started our computer simulation studies on polymer crystal growth in 1997 [20]. We first studied a two-dimensional (2D) model by assuming that a long chain C_{500} was strongly adsorbed on a growth surface of the crystal. By use of a simple bead-spring model, we could reproduce the crystallization of the chain into a neat chain-folded lamella. Crystallization rate and lamella thickness were found to show large undercooling dependence, which conformed qualitatively to well-known experimental knowledge. We also observed pronounced lamella-thickening and large structural fluctuations during crystallization at higher temperatures.

Then we extended the 2D-model to a 3D one [21]. We considered crystallization of a single polymer chain C_{500} from a vapor phase onto a solid substrate, taking into account detailed interactions between the chain and the substrate. Though the polymer molecule in a vacuum was collapsed, like in a very poor solvent, under the influence of bare van der Waals interactions between atoms, the molecule was found to show quick adsorption and crystallization into a rather neat chain folded lamella.

Finally, we were led to the last stage of research where we treated the crystallization from the melt in multiple chain systems [22–24]. In most cases, we considered relatively short chains made of 100 beads; they were designed to be mobile and slightly stiff to accelerate crystallization. We could then observe the steady-state growth of chain-folded lamellae, and we discussed the growth rate vs. crystallization temperature. We also examined the molecular trajectories at the growth front. In addition, we also studied the spontaneous formation of fiber structures from an oriented amorphous state [25]. In this chapter of the book, we review our researches, which have been performed over the last seven years. We want to emphasize the potential power of the molecular simulation in the studies of polymer crystallization.

2
Molecular Models

Crystal growth is a process where the polymer chains are continuously adsorbed to the growth surfaces. We must model both the polymer molecule and the growth substrate. The polymer chains we consider here are composed of many beads (from 100 to 1000) that correspond to a methylene-group of mass 14. The beads are connected by harmonic springs, which have a force constant similar to that of the C – C bond; the potential energy of the spring is thus

$$U_b = \frac{1}{2} k_B (r - r_0)^2 , \tag{1}$$

where $k_B = 3.5 \times 10^{25}$ [J/m^{2}mol], and the averaged bond length $r_0 = 0.154$ [nm]. Each pair of beads, more than two bonds apart, is assumed to interact via van der Waals forces of the usual Lennard-Jones type;

$$U_{vdW}(r) = 4\varepsilon \left[\left(\frac{\sigma}{r} \right)^{12} - \left(\frac{\sigma}{r} \right)^{6} \right] , \tag{2}$$

where the values of ε and σ are taken from the parameters for the CH_2 united atom ($\varepsilon = 598.64$ [J/mol], $\sigma = 0.3923$ [nm]), and the interactions are usually cut-off at $r = 2.5\sigma$ except where stated separately (Fig. 1a and Table 1).

The molecule is either fully flexible or semi-flexible. The fully flexible chains are generally harder to crystallize than semi-flexible chains [35]. In the latter part of the paper (Sect. 5), where we discuss crystallization from the melt, we consider a semi-flexible chain, the flexibility of which is adjusted to reproduce the characteristic ratio of real polyethylene. We there make the

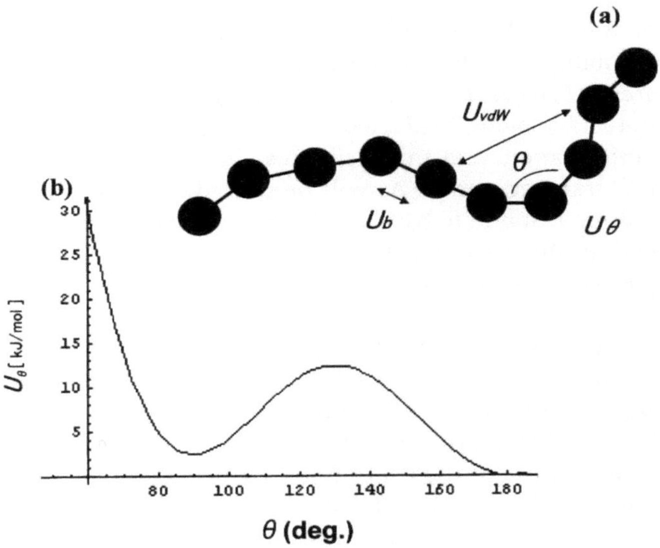

Fig. 1 **a** Model bead-spring chain interacting through bond potential U_b, bond angle potential U_θ, and van der Waals potential U_{vdW}, and **b** the form of the bond angle potential U_θ

Table 1 List of parameters and their values

Parameter	Values	Units
m	14×10^{-3}	kg/mol
k_B	3.5×10^{25}	J/m^{2}mol
r_0	0.154	nm
a	7.440×10^3	J/mol
b	2.297×10^4	J/mol
d	7.386×10^4	J/mol
θ_0	108.78	deg
ε	598.64	J/mol
$^*\sigma$	0.392	nm
λ	0.433	nm
d_s	0.375	nm
Z_c	0.229	nm

$^*\sigma$ is a unit length in reduced unit.

chain slightly stiffer by adding the potential for bond angle bending,

$$V_\theta = a - b(\cos\theta - \cos\theta_0) + d(\cos\theta - \cos\theta_0)^3 , \quad (3)$$

where θ is the C – C – C bond angle, and θ_0 is 108.78°. This potential was constructed to give the lowest energy of 0 kcal at $\theta = 180°$, the local minimum of 2.51 kJ/mol at $\theta = 90°$ (90° kink), and the energy barrier of about 12.54 kJ/mol at $\theta = 130°$ (Fig. 1b). These parameter values were selected to mimic the flexibility of polyethylene chain due to gauche bond generation. Thus our present polymer model is akin to polyethylene, with the dihedral angle potentials being neglected. Our present molecule prefers to take a straight conformation, and this facilitates chain diffusion along the chain axis as well as transverse to it. As will be described later, this model chain made of 100 atoms is found to have a mean-square end-to-end distance $\langle R_{100}^2 \rangle$ of about $120\sigma^2$ in the melt and the characteristic ratio $\langle R_{100}^2 \rangle / 100 r_0^2$ is then about 7.8; this value shows good correspondence with that of real polyethylene of about 6.7 [37].

There are several ways to model the substrate. The simplest would be to consider the substrate as a structureless attractive wall. However, since we want the polymer molecules to be parallel to each other on the substrate, we impose a directional force. In 2D crystallization, we took the substrate structure into account by use of the continuous substrate potential $U_{\text{sub}}^{2\text{D}}$, a sort of mean field potential that restricts the molecular motion on the substrate [20];

$$U_{\text{sub}}^{2\text{D}}(x, y) = U_0(1 - \cos(2\pi x/\lambda)) . \quad (4)$$

The substrate potential restricts the motion along the x-axis only, and it exerts no hindrance to the motion along the y-axis. The period $\lambda = 0.433$ nm is chosen so that the polymer chain pack closely on the surface; the value of λ is similar to the interchain separation on the {110} surface of the polyethylene crystal. On the other hand, the energy barrier $2U_0$ of the potential $U_{\text{sub}}^{\text{2D}}$ is chosen a little bit arbitrarily to have the same order of magnitude as the energy necessary for detaching the CH_2 atoms from the {110} surface of the polyethylene crystal ($U_0 = 2100$ [J/mol]). In this model, the beads making up folds or loops are located at the high energy hills of the substrate potential and are allotted excess substrate energy (Fig. 2).

In the latter part of this chapter where we study crystallization in 3D space [21–24], we consider the substrate-polymer interactions precisely by use of the method of Steele [38]. The substrate crystal is considered to be made of the same bead-spring chains closely packed, and at the lateral surface (the substrate) parallel chain stems are arranged at a spacing of $\lambda = 0.433$ nm (Fig. 2). The substrate is again similar to the {110} surface of the orthorhombic form of polyethylene, and the interlayer spacing of the {110} plane of the hexagonally packed chains d is $\sqrt{3}\lambda/2$. The substrate atoms only exert averaged forces, and no thermal motion of the substrate atoms is considered. Due to the periodic arrangement of the substrate atoms, the surface potential $U_{\text{sub}}^{\text{3D}}(x, y, z)$ between each chain atom and the substrate is a periodic function in x and y,

$$U_{\text{sub}}^{\text{3D}}(x, y, z) = U_{\text{sub}}^{\text{3D}}(x + \lambda, y + r_0, z)\,. \tag{5}$$

Following the treatment by Steele, we express the substrate potential as a Fourier series,

$$U_{\text{sub}}^{\text{3D}}(x, y, z) = \sum_q U(q_x, q_y, z) \exp(-2\pi i(q_x x + q_y y)) \tag{6}$$

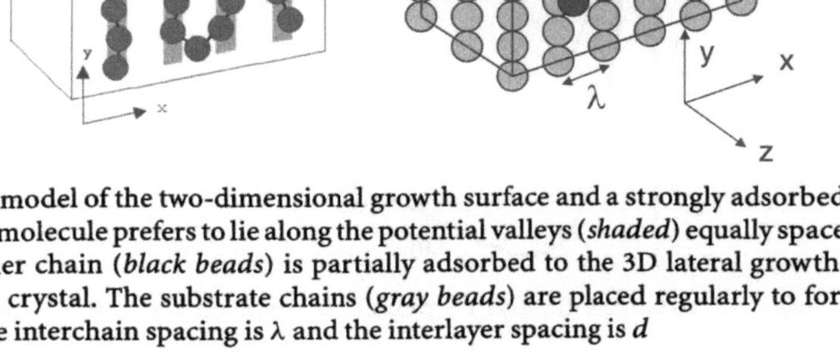

Fig. 2 **a** A model of the two-dimensional growth surface and a strongly adsorbed molecule, where the molecule prefers to lie along the potential valleys (*shaded*) equally spaced in λ, and **b** a polymer chain (*black beads*) is partially adsorbed to the 3D lateral growth-surface of a polymer crystal. The substrate chains (*gray beads*) are placed regularly to form ordered crystal; the interchain spacing is λ and the interlayer spacing is d

and

$$U(q_x, q_y, z) = \frac{2\pi\varepsilon}{\lambda r_0} \sum_{\alpha} \left\{ \frac{\sigma^{12}}{30} \left(\frac{\pi q}{z_\alpha} \right)^5 K_5(2\pi q z_\alpha) - 2\sigma^6 \left(\frac{\pi q}{z_\alpha} \right)^2 K_2(2\pi q z_\alpha) \right\} \tag{7}$$

where $q = (q_x, q_y)$ is a two dimensional reciprocal lattice vector of the surface structure, $q = |q|$, K_n is a modified Bessel function, and the sum over α represents the addition of the contributions from $\{110\}$ planes of different depths. The modified Bessel function K_n is a rapidly decreasing function of $2\pi qz$, and higher order q components make only very small contributions. Since $\lambda = 0.433$ nm is much larger than $r_0 = 0.154$ nm, the dominating contributions come from $q = (0, 0)$ and $q = (\pm 1, 0)$. We neglect higher order terms, and then the substrate energy is written as,

$$U_{\text{sub}}^{3\text{D}}(x, y, z) = U_0(z) + U_1(z) \cos\left(2\pi \frac{x}{\lambda}\right), \tag{8}$$

where

$$U_0(z) = \frac{2\pi\varepsilon}{\lambda r_0} \left(\frac{2\sigma^{12}}{5z^{10}} - \frac{\sigma^6}{z^4} - \frac{\sigma^6}{3d(z + z_c)^3} \right) \tag{9}$$

and

$$U_1(z) = \frac{4\pi\varepsilon}{\lambda r_0} \left(\frac{\sigma^{12}\pi^5}{30(z\lambda)^5} K_5\left(\frac{2\pi z}{\lambda}\right) - \frac{2\sigma^6\pi^2}{(z\lambda)^2} K_2\left(\frac{2\pi z}{\lambda}\right) \right). \tag{10}$$

In our study of 2D crystallization, we assumed that the two dimensional substrate potential is only a function of x as in Eq. 4. Here again we take the

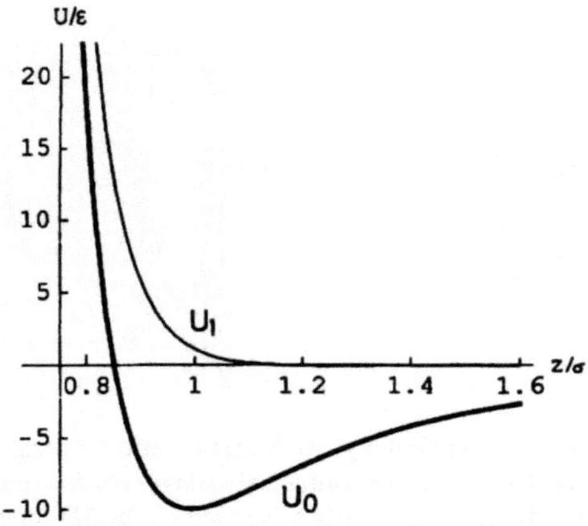

Fig. 3 The energy contributions U_0 and U_1 plotted vs. reduced separation z/σ from the substrate surface

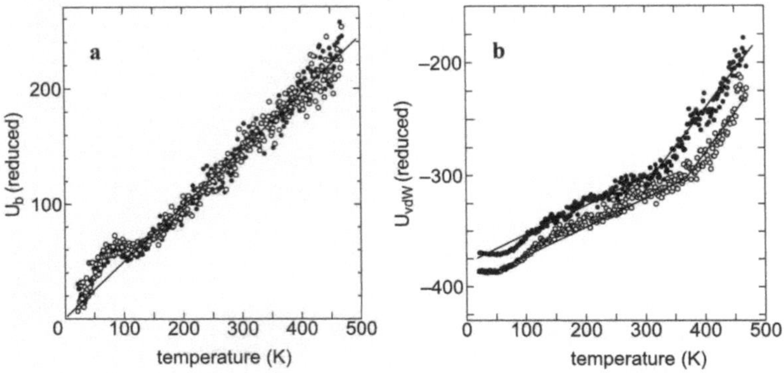

Fig. 4 Increases in the energies of **a** bond stretching and **b** van der Waals interactions during the melting process of the lamella of 20 bonds thick (•) and that of 50 bonds thick (∘); the heating rate for both 1.2 K/ps; the energies are expressed in Kelvin/bond (or atom). Marked changes in the van der Waals energy are observed, around 320 K for the thin lamella, and around 380 K for the thick lamella

the temperature increase first flows into the bond-stretching energy U_b, and is then redistributed into both van der Waals energy U_{vdW} and substrate energy U_{sub}^{2D}. Figure 4a shows the changes in U_b vs. temperature. The bond-stretching mode has the shortest relaxation time, and indeed the approximate thermal equilibrium value $k_B T/2$ is attained, except the initial states up to 100 K where the bond stretching modes are superheated.

By the temperature increase, the interchain packing and the molecular fit to the substrate become more and more disturbed resulting in marked increases in U_{vdW} (Fig. 4b). Both the van der Waals energy and the substrate energy increase linearly with temperature, until they show slight upswings around 300–400 K suggesting the onset of melting of the lamella. The calculation did not show any appreciable energy jump characteristic of the first order transition obviously due to a fast heating rate; definite energy differences between crystalline and random coil states will be evident in the discussion in the next section on crystallization by slow cooling. Direct inspection of the chain conformation tells us that the molecule starts large translational motions around 300–400 K leading to the disruption of the lamella structure and eventually to melting. Though a slight superheating may be present, the equilibrium melting temperature of a sufficiently thick lamella will be around 400 K.

3.2 Crystallization

Molecular processes at the growth surface of the crystal are one of our greatest concerns. By melting a chain-folded lamella at 600 K for 200 ps, we prepared a 2D random-coil of the molecule. The random-coil was then instantaneously quenched to each crystallization temperature $T_c(20\,\mathrm{K} \leq T_c \leq 320\,\mathrm{K})$, and the molecular processes of crystallization that followed were monitored. The crystallization was generally faster at lower T_c resulting in thinner lamellae. But at the lowest temperature $T_c = 20\,\mathrm{K}$, the system was trapped in a metastable state and remained as small separated clusters even after 10 ns. At higher temperatures, on the other hand, the system rapidly transformed to a single lamella within 10 ns.

The potential energies showed pronounced decreases by crystallization. Figure 5 shows the changes in U_{vdW} vs. time at various T_c. At each T_c the system approaches a stationary state, either stable or metastable, within a few ns. The crystallization process, the decreases in the energy, can be divided into three stages for convenience. The initial stage is up to about a few ps, where the energy shows only slight decreases. Then, up to about a few tens of ps, follows the intermediate stage characterized by a marked decrease in the energy. The late stage that follows is the final ordering process to complete the lamella structure.

In the initial stage of crystallization, the molecule prepares for subsequent ordering into the lamella structure. Within a few ps, the random coil molecule is locally straightened by partial settling down to the substrate valleys (Fig. 6). A single straight stem is not always stable and easily desorbed. The adsorbed stems are stabilized when they make chain folded clusters (hairpins) by reeling-in adjacent segments. The clusters connected by chain loops exert pulls and tend to collapse further. Unlike the usual collapse in free space, the molecule is now locally stuck on the substrate, and the molecular mobility is largely restricted resulting in slow collapse.

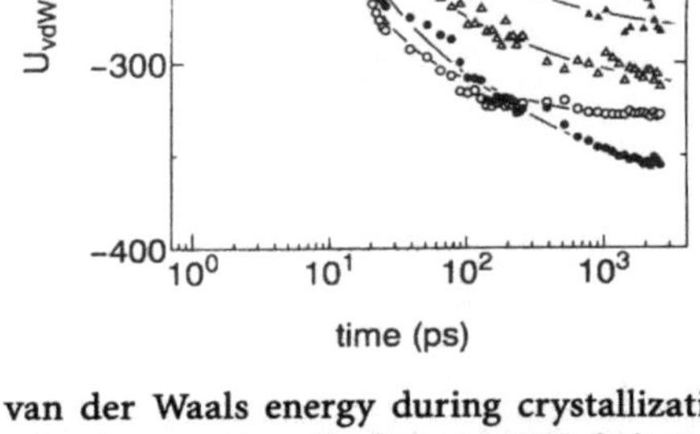

Fig. 5 Changes in the van der Waals energy during crystallization by quenching from 600 K to each crystallization temperature T_c: (▲) at 300 K, (△) at 200 K, (•) at 100 K, and (○) at 20 K. The energies are expressed in Kelvin/atom

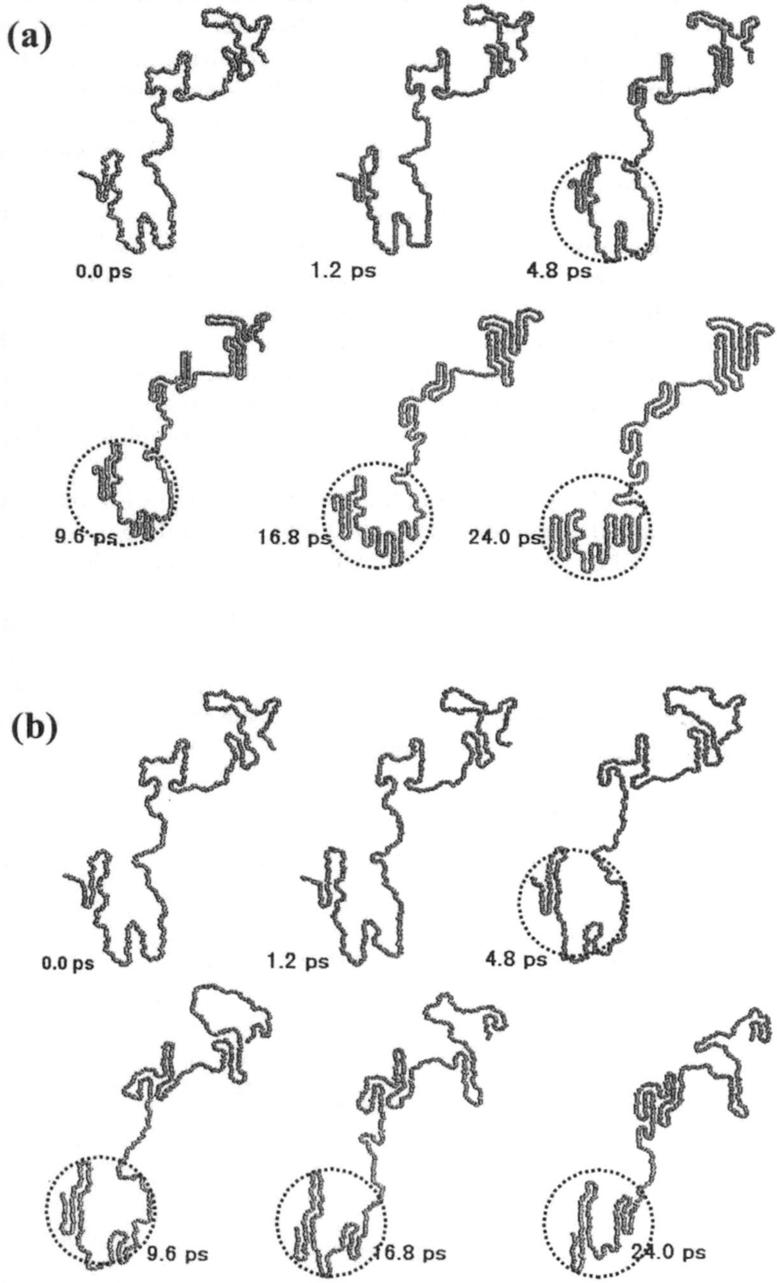

Fig. 6 Snapshots at indicated times during the initial to the intermediate stages of polymer crystallization **a** at 50 K and **b** at 300 K, and in the intermediate to the late stages

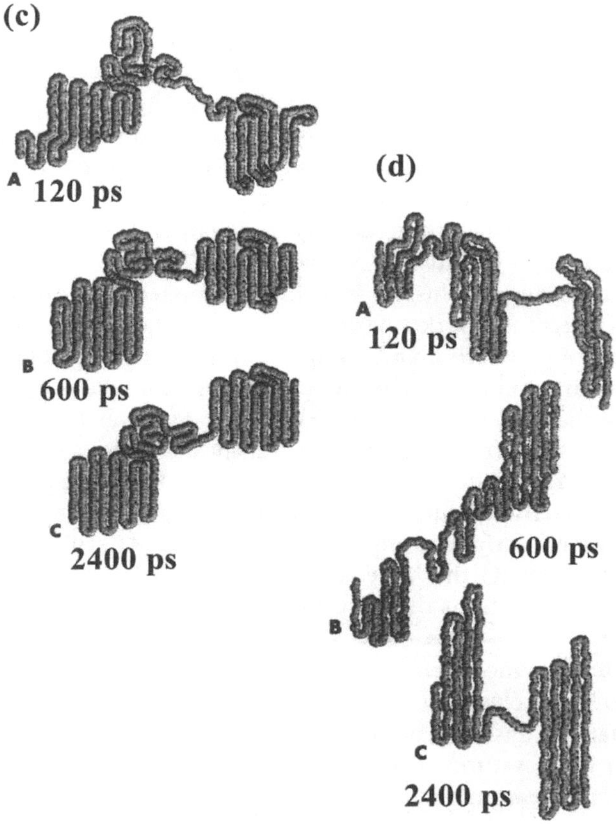

Fig. 6 c at 50 K, and **d** at 300 K

According to the conventional theory of Lauritzen–Hoffman, the very onset of ordering on the growth surface, the secondary nucleation, is the deposition of a single straight stem, which triggers the subsequent chain folding onto the stem. However, Fig. 6 suggests that the single adsorbed stem is not always stable and it is often destroyed or desorbed from the potential valley. On the contrary, the pair of stems that look like hairpins seem to be much more stable and leads to further growth into a cluster. Systematic investigations by Doye and Frenkel showed that the first stem adsorption does not give a free-energy barrier but the stem pair could be a nucleation barrier [29], the original concept suggested by Point [11].

A comparison of Figs. 6a and 6b also shows that the stem lengths of stable hairpins or clusters are appreciably smaller at 50 K than those at 300 K. The clusters of short stems are unfavorable at higher temperatures, and the stems lengthen considerably in order to make stable clusters. The stability of longer stems at higher temperatures correspond to the traditional Lauritzen–Hoffman picture of polymer crystallization, where longer chain stems are necessary at higher temperatures in order to compensate the excess energy at the folds.

In the intermediate stage from a few ps to several tens of ps, the energy of the system decreases greatly; the energy decreases come from the gathering and coalescing of small clusters into larger ones (Figs. 6c,d). Such large-scale motions of clusters are only facilitated in relatively small-sized clusters. During the intermediate stage, there emerge neat lamellae composed of parallel chains of similar stem lengths.

The local ordering into clusters is nearly completed in about a hundred ps, and the energy decrease markedly slows down. The late stage, after a hundred ps, is a process of generation and completion of a single lamella through the coalescence of larger clusters. The completion of a lamella is a very slow process requiring the coalescence of large clusters. At lower temperatures (20 K and 50 K) the system froze into an apparently unstable state that consisted of separate clusters and no further aggregation was observed even after 6 ns. At slightly higher temperatures (100 K and 200 K), the formation of a single lamella was rather fast and completed within 2 ns. At still higher temperatures (250 K and 300 K) the lamella completion again slowed down due to large structural fluctuations during crystallization; the completion of the single lamella took about 10 ns. It is very suggestive that at high temperature (300 K) the clusters once formed are often destroyed in the course of reconstruction into larger clusters (Fig. 6d). This is evidently the origin of the large energy fluctuations observed at higher temperatures in Fig. 5. The crystallization at higher temperatures is not a simple downhill process of energy but it accompanies the creation and annihilation of local order.

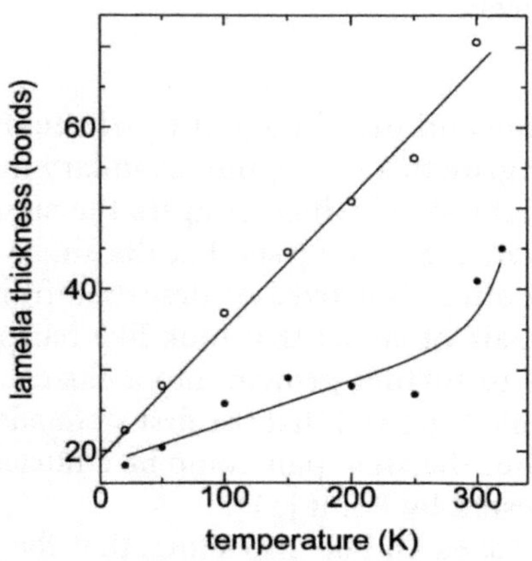

Fig. 7 Lamella thickness 2.5 ns after quench to each crystallization temperature (•), and that after annealing 6.4 ns (○). At 250 K and 300 K, the lamella is still thickening even after 6.4 ns by an appreciable rate

Another pronounced feature is considerable lamella thickening during the intermediate and the late stages. The thickening was faster and persisted longer at higher temperatures. Such marked thickening may not be a general phenomenon but only observed in systems with large molecular mobility. We expect, however, that the polymer molecules at the growth surfaces are highly mobile as suggested by our previous MC simulation [16, 17] and described later in this chapter. If this is the case, the polymer crystallization will generally involve such lamella thickening at the growth surfaces.

At each crystallization temperature T_c, well-defined lamella structure developed within a few ns. We found that the thickness of the lamella depends sensitively on T_c. The lamella thickness (the average stem length) observed at 2.56 ns after quenching is plotted vs. T_c (Fig. 7). The lamella thickness is shown to increase steeply near the estimated melting temperature $T_m = 400$ K. Such behavior resembles the observed dependence of the lamella thickness ℓ on undercooling $\Delta T = T_m - T_c$. The lamella shows further thickening even after completion of a single lamella, especially at higher temperatures, until it reaches an apparent limiting thickness.

3.3 Lamella Thickening

The thickening of two dimensional chain-folded clusters on the growth surface was shown to be an important molecular process in polymer crystallization. Besides the significance in polymer crystallization, the molecular processes of thickening are also very interesting since they are specific to polymers requiring highly cooperative diffusion along the chain contour. We made separate simulations of lamella thickening in regularly-folded lamella in 2D.

The lamella thickening depends sensitively on the initial lamella thickness as well as on the annealing temperature. We first considered the thickening of very thin lamella of about 18-bonds thick; this thickness nearly corresponds to that of the lamella crystallized at 0 K (Fig. 7). The temperature of annealing T_a was taken between 20 K and 150 K; a quick jump to a higher temperature resulted in partial melting and re-crystallization and a continuous thickening process could not be observed.

Between 20 K and 150 K, the lamella showed remarkable thickening within 14 ns. Figure 8a shows the changes in lamella thickness averaged over three runs. Initially the lamella thickens very rapidly but soon levels off around 5 ns. Such saturation of the thickening has not been observed in real 3D crystals of polymers; the thickening is generally very slow and is believed to follow the log(t)-rule. We cannot draw a definite conclusion as to whether the lamella thickness approaches the limiting value or still increases very gradually. In Fig. 8, the data are shown to fit well to a formula obtained through a phenomenological argument, similar to that of Sanchez [39], which leads to a presence of the limiting thickness. Typical snapshots at 50 K are shown in Fig. 9. At the early stage of thickening, the lamella takes lens-like shape at lower T_a (20 K and 50 K), and both chain ends have marked preference to sit on the fold surfaces giving integer folds.

It is well acknowledged in real polymer crystals that the thickening is pronounced when the lamella is annealed at temperatures higher than T_c at which the lamella was grown. A thicker lamella, which is obtained at higher T_c, must be annealed at still higher temperatures in order to observe considerable thickening. We studied the thickening of a lamella of about 40-bonds thick; we should remember that our model system has given the lamella a thickness of about 40 bonds when crystallized around 100 K, $T_c = 100$ K (Fig. 7). The thick lamella behaved quite differently. Below 100 K, the lamella showed no appreciable thickening, while it thickened considerably above 100 K (Fig. 8b). The presence of such a temperature threshold for the onset of lamella thickening corresponds to the experimental result mentioned above. At 150 K, the lamella thickness approached a limiting value of 45 bonds. We remember that the thin lamella has also thickened to a similar limiting thickness of about 46 bonds when annealed at 150 K. Thus, our model lamella seems to attain the same limiting thickness irrespective of the initial thickness.

Rapid lamella thickening was observed clearly in such a highly mobile phase of polymer crystal. The elementary process of thickening is diffusive translation of the molecule along the chain. It is a highly cooperative motion of the whole chain. What is a fundamental driving force for the lamella thickening? In the phenomenological theory, the thickening is considered as driven by the decrease in the total surface free energy, the decrease in the fold

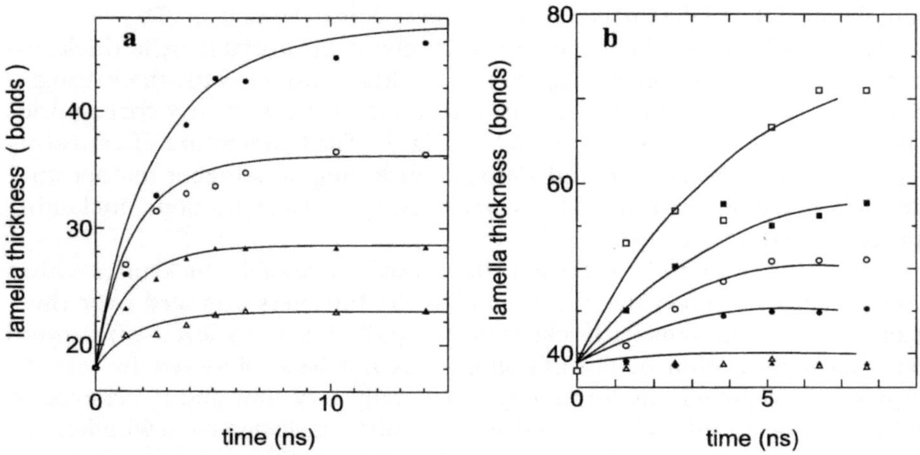

Fig. 8 Increases in the average lamella thickness by annealing **a** a thin lamella, at (△) 20 K, (▲)50 K, (○) 100 K, and (•)150 K, and **b** a thick lamella at (▲) 20 K, (△) 100 K, (•) 150 K, (○) 200 K, (■) 250 K, and (□) 300 K. The average was taken over three runs with different initial velocity distribution

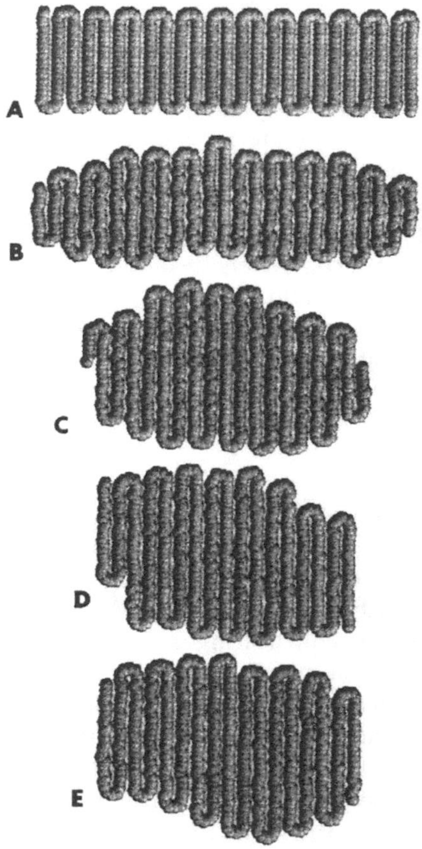

Fig. 9 Typical molecular process of thickening in the thin lamella at 50 K: **A** the initial lamella, **B** after 0.13 ns, **C** after 2.56 ns, **D** after 5.12 ns, and **E** after 10.24 ns. The thickness of the initial lamella is about 18 bonds

surface energy at the expense of the energy at the side surfaces. We calculated the decreases in the van der Waals energy, Eq. 2, and the substrate energy, Eq. 4, both averaged over 3.2 ps at each stage of the annealing; the former term reflects the changes in the long-range attractive potential due mainly to the overall changes of lamella shape, while the latter comes from the decreases in the number of folds where excess substrate energy is accumulated in surmounting hills between valleys. We found that both the energy contributions show large decreases by lamella thickening, but the van der Waals energy decreases considerably more than the substrate energy; the lamella thickening of our model system is mainly driven by the van der Waals attraction between atoms.

4 Three-Dimensional Crystallization of a Single Chain from Vapor

Real polymer processes involved in polymer crystallization are those at the crystal-melt or crystal-solution interfaces and inevitably 3D in nature. Before attacking our final target, the simulation of polymer crystallization from the melt, we studied crystallization of a single chain in a vacuum: adsorption and folding at the growth front. The polymer molecule we considered was the same as described above: a completely flexible chain composed of 500 or 1000 CH_2 beads. We consider crystallization in a vacuum or in an extremely poor solvent condition. Here we took the detailed interaction between the chain molecule and the substrate atoms through Eqs. 8–10.

4.1 Melting Point of the System

Here again, the estimation of the melting point of the present model is a prerequisite for discussions. We first studied the melting of a patch of chain-folded lamella on the growth surface by heating at a constant rate of 1.2 K/ps. Like the argument described earlier, a precise melting point is blurred by the lamella thickening during slow heating or by an appreciable superheat during rapid heating, so we estimated T_m only roughly. It was noticed that the chain adsorption was so strong that the chain did not easily detach from the surface within the present temperature range. Changes in chain conformation and energy clearly indicated melting of the lamella around 300 K. The melting point T_m of the present system is considerably lower than the above 2D case; since the chain flexibility is the same, the difference in T_m must be due to the smaller energy barrier 2U1 for the lateral translation of the chain.

4.2 Chain Conformation before Adsorption

The chain conformation or the degree of chain expansion before adsorption will have significant effects on the molecular mode of crystallization. In the case of crystallization of a single chain, a more expanded conformation has less surface segment density and a lower degree of chain entanglement. Less segment density necessitates longer range diffusion making crystallization more sluggish while less entanglement facilitates the formation of an ordered lamella structure. An example we have is the crystallization in 2D-space investigated above; here the chain was well expanded with no chain entanglement at all and it crystallized into a rather neat chain-folded lamella with predominantly adjacent reentry folds. In the solution or melt, the van der Waals attractions are screened by the intervening media, and the molecule

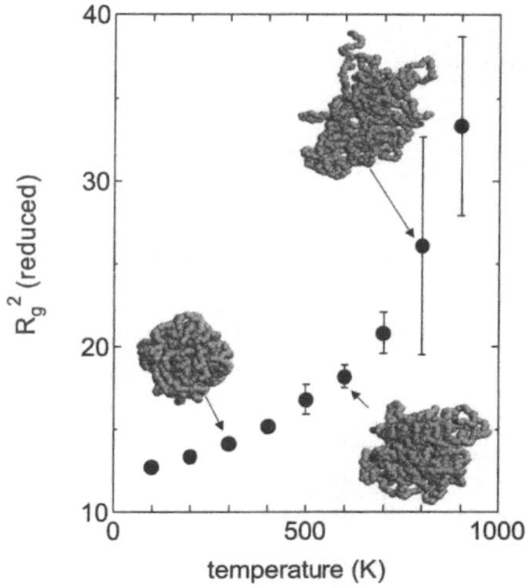

Fig. 10 Equilibrium radius of gyration of a molecule plotted as a function of temperature; the molecule is composed of 1000 beads. The radius of gyration shows a steep increase and a large fluctuation above 700 K. The insets show typical chain conformations at indicated temperatures. Note that the ideal random coil state of this fully flexible chain should have the mean-square radius of gyration $R_g^2 = 1000 \times (1.54/3.92)^2/6 = 25.7$, the value is around 800 K

has an expanded random-coil conformation. In the present model, however, the atoms are subjected to bare van der Waals interactions, and therefore the chain is expected to be collapsed. We simulated the conformation of the isolated chain in a vacuum at various temperatures. In Fig. 10, the equilibrium radius of gyration R_g^2 is plotted vs. temperature; the insets are typical chain conformations at given temperatures. An abrupt increase in R_g is readily seen around 700 K, indicating a transition from a compact globule to an expanded coil. The molecule therefore remains a globular strongly collapsed state within the temperature range of interest: $0 < T < 300$ K.

4.3
Adsorption and Ordering of the Globular Chain

When the globular molecule is placed near the crystal surface, the molecule begins to attach to the surface due to strong attraction, and at the same time the molecule is forced to settle into the potential valleys on the surface. Figure 11 shows the changes in the atomic distributions, which is plotted vs. distance z from the surface, with time and temperature. By progressive adsorption, the initial broad distribution changes to the distribution with distinct peaks. After a sufficiently long time, the polymer molecule forms a layer structure: multilayered at lower temperatures and monolayered at higher temperatures. The progress of adsorption can be described by use of the surface normal component of the radius of gyration;

$$\langle R_z^2 \rangle = \left\langle \frac{1}{N} \sum_j^N \left(z_j - z_g \right)^2 \right\rangle , \tag{11}$$

where z_j and z_g are the surface normal coordinates of the j-th atom and that of the center of mass of the molecule, respectively. At each temperature, the squared radius $\langle R_z^2 \rangle$ normal to the substrate was found to shorten quickly,

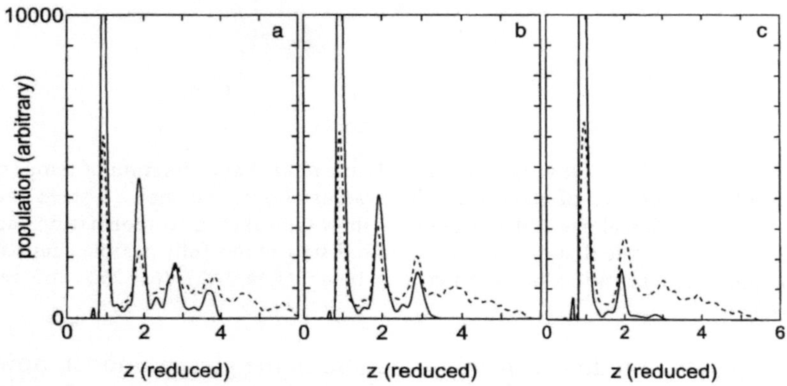

Fig. 11 The number density of chain atoms vs. distance z from the substrate surface, at the early stage of 12.8 ps (*dash*), and at the late stage of 1280 ps (*solid*). The data were obtained at **a** 50 K, **b** 100 K, and **c** 250 K. The layer structures are readily noticed at the late stage irrespective of temperature

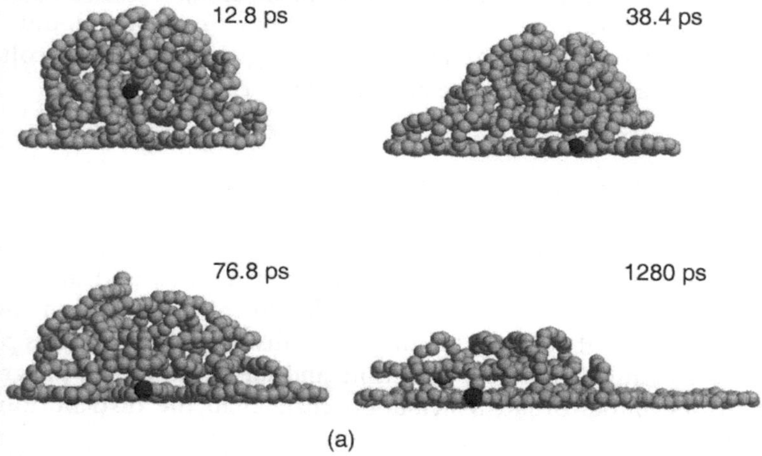

Fig. 12 Typical trajectory at 100 K, **a** side views showing adsorption during the initial stage of adsorption and at the late stage of 1280 ps, and

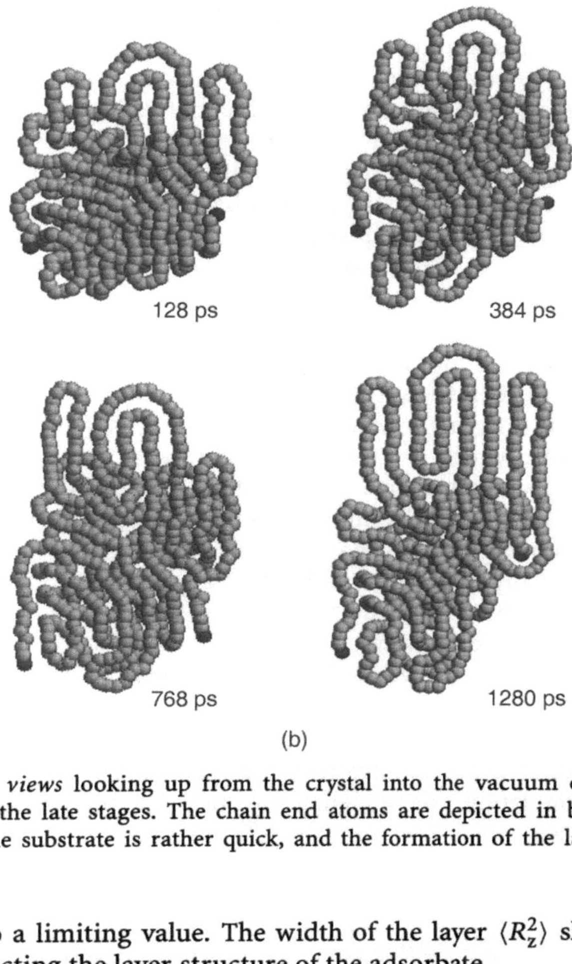

Fig. 12 b *bottom views* looking up from the crystal into the vacuum during from the intermediate to the late stages. The chain end atoms are depicted in black. The chain adsorption to the substrate is rather quick, and the formation of the layer structure is readily noticed

within 1 ns, to a limiting value. The width of the layer $\langle R_z^2 \rangle$ shows stepwise decreases reflecting the layer structure of the adsorbate.

By adsorption, the shape of the globular chain transforms from a hemisphere to a flat disk. Figure 12 shows typical initial trajectories during adsorption. When the globule comes to touch the substrate, the peripheral atoms begin to be adsorbed to form a layer. Since the molecule is strongly collapsed, the density of the adsorbed atom on the substrate is very high. Therefore, the chain ordering on the substrate does not proceed sequentially from the chain end as conceived in the traditional picture of polymer crystallization, but it is a highly cooperative process.

With progress of adsorption, the molecular order in the adsorbed layer gradually grows. The chain entanglements, however, hinder the development of order, and cause the persistent amorphous overlayers. At low temperatures, the molecule does not completely spread over the surface leaving a large amount of amorphous segments. At higher temperatures, on the contrary, the molecule spreads easily and forms an approximate monolayer. The chain entanglements are removed considerably during the initial spreading, and the formation of long loops across many potential-valleys is restricted by the presence of the crystalline field $U_1(z)$. All these are considered to be origins of the formation of a rather neat chain-folded lamella (Fig. 13).

In order to investigate the competition between adsorption and ordering, we define the adsorption-ratio p and the average stem-length l_{av}, where p is the number of atoms in the first adsorption layer divided by the total number of chain atoms, and l_{av} is defined as a weight-averaged length of straight-stems in the first adsorbed layer. We can think of two extreme molecular pictures of adsorption and ordering. One is that the chain is first adsorbed onto the substrate then followed by the ordering into a chain folded crystal, in which case the ordering process has essentially a two-dimensional character. The other one, which is a traditional picture of polymer crystallization, is that the polymer chain segment of constant length is deposited to the side edge (kink site) of the lamella; in this case the adsorption ratio increases with the stem length being kept constant. We monitored p and l_{av} during the initial process of adsorption, and found that at any temperature the increases in p and l_{av} keep pace with each other(Fig. 14). The adsorption and the ordering are thus found to be cooperative.

We also found that the average stem length after a sufficiently long time, which corresponds to the thickness of the adsorbed lamella, shows a pronounced dependence on crystallization temperature (Fig. 15). Though the detailed molecular processes involved are quite different from those in our previous 2D simulation, the tendency for thicker lamella at higher T_c is here again reproduced.

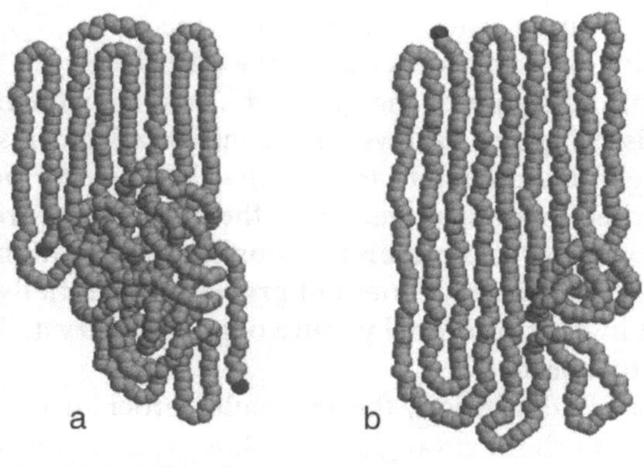

Fig. 13 Asymptotic chain conformations (*top views*) obtained at 2.56 ns: **a** at 100 K and **b** at 250 K. On the ordered layer remain entangled amorphous segments. The amorphous segments are richer at lower temperatures

When we simulate crystallization at some T_c, we have to know the substrate thickness (lamella thickness) at that T_c, while the lamella thickness (the average stem-length) depends on T_c and should be determined from the crystallization process itself; the substrate thickness and the resulting average stem-length must be self-consistent. In the present calculation, however, we consider the crystallization on the substrate of arbitrary chosen thickness that corresponds to 30 bonds. Therefore, the argument is not self-consistent. Below we discuss the effect of a finite substrate on the molecular process of chain folding. Special interests are the resulting chain conformation in the amorphous phase and the fate of chain entanglements.

Since the interaction energy $U_{\mathrm{sub}}^{\mathrm{3D}}$ with the finite substrate is not easy to handle due to the loss of two dimensional periodicity assumed in the derivation of Eq. 5, we simplify the substrate interaction. We considered that the substrate was a sandwich of the amorphous and the crystalline layers, and the attractive potential $U_0(z)$ works at any point (x, y), while the translational barrier $U_1(z)\cos(2\pi x/\lambda)$ only works on the crystalline substrate. The implicit assumption is that the atomic densities of the crystal and the amorphous are not so different.

Figure 16 shows early crystallization at 250 K. The molecule is adsorbed uniformly, irrespective of whether on the crystal substrate or on the amorph-

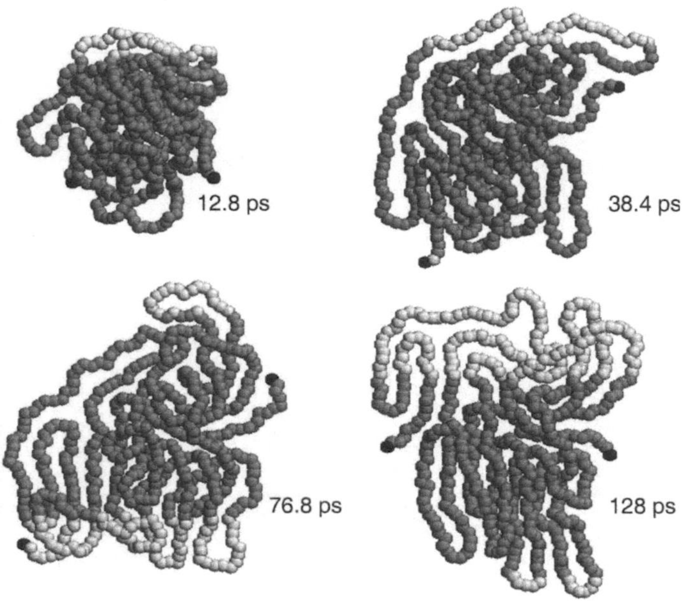

Fig. 16 Molecular trajectory in the early stage of chain adsorption onto the thin substrate. The chain atoms on the crystalline substrate and the amorphous substrate are depicted in *dark gray* and in *white*, respectively. The presence of long loops in the amorphous region is *quite* pronounced

ous substrate, caused by the common attractive potential $U_0(z)$. We should remember that the distinction of the amorphous substrate from the crystalline one is, in the present model, only in the absence of the surface ordering potential $U_1(z)$. Similar to the previous case mentioned, adsorbed segments quickly align along the y-axis to form parallel bundles of stems. At lower temperatures, the development of order is restricted and considerable amorphous overlayer remains, while at higher temperatures large molecular mobility enables the chain to spread over the substrate very quickly. Due to limited thickness of the crystalline substrate, the chain ordering grows preferentially to the lateral direction. Contrary to the usual picture of crystallization and mainly due to the adsorption from vapor, the lateral growth of the crystallite proceeds not by a sequential addition of the segments to the side edges of the grown lamella, but the chain can even squeeze into the ordered parallel stems of the grown lamella.

In contrast to the case of the infinite crystalline substrate, a presence of long loops in the fold surface is readily noticed. Since there is no in-plane potential $U_1(z)$ in the amorphous regions, chain segments are allowed to make a long traverse in the x-axis direction during the initial adsorption, and such chain segments give rise to long loops in the final lamella crystal. Further-

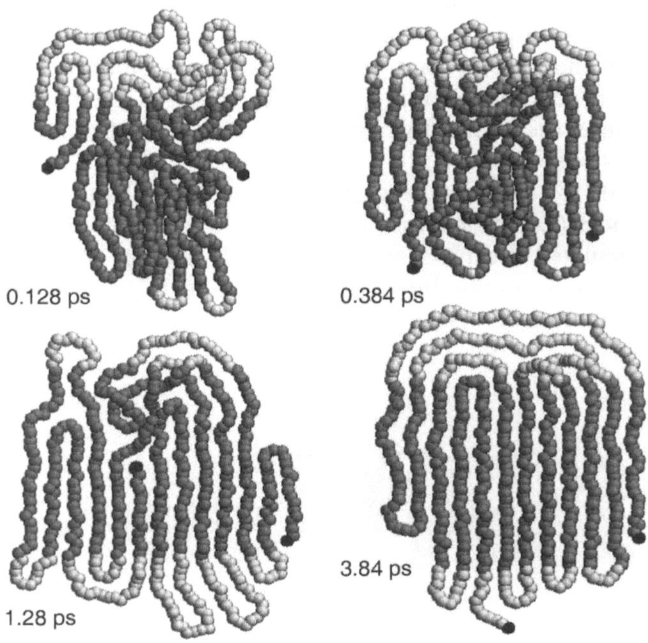

Fig. 17 Molecular trajectory in the late stage of chain adsorption. The chain atoms on the thin crystalline substrate and the amorphous substrate are depicted in *dark gray* and in *white*, respectively. The chain entanglements are pushed out of the crystalline substrate

more, the residual chain entanglements in the amorphous overlayer find easy paths toward the amorphous regions creating rich entangled loops in the amorphous (Fig. 17).

5 Polymer Crystallization from the Melt

Through the studies described above, we are now at the last stage and can attack our final target, the simulation of polymer crystallization from a dense melt. Crystallization from the melt is a much more complex process. The systems are dense and inevitably we have to treat much larger systems. We considered that the polymer melt was sandwiched between two parallel substrates, and we adopted free-boundary conditions in the x- and y-axis directions in order to reduce computational load and to avoid interactions of the chains with their periodic images (Fig. 18). The sizes L_x, L_y, and L_z of the present system were all the same and either 15σ or 30σ for systems of 8000 atoms and 64 000 atoms, respectively. The average density of the initial melt was about 0.86 [g/cm^3]. The interactions with the substrate were calculated by Steele's method taking only the lowest order term of the potential. In the ideal melt state, chains should have an extension of about the radius of gyration R_g, the mean-square of which should be proportional to the number of atoms in the chain. If we want to treat a very long chain, we must prepare a very large MD cell to accommodate bigger chains without serious boundary effects. Therefore, we mainly consider relatively short chains (C_{100}) in the present discussions.

The polymer we consider here is a semi-flexible chain which has some bending stiffness (Eq. 3). We first estimated the chain conformation in the melt. The calculated mean-square end-to-end distance R_n^2 between atoms n-bond apart has shown that the chains have an ideal Gaussian conformation; R_n^2 is a linear function of n (see Fig. 35 given later). The value of R_n^2 for $n = 100$

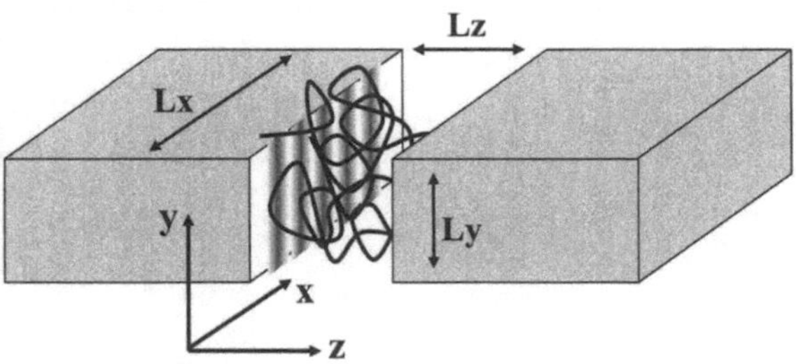

Fig. 18 Schematic picture of the system for simulating polymer crystallization from the dense melt. Polymer chains that should be crystallized are sandwiched between parallel side surfaces of the lamellae made of the same polymer chains. The z-axis is taken normal to the substrate, while the y-axis is along the chain direction of the substrate crystals

was about $120\sigma^2$, which means the characteristic ratio C = 7.8. This is in good agreement with experimental values of C around 6.7 in polyethylene [37]. By holding the system at 600 K, well above the melting point ($T_m \simeq 390$ K) for about 1 ns, we generate the initial isotropic melt. The melt is then quenched to various crystallization temperatures ranging from 250 K to 400 K, and we examine the crystallization process up to about 20–40 ns at each temperature.

5.1 Layering

The structures of the solid-melt interface and the melt confined within a narrow gap are of great significance in diverse areas of research such as lubrication, adhesion, or in future nanometer science. It is well recognized that the melt of *n*-alkanes, and other simple molecules show anomalous oscillations in density, viscosity, etc. vs. depth from the surface showing the presence of marked layer structures in the melt [40]. Even in polymer melts similar layering phenomena were suggested near the solid surface [41], but no pronounced ordering or the onset of crystallization were reported.

We first examined the density distribution near the solid-melt interface vs. depth from the surface. Figure 19 shows typical density profiles in a relatively small system of 8000 atoms (80 chains of C_{100}). It is readily noticed that even at 500 K marked density oscillation is present near the solid surface, though

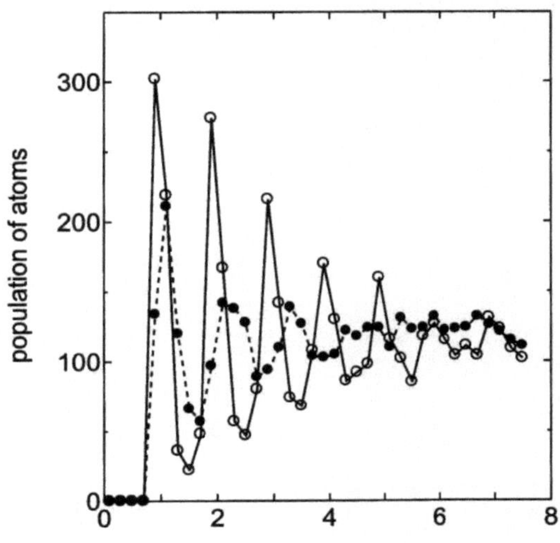

Fig. 19 Profiles of the atomic densities averaged in the *x*-*y* plane, at 500 K (•) and at 300 K (○), both after 1.28 ns. Even in the melt at 500 K, marked layer structure in the density distribution is quite evident near the solid-melt interfaces. The increase in the density oscillation at 300 K is an indication of the onset of crystallization

the oscillation readily dampens showing only a few layers near the surface. The interlayer spacing is approximately σ, which is the characteristic length of the van der Waals interactions. The attractive potential of the substrate is so deep (of the order $10\varepsilon = 720$ K) that the molecules near the substrate are forced to align in layers.

With decreasing temperature, the density oscillation becomes very pronounced and grows into a deeper melt region. At 300 K, for example, we can see at least 5 layers after 1.28 ns. Within the layers, as will be shown later, definite order in chain orientation and chain packing is observed suggesting the growth of polymer crystals.

5.2 Chain Order Within Each Layer

Successive layers were formed near the solid-melt interface when we lowered the temperature. Figure 20 shows typical snapshots of the molecular arrange-

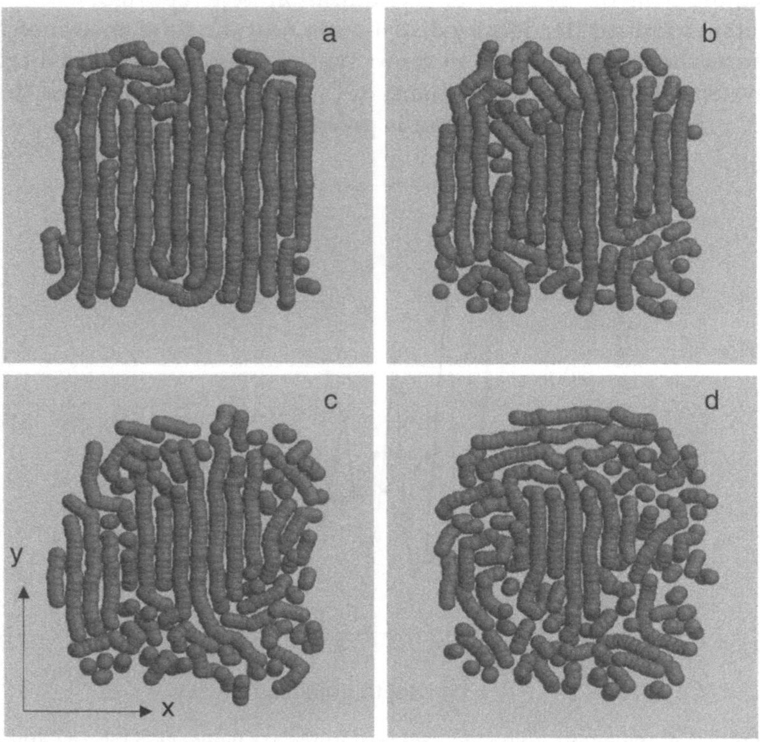

Fig. 20 Pictures of atoms that belong to **a** the first, **b** the second, **c** the third, and **d** the fourth **e** layers, which show the in-plane ordering of the chains in each layer at 300 K after 1.28 ns. Especially marked is an ordering in the first layer nearest to the substrate

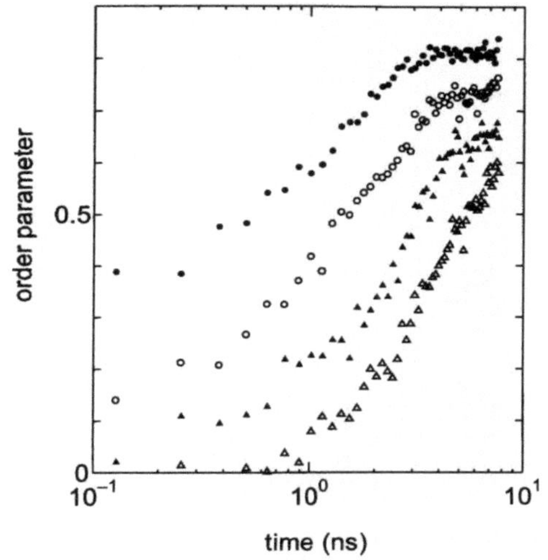

Fig. 21 Time evolution of the orientational order at 350 K, in the first (•), the second (○), the third (▲), and the fourth (△), layers. The ordering is faster in layers nearer to the substrate

ment in the layers after 1.28 ns of simulation at 300 K in the same system as Fig. 19 (8000 atom system). We can readily notice that the chains have a marked tendency to lie parallel to the *y*-axis, the chain direction of the substrate. The orientational order is the highest in the first layer and is gradually lost in deeper layers until a fully disordered melt structure is recovered. The development of chain orientation within each layer can be described in terms of the order parameter given as follows,

$$P_l = \left\langle (3\cos^2\theta - 1)/2 \right\rangle_l , \tag{12}$$

where P_l is the orientational order within the *l*-th layer, θ is the angle between the C – C bond and the *y*-axis, and the average is taken over all bonds that belong to the *l*-th layer. The time evolution of the order parameter in several layers at 350 K is shown in Fig. 21. The first layer rapidly attains the equilibrium value of about 0.8 within 3 ns, while at deeper layers the ordering is much slower but seems to be aiming at similar asymptotic values.

5.3 Stationary Growth of the Chain Folded Lamellae

The development of order in each layer is actually the growth of crystalline lamellae. We show in Fig. 22 a typical snapshot of the system of 640 chains of C_{100} (64 000 atoms) at 12.8 ns after quenching from 600 K down to 350 K. It is clearly noticed that the stacked lamellae grow from both side substrates. For further analysis of the data we need to extract crystalline regions. In order to

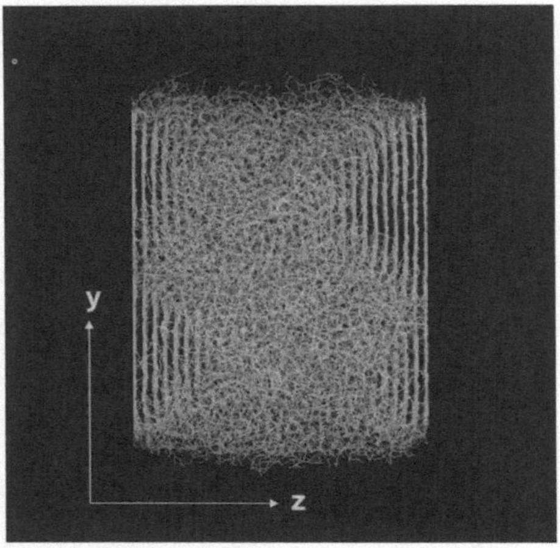

Fig. 22 A representative snapshot, obtained at 12.8 ns after quenching to 350 K, of the system made of 640 chains of C_{100}. We readily notice crystalline domains growing toward each other from both substrates

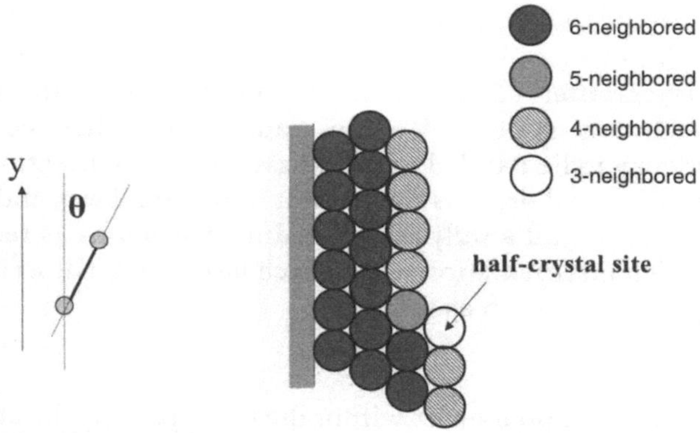

Fig. 23 The bonds that constitute crystalline domains must lie nearly parallel to the y-axis with an angle θ of less than 20°. Furthermore, the bonds must have at least three neighbors that satisfy $0.7\sigma < \sqrt{r_x^2 + r_z^2} < 1.3\sigma$ and $|r_y| < r_0/2$. Note that the crystalline stems deep inside the crystal (*black spheres*) have six neighbors, while those on the free surfaces (*hatched spheres*) have four neighbors. The stems at the half-crystal site, or at the kink site, (*white sphere*) have three neighbors. Stems attached on the free surface, and those floating in the melt phase have less than three neighbors

demarcate the crystalline regions, we define the crystal domains as follows. The crystal is made of well-aligned and closely-packed C – C bonds which are nearly parallel to the y-axis with angles θ less than 20° (Fig. 24), and in addition the crystalline stems must have at least 3 neighbors (Fig. 23). We consider that the bonds are neighboring when the vector $\boldsymbol{r}$ connecting the centers of the bonds satisfies $0.7\sigma < \sqrt{r_x^2 + r_z^2} < 1.3\sigma$ and $|r_y| < r_0/2$.

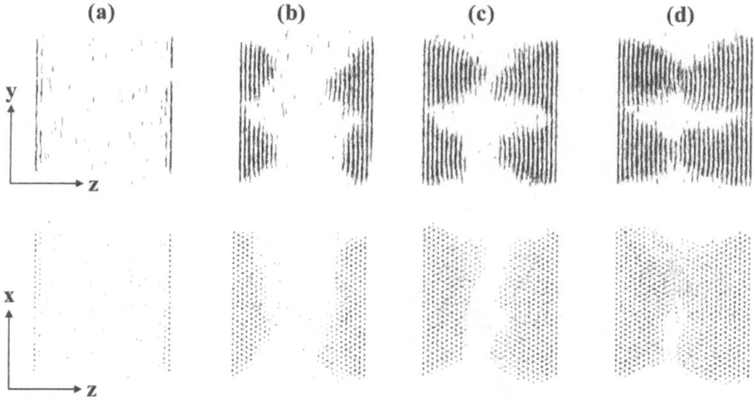

Fig. 24 Growth of lamellae at 330 K, viewed along the x-axis (*above*), and along the y-axis (*below*); **a** at 0.128 ns, **b** at 6.4 ns, **c** at 12.8 ns, and **d** at 19.2 ns. We notice the tapered growth fronts and their advances in the normal (the z-axis) direction together with the lamella thickening along the chain axis (the y-axis direction)

A typical growth process at 330 K of the crystal domains defined in this way is shown in Fig. 24; the lower pictures are viewed along the y-axis and show the hexagonal packing of the chains, while the upper pictures viewed along the x-axis clearly demonstrate the growth of stacked lamellae. All lamellae have a marked tapered shape and show thickening growth along the chain axis as well as the usual growth perpendicular to it. The lamellae come to collide with those grown from the opposite substrate around 10 ns and merge into larger lamellae around 20 ns.

The overall rate of crystallization is evaluated through crystallinity. The crystallinity χ is estimated as the fraction of atoms that belong to the crystal domains. Figure 25 shows the increases in χ at various crystallization temperatures. At each temperature, the crystallinity χ initially increases up to around 10%; this fraction is not a consequence of crystallization but comes from the surface layering near the substrate. After the initial increases, the crystallinity shows an approximately linear increase due to the usual crystal growth, until it begins to slow down around 10 ns. The growth rate saturation may be due to lamella collisions but can also be due to other factors yet unknown. The crystallization rate depends on T_c very sensitively. The crystallization is very slow at higher temperatures around T_m; it is actually too slow to be detected around $T_c = 370-390$ K. The crystallization rate increases markedly with decreasing temperature, and it shows an apparent maximum around $T_c = 330-320$ K. At lower temperatures below 320 K the crystallization rate again decreased, which was considered to be an indication of restricted chain mobility at lower temperatures [22].

Though the overall crystallinity showed a monotonic increase until around 330 K with lowering temperature, the detailed growth rate of each lamella

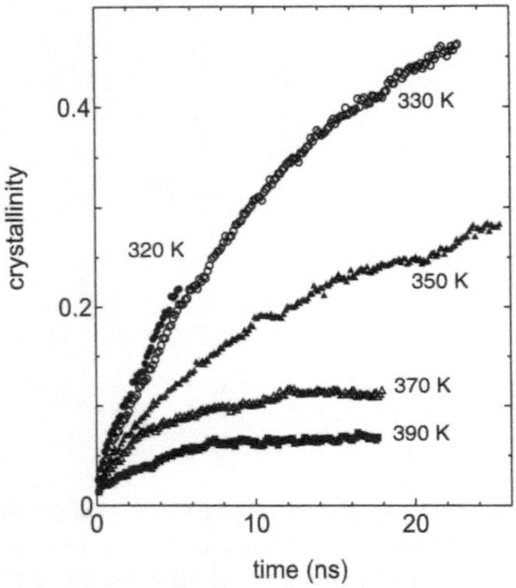

Fig. 25 Increases in the overall crystallinity with time, at 390 K (■), at 370 K (△), at 350 K (▲), at 330 K (○), and at 320 K (•). The crystallinity is defined as the fraction of chain atoms that belong to the ordered domains

shows quite large fluctuation. Figure 26 shows the changes in the sizes of each lamella, the number of atoms within each lamella, (a) at 330 K, and (b) at 350 K. It is clearly seen that the lamellae do not always grow keeping pace with each other. The irregular growth rate is especially marked at higher temperatures (Fig. 26b), where the four lamellae grow in quite distinct rates.

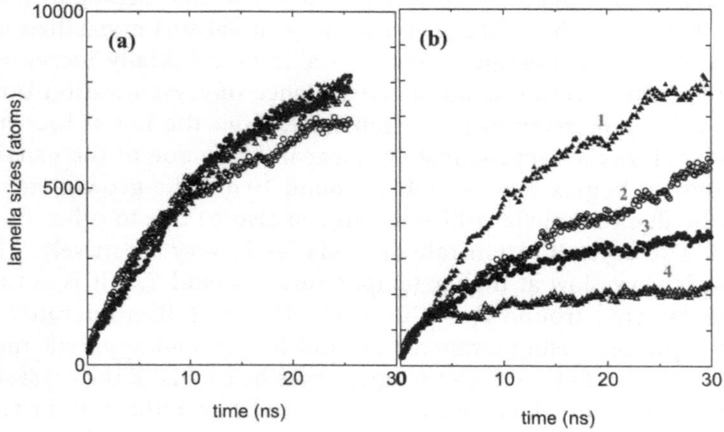

Fig. 26 Growth of each lamella, the number of atoms in the lamella, vs. time, **a** at 330 K and **b** at 350 K. Please notice that the lamellae show quite different growth rates especially at higher temperatures

These anomalous growth rates may be simply due to statistical fluctuations in small systems; lower probability of surface nucleation at higher temperatures will give larger statistical fluctuations. But similar incoherent lamella growth was also observed experimentally in a recent AFM investigation of the shish-kebab structure [42]. We suspect some unknown but very interesting mechanisms are operative, for example chain entanglements in the melt, stress concentration at the fold surfaces, etc.

5.4 Structures of the Growth Surfaces and the Fold Surfaces

Structures of the interfaces between the crystal domain and the melt are very interesting and in fact are very important in considering polymer crystallization. The structure of the lateral growth surface, whether it is rough or smooth, is of prime importance in constructing the molecular theory of crystal growth. Since the crystal-melt interfaces are extremely difficult to investigate by experiments, there has been no definitive information. We will now examine the structure of the growth surfaces and the molecular process of crystal stem deposition onto it.

Figure 27 shows the instantaneous hexagonal crystalline stems at 330 K after 6.4 ns, viewed along the chain axis (gray spheres). The growth surfaces are relatively smooth, though not that wide, and correspond to the $\{100\}$ planes of the hexagonal lattice. The stepwise addition of the crystalline stems onto the growth surface, just like the Kossel mechanism in the crystal growth of simple molecules, is thus suggested. We can verify the idea by monitoring the stem adsorption processes. Also shown in Fig. 27 are the positions of crystalline stems that are newly added to the surface during the next 0.128 ns (black spheres). New crystalline stems are preferentially added on the smooth surfaces near the kinks. Detailed stem addition processes over a shorter time scale (0.0128 ns intervals) are given in Fig. 28. It is clearly seen that the crystalline stems tend to attach at kink sites, but frequent attachments and detachments on the growth surfaces are observed resulting in average advance of the kink positions; only after time average can we observe preferential addition and in-plane growth from the kink sites. Thus, surface nucleation and growth mechanisms are strongly suggested, though at present we could not catch the surface nucleation process in the act. The surface is very rough, which might correspond to the multiple nucleation regime [9].

The structure of the fold surface has long been a most controversial topic (ever since the finding of the chain folded lamellae). What kind of fold structure will the direct MD simulation predict? Figure 29 shows: (a) the crystalline domains; and (b) the fold loops at 330 K after a sufficiently long time period of 38.4 ns; the crystallinity reaches about 52%. We noticed that most of the fold loops near the substrate are rather short. The presence of looser and longer loops and abundant cilium in the middle of the MD cell is obviously

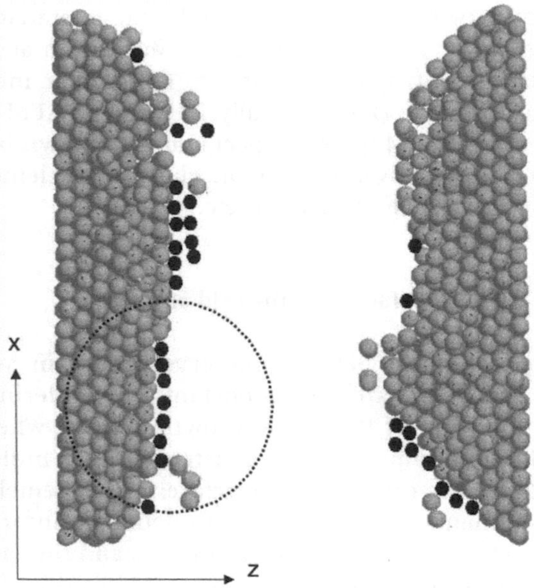

Fig. 27 Shape of crystalline domains at 330 K after 6.4 ns (*grey spheres*). The crystalline lamellae are found to have rather flat {100} surfaces. Also shown are newly added stems (*black spheres*) during the next 0.128 ns of simulation. The addition of the stems starts preferentially at kink sites

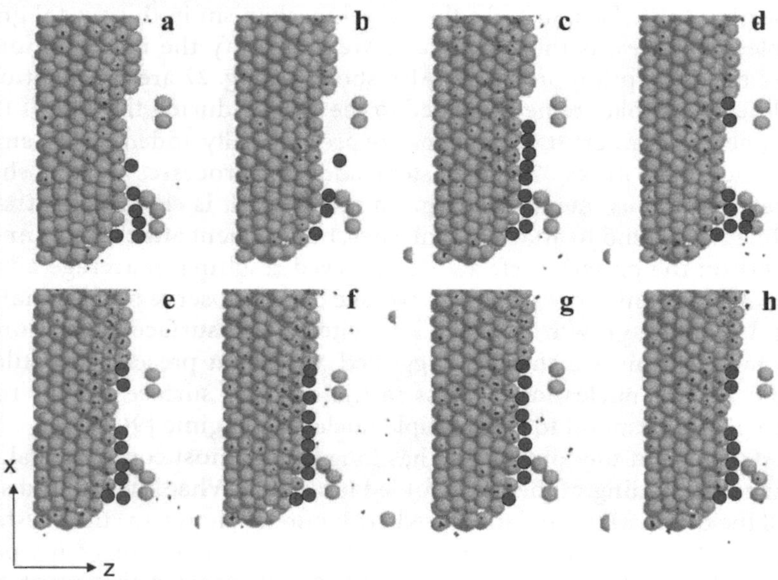

Fig. 28 Detailed step propagations at 330 K, from 6.4 ns in every 12.8 ps. Frequent attachment and detachment of the crystalline stems (*black circles*) are evident

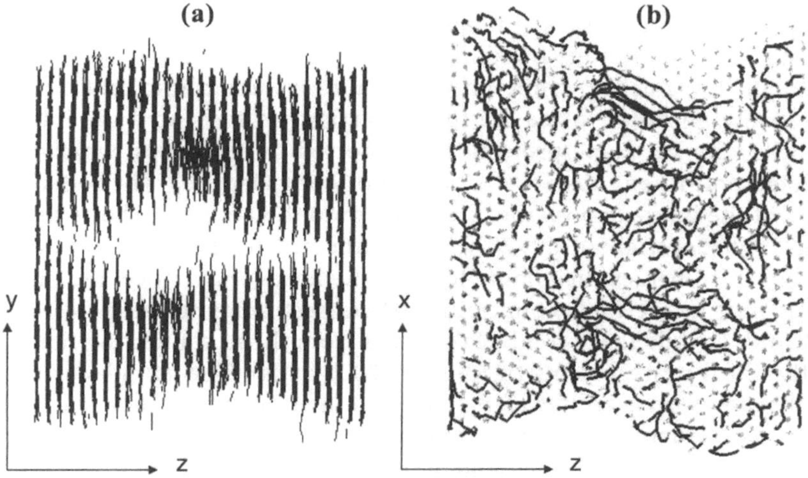

Fig. 29 **a** Typical crystalline domains at 330 K, and **b** polymer segments forming fold loops after a sufficiently long period of simulation of 38.4 ns

because the crystallization is not completed and many amorphous segments still remain to be crystallized near the center of the MD cell. Though the crystallization is still incomplete even after 38.4 ns, we analyzed the statistical distribution of the fold loops. The crystallinity is 52%, and the remaining

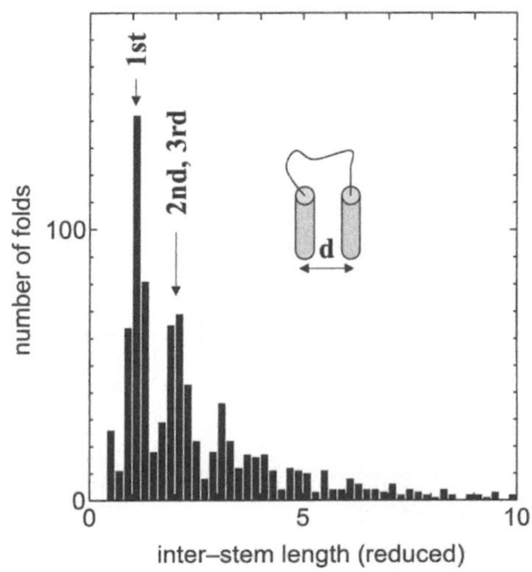

Fig. 30 Radial distribution of inter-stem vectors connecting stems linked by loops. The folds connecting stems separated by the ⟨100⟩, ⟨110⟩ (the nearest neighbors), and ⟨210⟩, ⟨200⟩ (the second and the third nearest neighbors) in the hexagonal lattice vectors are quite dominant

amorphous region consists of loops, cilium, or floating chains. Out of the remaining 48% amorphous fraction, about 27% are loops (whose chain ends are caught by lamellae) while only 17% and 4% are cilium and tie chains, respectively. In order to see the fold statistics, we measured the inter-stem distances (the x-z plane projection) connected by the loops, which are the projected end-to-end distances of the fold loops (Fig. 30). The radial distribution of the inter-stem vectors projected onto the x-z plane shows that most of the fold loops, about 60–70%, are connected to the nearest or the second and third nearest neighbor stems of the hexagonal lattice; the first peak corresponds to the folds connecting the nearest neighbor stems (⟨100⟩ and ⟨110⟩ folds), while the second peak comes from those connecting the second and the third nearest stems (⟨210⟩ and ⟨200⟩ folds). The abundance of short folds becomes even greater after longer simulation times. Detailed statistics of the fold structure are a very important subject of research, and indeed equilibrium Monte Carlo investigations have been proven to give important information [43]. Since our present lamellae are still growing with many melt chains waiting to be crystallized, full statistical analyses of the fold structure will remain a goal of our future projects.

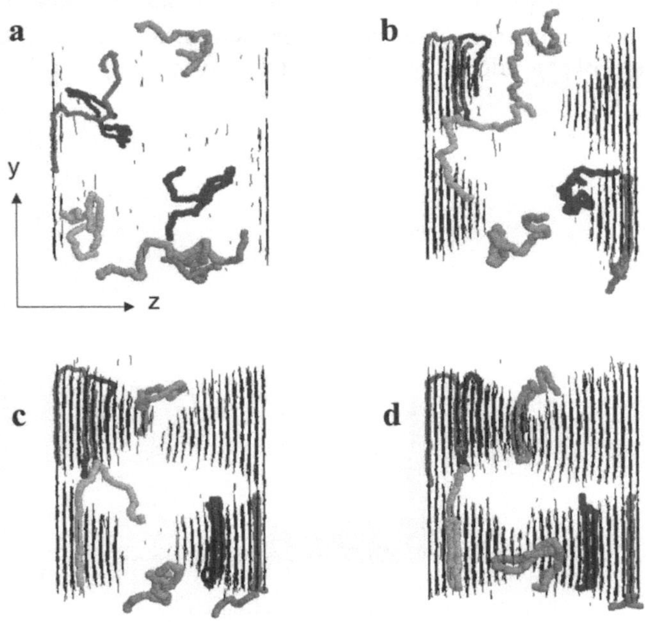

Fig. 31 Trajectories of seven crystallizing chains (*thick lines*) selected at random from 640 chains during crystallization at 330 K; **a** at 0.128 ns, **b** at 6.4 ns, **c** at 12.8 ns, and **d** at 19.2 ns. Also shown are the growing crystalline domains (*thin parallel lines*). Pictures are all side view along the x-axis

5.5
Chain Trajectories at the Growth Front

The molecular mechanism of the chains adsorbing to the growth front and folding to create a new crystalline layer is a central topic in polymer crystallization. We will monitor the trajectories of seven chains chosen at random. Figure 31 shows the crystallization processes of the seven chains (thick lines) at 330 K, superposed on the growing crystalline domains (thin parallel lines). The chains undergo active diffusion within the melt phase, though the center of mass of the chains does not move much, until they come into contact with the growth fronts of the lamellae. Also seen are the chains that once adsorbed to the growth fronts but soon detached into the melt phase region (not shown here). The crystallizing chains attach to the growth surface of the lamellae not necessarily from the chain ends but often in the middle part of the chains. Then the lengths of the adsorbed stems increase until they come to attain the final length, that is the lamella thickness at the growth front. Some chains form rather compact adjacent reentry folds, while others have long loops.

5.6
Melting of Lamellae and their Equilibrium Shape

Marked tapered shapes were noticed in the growing lamellae. Similar tapered edges, though of much smaller angles, are observed in the case of crystallization into highly mobile phases such as the high-pressure phase of polyethylene. The growth of very thin polymer lamellae is usually governed by kinetic factors. Therefore, we initially considered the tapered shape was also a kinetic form and examined how the tapered edges change their shapes during melting or when being held near the melting temperature. The melting of the lamellae at 380 K by quickly raising the temperature from 330 K is shown in Fig. 32. Quite unexpectedly, the lamellae showed a reverse process,

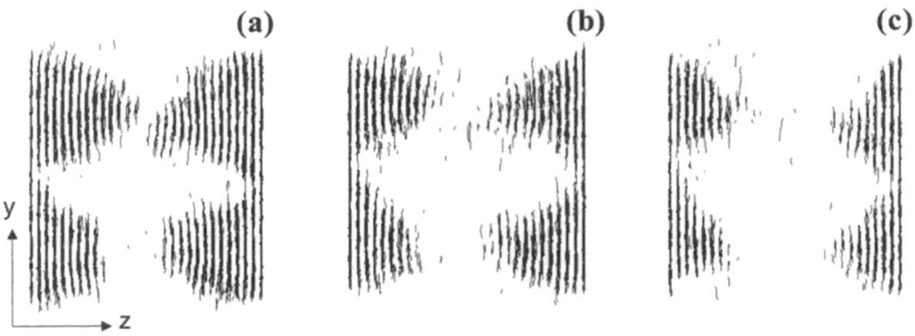

Fig. 32 Changes in the lamella profiles viewed along the x-axis, during melting at 380 K; **a** at 0.0 ns, **b** at 0.128 ns, **c** at 0.64 ns. We can see that the tapered growth fronts retreat at 380 K maintaining the tapered shape at the edges

with rapid retreat of the growth fronts while maintaining the tapered shapes. When the same initial lamellae were quickly heated and held at 360 K, the lamellae showed neither growth nor melting at least within 10 ns, but again the tapered shape of the lamellae was preserved at least up to about 9 ns. All these observations suggest that the tapered shape is not a kinetic form but rather an equilibrium form.

The tapered shape of the lamellae indicates the preference of the fold surfaces to have an inclination from the chain-axis normal. It is well known that the density difference between the amorphous and the crystalline phase results in the chain tilting. We made a rough estimation of the tilt angle by assuming that all the crystalline stems emanating from the fold surface join to form the amorphous phase; the fold loops are assumed not to be very tight even if the chains fold back to stems near the original stems from which the folds emanate. Indeed, the conformation of the chains near the growing edges (Fig. 33) shows that the growth fronts are composed of stems and many loose loops. Rich amorphous loops can cause considerable chain tilt to compensate for the density gap between the crystal and amorphous phases. The densities of the crystalline and amorphous phases are approximately 0.86 [g/cm^3] and 0.93 [g/cm^3], respectively, and then the resulting tilt angle is estimated to be about 20 deg, which is close to that observed by the present simulation.

As a final comment, we would like to add that the tapered shape of the growing lamella observed in our simulation of 3D crystallization from the melt can be a transient form of the growing lamellae. The advancing tapered

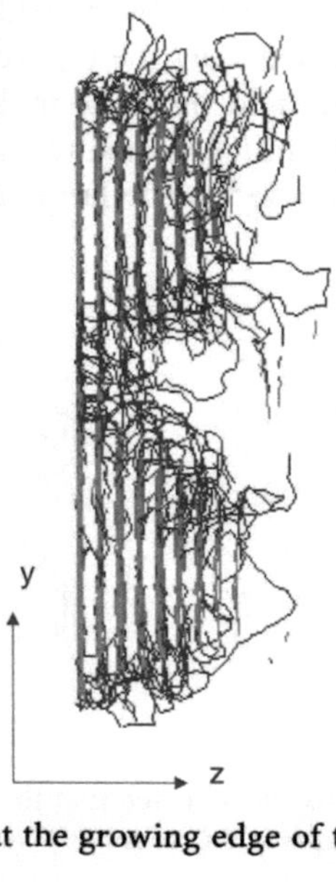

Fig. 33 Rich amorphous chains at the growing edge of the lamella

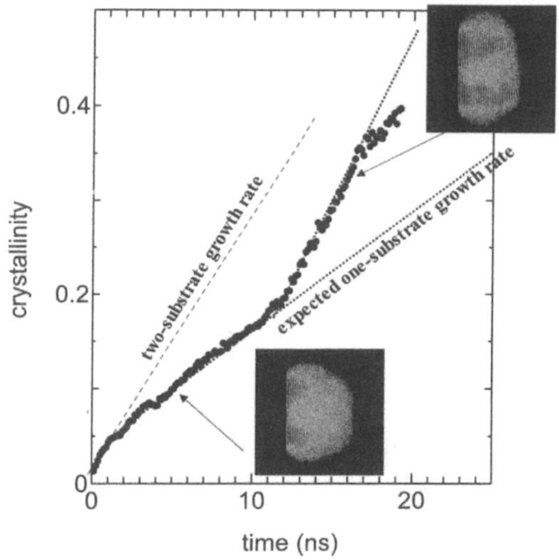

Fig. 34 Lamella crystals growing at 330 K from the left substrate only. The initial rate of crystallinity increase is about half of that of Fig. 25 as expected, but it unexpectedly accelerates around 12 ns suggesting a changeover in the crystallization mode

trapezoidal lamellae with their bases of the trapezoids at the substrate of limited height must have progressively smaller tilt angles. However, if the tapered form and the resulting chain-tilt are due to the overcrowding amorphous chains emanating at the fold surface, the progressively smaller tilt angle must result in the accumulation of surface stress. We have recently simulated the lamella growth in a larger space where the growing lamellae do not collide with those growing from the opposite substrate (Fig. 34). Since crystal growth from the right substrate is prohibited, the rate of crystallinity increase (dotted line) is expectedly about half that of the previous studies (dashed line). Quite unexpected is that the increase in the crystallinity does not slow down but rather it doubled around 12 ns; the growth rate in the normal direction is kept nearly constant in spite of such changes in the growth mode. Inspection of the snapshot suggests that the growing lamellae have changed shape from the tapered form to the usual constant thickness form. However, this is rather preliminary data and we are studying much larger systems under periodic boundary conditions.

5.7 Changes in Chain Extension

During crystallization, the molecules change their conformation from the random coil to chain folded. The chain conformation in the crystalline state has been the subject of great discussion, since it reflects the path that molecules have followed during crystallization. In the case of sufficiently long

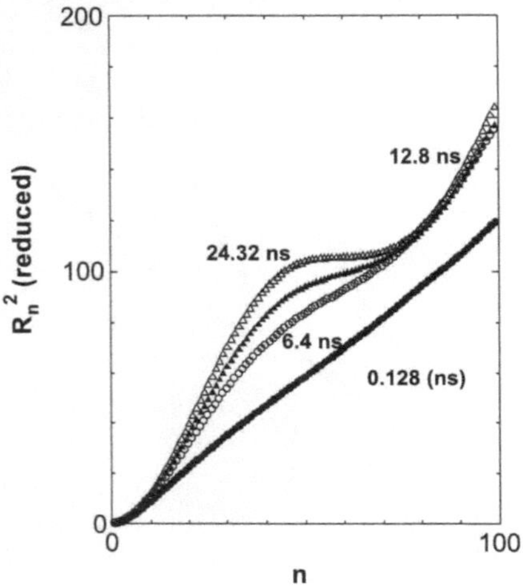

Fig. 35 Changes in R_n^2, the averaged distance between atoms n-bonds apart along the chain, during crystallization at 330 K. In the initial state of undercooled melt (•), R_n^2 depends linearly on n showing the random coil nature of the melt chains. With the onset of crystallization, the functional form of R_n^2 changes considerably; (○) after 6.4 ns, (▲) after 12.8 ns, and (△) after 24.32 ns

polymers crystallized from the melt, it is often indicated that the chains do not change their radius of gyration by crystallization and that the overall chain extension in the melt is maintained in the crystal.

The chain conformation, or extension, can be described well by the squared separation R_n^2 between atoms n-bonds apart within the chain,

$$R_n^2 = \left\langle \left(\boldsymbol{r}_{\mathrm{m}} - \boldsymbol{r}_{\mathrm{n+m}} \right)^2 \right\rangle , \tag{13}$$

where the average is taken over m-atoms in the chain and over all chains in the system. Shown in Fig. 35 is the change in R_n^2 during crystallization at 330 K. A linear increase in R_n^2 with n in the initial melt indicates that the chains are in the random coil state. With progress of crystallization, R_n^2 begins to show parabolic increases at smaller n, indicating that the chains are stretched locally by crystallization. Due to the chain folding, however, the increase in R_n^2 starts to decrease around $n = 40$. The final chain extension, the end-to-end separation R_{100}^2 is only slightly larger than that in the melt.

5.8 Local Structure of the Undercooled Melt

The microscopic structure of the undercooled melt has been a subject of great interest in studies of polymer crystallization. There have been long arguments in favor of the presence of mesoscopic local order in the melt or at the crystal-

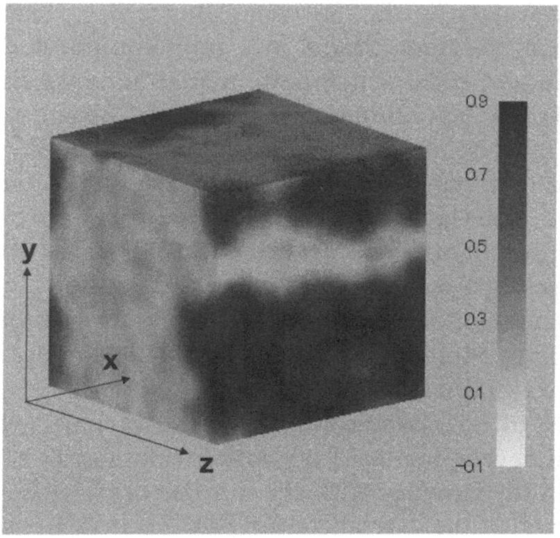

Fig. 36 Three-dimensional map of the local bond orientational order. The bond order is expressed by the density-scale given in the legend to the right

melt interface. We examine the degree of melt order in the chain orientation by use of the local order parameter $P(\boldsymbol{r})$ which is defined as follows:

$$P(\boldsymbol{r}) = \left\langle \left(3\cos^2\theta_{i,j} - 1\right)/2 \right\rangle , \tag{14}$$

where $\theta_{i,j}$ is the angle between the i-th and the j-th C – C bonds whose centers are both contained within a small cubic box of size 3σ (about 10A cubic) located at $\boldsymbol{r}$; each small cubic box contains approximately 60 C – C bond centers. The average is taken over all pairs of bonds whose centers are within the box. The order parameter P should have a large value where the neighboring chains have a definite tendency to be parallel, while P is very small for random bond orientation.

Figure 36 is a three dimensional representation of the order parameter P at 350 K after 19.2 ns of simulation, where about 25% of the system has transformed into the crystalline state. The black regions near both side surfaces correspond to the crystalline domains with higher P values, while the white regions are in a completely isotropic state of $P \cong 0$. Detailed inspection of these data has shown that no appreciable order is present in the melt. A simple interface model between the crystal and the isotropic melt seems to be more plausible at least in this case of short chain C_{100}.

5.9
Crystallization of Much Longer Chains from the Melt

The discussions have so far been concerned with relatively short C_{100} chains. We want to extend our simulations into much longer chain systems. We present some of our data on the ten-times longer C_{1000} chains. But we must begin with a caveat. Polymer chains in a fully equilibrated melt should be ideal Gaussian random coils, which have spatial dimensions proportional to the square root of their degrees of polymerization. The end-to-end distance of C_{1000} in the melt must be about three times longer than that of C_{100}. We previously confirmed that the present MD-cell size was large enough for the C_{100} chains to show the ideal Gaussian dimension. In order to provide a space large enough for the C_{1000} chains, we must adopt an MD-cell whose linear dimension is three-times larger containing as many as 2-million atoms. However, we had to deal with a system of the same MD-cell size as that for the C_{100} chains, 30σ cube, containing 64 chains of C_{1000}; we must, therefore, expect serious size effects in the crystallization processes.

The initial melt of C_{1000} was cooled down to various crystallization temperatures. Typical developments of crystalline domains at 370 K are shown in Fig. 37; here again the lamellae with the marked tapered shape are observed. Quite surprising is that the crystallization of C_{1000} is rather fast in spite of the much longer chains. Any growth of lamellae of C_{1000} at the temperature of 370 K, where C_{100} did not show any appreciable growth, will be an indication of the molecular weight effect.

Figure 38 is a molecular trajectory at 370 K of a chain selected at random. When a sufficiently long random coil chain approaches the growth front, a part of the chain is pulled out and adheres to the growth front. The resulting chain conformation at the growth surface is often a hairpin. The hairpins climb down or up the growth surface to complete the crystalline stems, followed by further folding of the tails connected to the hairpin. In the initial step of the primary nucleation and the secondary nucleation [11, 28, 29, 31],

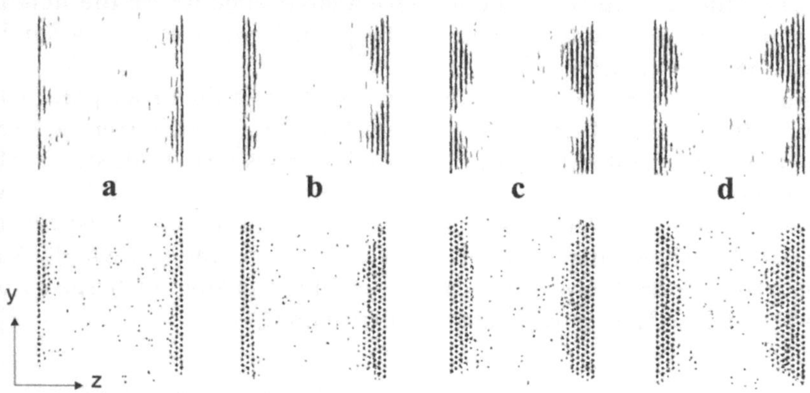

Fig. 37 Three dimensional pictures of growing lamellae of C_{1000} at 370 K **a** at 0.128 ns, **b** at 6.4 ns, **c** at 12.8 ns, and **d** at 19.2 ns. We again see the tapered growth fronts and their advances in the normal (the *z*-axis) direction together with the lamella thickening along the chain axis (the *y*-axis direction)

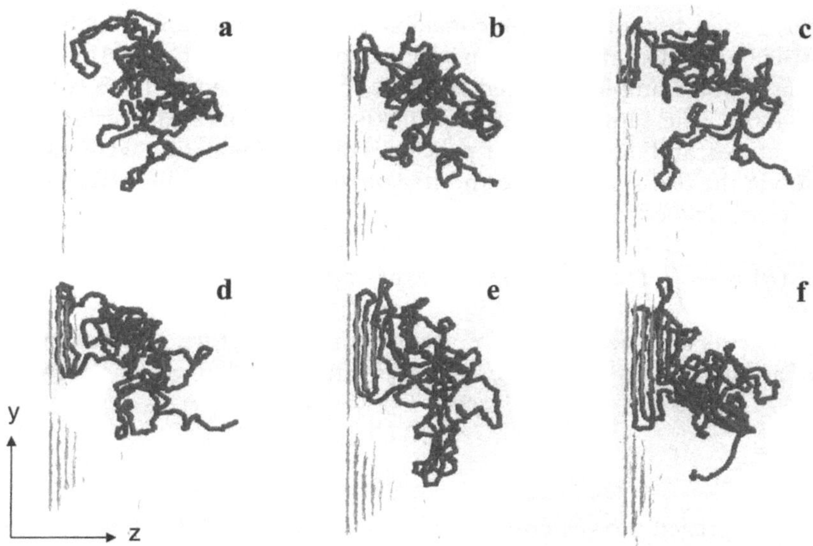

Fig. 38 Molecular trajectories of a crystallizing chain (*thick line*) selected at random from 64 chains during crystallization at 370 K; **a** at 0.128 ns, **b** at 0.64 ns, **c** at 1.28 ns, **d** at 2.56 ns, **e** at 6.40 ns, **f** 12.8 ns. Also shown are the growing crystalline domains of Fig. 12 (*thin parallel lines*). Pictures are all side view along the *x*-axis

the significance of such a stem pair or hairpin has already been suggested in previous papers. Here again, we have arrived at a similar conclusion.

We also examined the fold statistics in this C_{1000} system. The distribution of the inter-stem vectors connecting stems linked by the loops, and their radial distribution function again indicated that about 60–70% of the folds are short loops connecting the nearest or the second and third nearest stems, though the crystallization did not complete. The presence of local order in the undercooled melt in the present C_{1000} system is also examined through the same local order $P(\mathbf{r})$ parameter, the degree of bond orientation as a function of position $\mathbf{r}$, but again we did not detect any appreciable order in the undercooled melt.

6 Crystallization from an Oriented Amorphous State

Polymer crystallization under flow or under highly oriented states is of prime importance in industrial polymer processing. We expected that the crystallization would be highly accelerated when the initial amorphous chains were highly orientated. Therefore, we dared to use a realistic molecular model of

PE, the united atom model. We considered a sufficiently long PE chain made up of 5000 united atoms under periodic conditions in each direction. The initial amorphous sample prepared at 600 K was quenched to 100 K and drawn up to 400%. The sample was then quickly heated to various crystallization temperatures, and the molecular processes of fiber formation were monitored in situ via the real-space image and its Fourier transform, the structure function $S_{3D}(\boldsymbol{q})$ defined as follows:

$$S_{3D}(\boldsymbol{q}) = \frac{1}{N} \int \int \mathrm{d}\boldsymbol{r}\mathrm{d}\boldsymbol{r}' \langle \rho(\boldsymbol{r})\rho(\boldsymbol{r}') \rangle \exp(-\, i\boldsymbol{q}(\boldsymbol{r} - \boldsymbol{r}')) , \tag{15}$$

where $\boldsymbol{q}$ is the wave vector, $\boldsymbol{r}$ and $\boldsymbol{r}'$ are the position vectors, N is the total number of atoms, and the atomic density at position $\boldsymbol{r}$ is defined by the relation

$$\rho(\boldsymbol{r}) = \sum_{j=1}^{N} \delta(\boldsymbol{r} - \boldsymbol{r}_j) . \tag{16}$$

Here we averaged the structure function $S_{3D}(\boldsymbol{q})$ around the draw axis to give $S_{2D}(q_{||}, q_{\perp})$, where $q_{||}$ and $q_{\perp}$ are the components of the wave vector along the draw axis and perpendicular to it, respectively.

Typical snapshots and the structure functions at 350 K are shown in Fig. 39. In the initial state at 0.1 ns, we see two intense peaks on the meridian at $q_{||} = 52\,\mathrm{nm}^{-1}$ and on the equator $q_{\perp} = 14\,\mathrm{nm}^{-1}$, where the peaks correspond to half the fiber period of PE and the nearest neighbor inter-chain

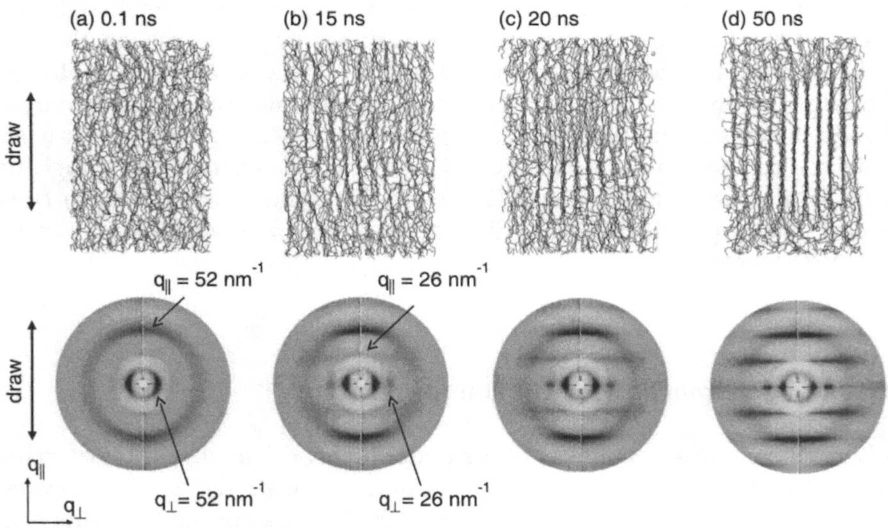

Fig. 39 Snapshots (*upper*) and the structure functions averaged around the draw axis, $S_{2D}(q_{||}, q_{\perp})$, (*lower*) at 350 K

spacing, respectively. Around 15 ns, highly ordered domains become visible (Fig. 39b); in the domain partially stretched molecular segments are aligned parallel to each other. The structure function appreciably changed from that at 0.1 ns. The broad layer streaks emerge along the layer line at $q_{||} = 26\ \mathrm{nm}^{-1}$, and the reflections at $q_{\perp} = 29\ \mathrm{nm}^{-1}$ on the equator become slightly sharper and stronger. In Fig. 39c, the layer streaks at $q_{||} = 26\ \mathrm{nm}^{-1}$ become much more evident, and the equatorial reflections around $q_{\perp} = 29\ \mathrm{nm}^{-1}$ split into independent reflections. Such a diffraction pattern is quite similar to that observed in the hexagonal phase of PE [44]. In the snapshot in Fig. 39c, we can see a crystalline nucleus emerged after 15 ns and still continued growing until its size becomes comparable to the MD cell dimensions (Fig. 39d). At the same time, the intensity contrast in the streak at $q_{||} = 26\ \mathrm{nm}^{-1}$ becomes pronounced. Even so, well ordered crystalline patterns, with high longitudinal order, could not be attained in the present simulation.

7 Summary and Discussions

7.1 Summary of the Present Studies

Through the stepwise revisions of the molecular model, we have achieved realistic modeling of 3D crystallization from the melt. But each step of the modeling has also revealed very interesting aspects of the crystallization. The simple 2D model of crystallization, though it was very simplified and seemed unrealistic, presented very clear results with respect to the molecular mechanism of chain folding. The crystallization of the random coil, which was highly expanded and disentangled, was shown to give rather neat chain folded lamella. The origin of such ideal chain folding was the local clustering. The subsequent reeling-in of the slack chains connecting separated clusters inevitably led to the regularly chain-folded lamellae. In addition, this model brought out the mechanism of lamella thickening. The lamella thickened continuously, with short pauses when the chain ends came close to the fold surface. The thickening and the underlying polymer diffusion along its contour showed a strange temperature threshold, and considerable chain motions were excited only above this threshold. The behavior was just like the onset of glass transition where the large scale molecular motions are switched on.

The crystallization of a single polymer chain in a vacuum onto the crystalline substrate was also insightful. Though the chain was highly collapsed, it transformed to the neat chain folded lamella through cooperative adhesion and ordering on the growth surface. The crystallization on the thin crystalline substrate demonstrated that the chain entanglements were pushed out of the crystalline substrate into amorphous regions leading to the accumulation of the chain entanglements in the interlamella regions.

Our final modeling, of the crystallization from the melt, was the most fruitful one. Thanks to the rapid development of powerful PCs and PC clusters, such large scale simulations became feasible. Though we began this work without confidence, we unexpectedly observed the rapid crystallization into chain folded lamellae. We observed the growth of lamellae that had a marked tapered shape; the lamellae showed thickening growth along the chain axis as well as the usual growth perpendicular to it. The crystallization was found to be very sensitive to temperature; the crystals nearly stopped growing around 370 K in C_{100} and at slightly higher temperatures in C_{1000}. The inspection of the growing lamellae showed that they did not always grow in concert, especially at higher temperatures near T_m. We also found that the lateral growth surfaces were locally flat, and that similar molecular processes like those in the Kossel mechanism seemed to be operative in the polymer crystal. The folds at the surfaces were found to be relatively short, at least 70% of which were connected to the nearest or the second and third nearest neighbor crystalline stems. Last but not least, our computer simulation did not indicate any sign of local order in the chain orientation within the undercooled melt of both C_{100} and C_{1000}.

7.2 Reconsideration of the Problems

Crystallization in polymers has long been one of the most difficult problems in polymer science. It was to our great surprise that the computer simulations proved very useful in studying this historical problem, if we properly devised the molecular models and the crystallization conditions. But I am aware that there are many problems in the present simulation. Major criticisms will be why the crystallization is so fast, what kind of relevance the present model has to real polymer systems, and how we can bridge the space and time gaps between the present model and real polymers.

The molecular process of crystallization from the melt was, in the present model, extremely accelerated due to following factors: (1) The polymers we considered were rather short, or highly disentangled. The short polymers have much larger mobility enabling quick response to the driving force toward crystallization, while highly disentangled chains do not need large scale diffusion to the growth surfaces; (2) the most conspicuous character of our model molecule is its molecular structure. The chain is like PE but the planar zigzag conformation is lost, which must have facilitated chain translation along the chain contour and transverse to it. The geometrical simplicity of the molecule will be also an origin of easy chain alignment and quick ordering. In our previous MD simulation of crystallization in *n*-alkanes [45, 46], the crystallization in the realistic planar zigzag chains was much slower than that in simple bead-spring chains; the former system was at least one order of magnitude slower in crystallization. Such high sensitivity of the characteristic time to the detailed molecular structure was also noticed in the melt rheology of polymers [47]. The crystallization is expected to be much more sensitive than melt rheology to small structural changes, which will lead to extremely slow dynamics in crystallization of real polymers. (3) The other possible reason for the very fast crystallization is that we employed very large undercooling. The present crystallization from the melt showed the fastest rate at 330 K, which was undercooled as much as 60 K. A similar aspect of the model is that the systems were very small in comparison with those of real polymers. The lateral dimensions were small, only 12 nm, and the kinks could only be propagated along a very limited length; the growth surface was effectively "rough" and led to rapid normal growth perpendicular to the growth front. (4) The last factor that accelerated crystallization was the chain orientation in the melt. As observed experimentally in very high speed spinning of fibers, it is quite reasonable that the crystallization can be very fast under large deformation.

In the world of molecular simulation, it would be more conventional to consider that the present model is a coarse grained model of real polymers, where the real time-scale is much longer than that of the present MD simulation time-scale. However, we did not intend to make a coarse grained model. The crystallization of polymers was shown to be rather universal. Various kinds of polymers, either fast crystallizing or slow crystallizing, were known to follow the same scheme with respect to the molecular mechanism of crystallization. So we studied this simple model expecting that the present model would also follow the same crystallization scheme and show the general molecular mechanisms of polymer crystallization.

7.3
Limitation of the Present Model and Future Problems

Within the present model, we have many unsolved problems. Most of the present studies on 3D crystallization from the melt deal with the relatively short C_{100} chain. The study of the much longer C_{1000} chain is still preliminary; we want to clarify more polymer-like behavior such as the reeling-in process of the chains. Since the polymers in the ideal melt are the ideal Gaussian and highly entangled, we need a much larger MD cell to accommodate such large polymers.

Another shortcoming of the present model is that the MD cell size in the y-axis direction, the direction parallel to the crystalline chain axis, was rather small. There must be serious size effects along this direction, even if we adopt the periodic condition in this direction. Here again we need a much larger MD cell in order to reproduce the lamella thickness vs. temperature relation in crystallization from the melt.

Even greater problems for the future include simulations of more realistic and complex polymers including helical polymers. Various interesting and fruitful subjects of future research are available, such as the molecular design of polymers having desired mesoscopic structures, elucidation of the very complex processes involved in the helical ordering at the growth front, a sort of molecular recognition at the crystal surfaces. Simulations of the crystallization under flow or extreme deformation, which we have already attempted, will become a very interesting field of research in the future. Even more exciting, and not too far from the present state of the art, are the investigations of structure formation in various nano-scale structures.

Acknowledgements The present work was supported by the Grant-in-Aid of Scientific Research on Priority Areas, "Mechanism of Polymer Crystallization" (No.12127206), from the Ministry of Education, Science, and Culture, Japan.

References

1. Wunderlich B (1976) Macromolecular Physics. Academic Press, New York, 1–2
2. Hoffman JD, Davies GT, Lauritzten JI (1976) Treatise on Solid-state Chemistry. Plenum, New York, 3
3. Armistead K, Goldbeck-Wood G (1992) Adv Polym Sci 100:219
4. Imai M, Mori K, Mizukami T, Kaji K, Kanaya T (1992) Polymer 33:4451
5. Lee CH, Saito H, Inoue T (1996) Macromolecules 29:7034
6. Fukao K, Miyamoto Y (1997) Phys Rev Lett 79:4613
7. Tashiro K, Imanishi K, Izumi Y, Kobayashi M, Kobayashi K, Satoh M, Stein R (1995) Macromolecules 28:8477
8. Mandelkern L (2002) Crystallization of Polymers. Cambridge University Press, Edinburgh
9. Hoffman JD, Miller RL (1997) Polymer 38:3151
10. Sadler DM, Gilmer GM (1984) Polymer 25:1446
11. Point JJ (1979) Faraday Discuss Chem Soc 68:167
12. Hokosaka M (1990) Polymer 31:458
13. Keller A, Hikosaka M, Rastogi S, Toda A, Barham PJ, Goldbeck-Wood G (1994) J Mater Sci 29:2579
14. Strobl G (2000) Eur Phys J E 3:165
15. Cheng SZD, Noid DW, Wunderlich B (1989) J Polym Sci, Part B 27:1149
16. Yamamoto T, Hikosaka M, Takahashi N (1994) Macromolecules 27:1466
17. Yamamoto T (1995) J Chem Soc Faraday Trans 91:2559
18. Kavassalis TA, Sundararajan PR (1993) Macromolecules 26:4144
19. Fujiwara S, Sato T (1997) J Chem Phys 107:613
20. Yamamoto T (1997) J Chem Phys 107:2653
21. Yamamoto T (1998) J Chem Phys 89:2356
22. Yamamoto T (2001) J Chem Phys 115:8675
23. Yamamoto T (2003) J Macromol Sci B 42:629
24. Yamamoto T (2004) Polymer 45:1357
25. Koyama A, Yamamoto T, Fukao K, Miyamoto Y (2002) Phys Rev E 65:050801
26. Chen CM, Higgs PG (1998) J Chem Phys 108:4305

27. Liu C, Muthukumar M (1998) J Chem Phys 109:2536
28. Welch P, Muthukumar M (2001) Phys Rev Letts 87:218302
29. Doye JP, Frenkel D (1998) J Chem Phys 109:10033
30. Doye JP, Frenkel D (1999) J Chem Phys 110:2692
31. Meyer H, Mueller-Plathe F (2001) J Chem Phys 115:7807
32. Meyer H, Mueller-Plathe F (2002) Macromolecules 35:1241
33. Waheed N, Lavine MS, Rutledge G (2002) J Chem Phys 116:2301
34. Hu W, Frenkel D, Mathot VBF (2003) Macromolecules 36:549
35. Miura T, Kishi R, Mikami M, Tanabe Y (2001) Phys Rev E 63:061807
36. Toda A, Kiho H (1987) J Phys Soc Japan 56:1631
37. Flory PJ (1969) Statistical Mechanics of chain molecules. Wiley, New York, 432 pp
38. Steele WA (1973) Surf Sci 36:317
39. Sanchez IC, Colson JP, Eby RK (1973) J Appl Phys 44:4332
40. Granick S (1991) Science 253:1374
41. Yoon DY, Vacatello M, Smith GD (1995) In: Binder K (ed) Monte Carlo and Molecular Dynamics Simulation in Polymer Science. Oxford Univ Press, Oxford, pp 433–475
42. Hobbs JK, Miles MJ (2001) Macromolecules 34:353
43. Gautam S, Balijepalli S, Rutledge GC (2000) Macromolecules 37:9136
44. Yamamoto T (1979) J Macromol Sci Phys B16:487
45. Shimizu T, Yamamoto T (2000) J Chem Phys 113:3351
46. Li HZ, Yamamoto T (2001) J Chem Phys 114:5774
47. Abraham CF, Kremer K (2002) J Chem Phys 116:3162

Adv Polym Sci (2005) 191: 87–135
DOI 10.1007/12_009

Published online: 25 October 2005

Pre-Crystalline, High-Entropy Aggregates: A Role in Polymer Crystallization?

Giuseppe Allegra · Stefano Valdo Meille (✉)

Dipartimento di Chimica, Materiali e Ingegneria Chimica "G. Natta",
Politecnico di Milano, Via L. Mancinelli 7, 20131 Milano, Italy
giuseppe.allegra@polimi.it, valdo.meille@polimi.it

Abstract We distinguish three main modes of crystallization for polymers with a relatively flexible main-chain, i.e., (i) usual lamellar crystallization occurring by cooling from the reference state (melt or solution) above the temperature T_0 down to $T > T_g$; (ii) crystallization from the glass; (iii) crystallization from a stable thermotropic mesophase. In all three cases we propose that structure development proceeds via high entropy pre-crystalline aggregates, which may influence features of the crystalline organization. Pre-crystalline structures characteristic of modes (i) and (ii) are identified with bundles, i.e., energy-driven hexagonal associations among chain segments. At $T < T_0$ the polymer solution is regarded as meta-stable, and in this state the bundle segments are essentially consecutive whereas in the melt and the glass bundles also comprise non-consecutive chain segments. The fold thickness L observed in lamellar crystallization, resulting from bundle aggregation and rearrangement, is basically controlled by the average fold length in the consecutive chain portions within bundles. For small values of $\Delta T(= T_0 - T_{\text{crystallization}})$ we obtain $L \propto 1/\Delta T$, in agreement with experimental data from polyethylene as well as with several simulation results; the proportionality factor appears to be the same for the solution and the melt. The bundle model appears to be consistent with indirect evidence such as segregation of short chains in the crystallization process and clustering of segments belonging to the same chain in the crystal. In mode (ii) it is plausible, at least in certain instances, that crystallization is preceded by bundle aggregation leading to phase separation. In the case of crystallization in mode (iii) we can identify the pre-crystalline high entropy state with the thermotropic mesophase itself. Such phases involve large domains of parallel, hexagonally packed, conformationally disordered chains, with a high propensity to fully extended macroconformations. They occur with polymers with a large persistence length of entropic (i.e., elastic) origin, mainly due to conformational disorder of the side groups. Folds and hairpins in these mesophases are energetically disfavored because adequate compensatory inter-stem attractions are missing. Finally, it is shown that crystallization of helical non-chiral polymers into crystalline modifications comprising isochiral helices only, may in certain cases be accounted for on the basis of hexagonal pre-crystalline intermediates like bundles and mesophases discussed in the present contribution.

Keywords Bundles · Chain-folding · Chiral crystal polymorphs · Mesophases · Polymer crystallization · Pre-crystalline order

Abbreviations

β^2	mean square contraction of an otherwise unperturbed chain with bundles
T_g	glass transition temperature
T_0	ideal dissolution temperature, or melting temperature
T	actual temperature
ΔT	$T_0 - T$
Z	partition function of the chain
E_i	energy of the i-th conformational state of the chain
E	melting energy per chain atom
Ω_i	multiplicity of the i-th state
L	lamellar thickness
$n_{\text{bundle}}, n_{\text{stem}}, n_{\text{loop}}, n_{\text{bridge}}$	no. of chain atoms in a bundle, a stem, a loop, a bridge
p	no. of stems per bundle
P	persistence length of the polymer chain
D	chain diameter
ΔL	length of the independent chain element
C_∞	characteristic ratio of a polymer chain
$C_{\infty B}$	characteristic ratio of a chain with all side groups replaced by hydrogens
l_0, l	skeletal bond length and its projection on the chain axis
C, M, L	crystalline phase, mesophase, and liquid phase of the polymer
H	hexagonality index

1
Introduction

Polymer crystallization [1, 2] is a complex process involving different steps and mechanisms [3–5] depending upon the specific molecular system and the conditions under which crystallization is carried out. Crystallization may develop from either a solution or a melt, which in turn may be either in a quiescent state or under flux strain [6], or it may proceed from a mesomorphic phase or again from the glass: under such different circumstances polymer chains are subject to widely different constraints. In turn copolymers or highly defective polymers may show behaviors differing substantially from highly stereoregular homopolymers. Chain stiffness is also likely to play a key role. In the present contribution our discussion will be limited to fairly flexible polymers most of which possess the ability to give rise to chain-folded crystals and exhibit T_gs typically below 100 °C.

Limiting ourselves to crystallization under quiescent conditions, we shall focus our structural-statistical analysis on states intermediate between the liquid—or dissolved—polymer and the crystalline polymer. Our basic hypothesis is that, since polymer crystallization occurs at significant degrees of supercooling, it is plausible that macromolecules in the crystallizing melt or in solution present some degree of structural organization that may be described in terms of a metastable configurational equilibrium. At $T < T_0$ (i.e., *the ideal dissolution, or melting temperature*) we envisage for polymers two distinct possibilities: (i) they comprise *bundles*, or relatively short, metastable folded chain portions; and (ii) they are organized in thermotropic mesomorphic domains involving parallel extended chains. Both these states involve high-entropy structures where parallel chain segments will be assumed for simplicity to organize in hexagonal arrays. Tetragonal or trigonal packing modes may in principle occur, especially with chains which deviate substantially from cylindrical shapes, but due to their lower probability in the case of flexible chains such alternative arrangements will be disregarded. We consider such intermediate states, with partially ordered chain aggregates, as *pre-crystalline* since they present degrees of order (or, more precisely, of structural organization) ranging between the fully amorphous chains and polymer crystals. In this context they will be considered as distinct thermodynamic states—metastable in the case of bundles and stable in the case of thermotropic mesophases—rather than mere stages in the process leading to polymer crystals.

The bundle model of polymer crystallization will be discussed first. This mean-field approach describes the metastable configurational equilibrium of the undercooled polymer chain in solution or in the melt, and we will summarize and update previous work of ours [7–9]. The concept of the bundle, i.e. an aggregate of a few parallel polymer segments or *stems* connected by *folds* (see Fig. 1; see also [10]) and stabilized by attractive crystal-like interactions, is central as some of the properties of the resulting chain-folded crystals (most notably the fold length) can be interpreted simply in terms of the fold length statistics in consecutive chain portions within bundles.

We will then examine other flexible polymer crystallization instances which may be interpreted, at least qualitatively, in terms of the bundle model. We will concentrate on crystallization occurring through metastable mesophases which develop by quenching polymers like isotactic polypropylene, syndiotactic polypropylene etc. In principle also hexagonal crystallization of highly defective polymers, and order developing in some microphase-separated copolymer systems could be discussed in a similar perspective but these two areas will be treated in future work. A comparison between the bundle approach and pertinent results of selected molecular simulation approaches follows.

In the subsequent discussion of enantiotropic mesophases we will follow the lines of a recent contribution [11] evidencing how relatively flexible polymers can develop significant stiffness by a self-compacting process. Such mesophases are basically stabilized by entropic factors as in Onsager's athermal theory of the liquid-crystalline state [12]. In their simplest versions these phases are characterized by intermolecular enthalpic interactions comparable to those in the melt, whereas entropy is maximized by the conformationally disordered chains adopting, on average, a straight, cylindrical structure. As a consequence the fold statistics found in the isotropic melt and in chain-folded crystals is lost and hexagonal, extended-chain phases tend to develop.

Before the Concluding Remarks the possible role of hexagonally packed pre-crystalline entities, i.e., bundles, bundle aggregates, hexagonal mesophases etc. in the crystallization of chiral polymer crystal structures will be commented expanding our previous work [13]. Instances where mechanisms involving nucleation by chiral surfaces may apply (e.g. [14]) will be briefly discussed or referred to. Conversely, where no such surfaces can be envisaged a key role of pre-crystalline entities appears highly plausible.

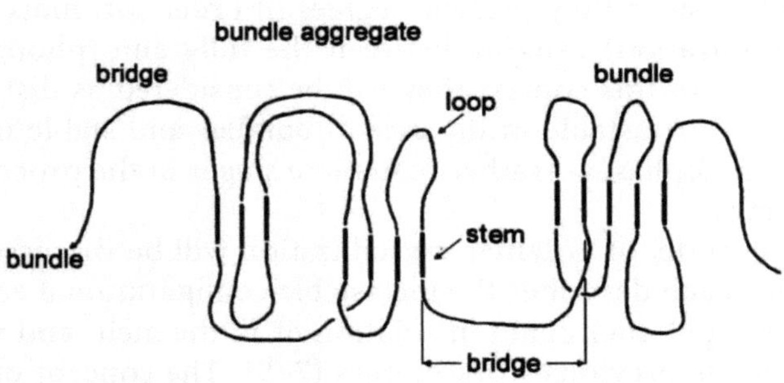

Fig. 1 Schematic representation of a bundle and a bundle aggregate in a polymer chain. Bridges, loops, and crystallized stems are evidenced

2
The Bundle Model

2.1
General Concepts

We shall consider the intramolecular equilibrium of an isolated, very long chain wherein the atoms may attract each other by means of crystal-like packing forces. The chain will be considered either in ideal solution or in the melt, and the resulting structure will comprise *bundles*, i.e., pre-crystalline objects built with parallel chain strands (*stems*) in a hexagonal arrangement. In the solution state the stems are consecutive and therefore fully connected by *chain folds*, see Fig. 1, whereas in the melt a bundle may comprise stems from distant chain portions or even from different chains. Under appropriate circumstances bundles may produce a lamellar-crystalline state with a free energy decrease and accordingly, we shall regard the polymer equilibrium, including the bundles, as metastable [7–9]. In the Appendix a concise account of the bundle statistical mechanical treatment is given [8, 9].

Regarding the chain as unperturbed, both in solution and in the melt the statistical contact probability for two chain atoms separated by m bonds is proportional to $m^{-3/2}$ if m is large enough, say approximately > 15–20 in the case of polyethylene [9]. In fact the interatomic root-mean-square distance is $\propto m^{1/2}$, whereas the contact probability p_m is inversely proportional to the mean corresponding volume V_m, i.e., to $V_m \propto m^{3/2}$. As a consequence the statistical probability of a chain fold decreases rather quickly with its contour length. If the temperature T is well below T_0, a large number of bundles is formed in a very long chain, each comprising consecutive chain strands, and the average fold length will be small. Conversely, *the number of bundles tends to zero with T approaching T_0 and their average length goes to infinity*, as $\langle m \rangle = \int_{\overline{m}}^{\infty} mp(m)\mathrm{d}m \to \infty$, $\overline{m}$ being a suitable lower limit. Another essential aspect of the bundle approach is that the lamellar crystal is formed by transfer of stems (see Fig. 1) from bundles to the lateral lamellar surface. Under the assumption that the number of stems does not change in the process, the lamellar thickness is equal to the sum of the contour lengths of the stem and of the fold, and goes to infinity as $T \to T_0$.

The above model was initially devised for polymer solutions, and in principle it cannot be applied as such to polymer melts, particularly at large undercoolings $\Delta T = (T_0 - T)$ [7–9]. In fact, in such systems the large friction forces and the strong kinetic barriers represented by chain entanglements—leading to the slowly diffusive chain reptation motion—give rise to chain relaxation times much larger than the half-lifetime of the hexagonal-packed structure of the bundle stems. Although we will not develop here the theoretical aspects of the resulting incomplete bundle equilibrium, we may conclude that bundles in the bulk will also comprise chain strands very distant along the chain sequence or even portions of different chains. The similarity between the lamellar thickness of solution- and melt-crystallized samples at the same undercooling is consistent with the prime importance of the attractive interactions between consecutive chain stems. It is qualitatively supported by the fact that chains in the melt and in solution present similar unperturbed dimensions and by the small relevance of large bundles and of bundle aggregates in the model calculations. It also suggests that results obtained from the bundle approach for solution crystallization may be partly extended to the melt; in particular, pseudo-hexagonal packing of the chains should be a common feature, in addition to the average lamellar thickness. Postponing further discussion of this issue, in the following we will generally use the same expression "bundle equilibrium" for both solution and bulk crystallization.

2.2
Bundle Crystallization: A Statistical Description

Unless stated otherwise, in this Subsection we shall deal with solution crystallization. After suitable concentration and re-organization, bundles may produce a sufficiently large crystalline aggregation to sustain spontaneous growth: we have *homogeneous crystallization*, resulting from the formation of a critical nucleus. Its further growth may result from deposition of either single or multiple strands on the crystal surface, with formation of a lamellar structure (*heterogeneous crystallization*, see Fig. 2). We remark that in principle the process may be regarded as reversible if the value of $-\Delta G$ for the chain deposition is sufficiently small. Heterogeneous crystallization will represent the core of our analysis; we will not attempt in this context to carry out a comparative discussion with the homogeneous case (i.e., the formation of the primary nucleus), in the assumption that substantial analogies may be envisaged in the two cases for the features under discussion. The parallel stems belonging to the hexagonal bundle arrangement (see Figs. 2 and 3) will be gradually adsorbed onto the crystal surface in a dynamic process by which some crystal stems will revert to bundles. *The net flux of stems from bundles to crystals yields the amount of crystal growth. Although the local conformation of the chain strands adjoining the stems transferred to the crystal will undergo a straightening process, some overall conformational aspects will persist, in particular the total number of stems.* This will substantially contribute to determine the initial lamellar thickness as well as some degree of adjacent re-entry of the folds.

In any state preceding the onset of crystallization at $T < T_0$ we assume that bundle stability is favored by localized attractive interactions between contacting (short) *stems*, some enthalpy advantage being balanced by a corresponding entropy loss (see Fig. 3). Depending upon the core structure of the crystalline stems, various bundle models were examined [8, 9]. In the present

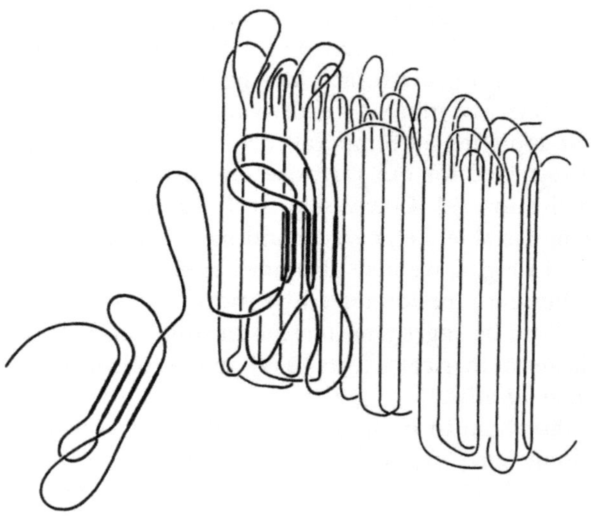

Fig. 2 Bundles adjoin the side surface of a lamellar crystal

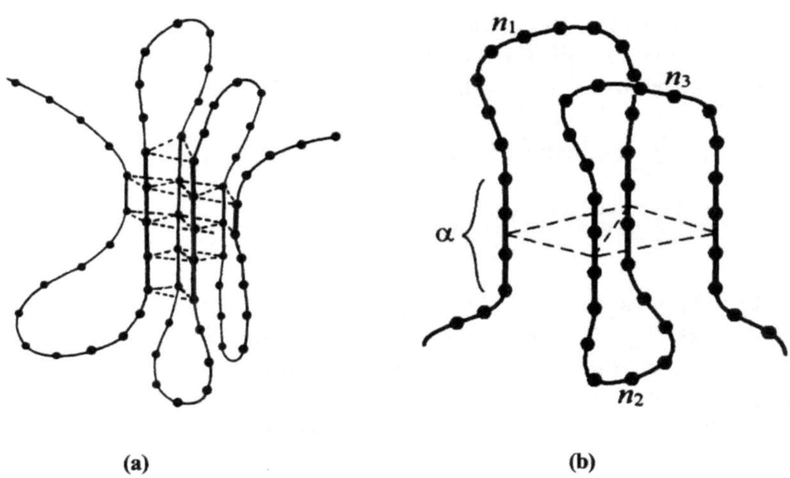

Fig. 3 **a** Scheme of a bundle (Model A [8, 9]). Chain repeating units are evidenced as *black dots*, crystalline packed stems are shown as *thickened straight portions* of the chain, *dashed lines* stand for crystal-like attractions. **b** A bundle comprising 4 stems, 3 loops (see Appendix). *Dashed lines* stand for energy attraction between crystal-like stems with α chain atoms; n_1, n_2, n_3 are numbers of chain atoms in the loops

review we will concentrate on Model A, where the existence of three longer, identical stems is assumed, together with shorter stems of variable length (see Fig. 3a). Model A allows for more degrees of statistical freedom than Models B and C, where all stems within each bundle are of equal length, and leads to values of the lamellar thickness in better agreement with experiment. Examples of localized interactions may be the insertion of a $-CH_3$ group in an appropriate "hole" of a growth face, the interaction between appropriately oriented phenyl rings and more generally van der Waals or dipolar attractions between organic moieties. At temperatures $T \geq T_0$ bundles are ineffective—and will be disregarded for our purpose—as they are not stable enough to produce crystallization [9]. Conversely, at $T < T_0$ the bundle distribution is assumed to be metastable, i.e., corresponding to a relative free energy minimum. As shown in Fig. 1, we may also have *bundle aggregates*, where two or more different bundles are connected by more than one loop. Since explicit consideration of bundle aggregates instead of simple bundles does not appear to entail significant numerical effects [8, 9], in the following *we shall limit our analysis to simple bundles.*

We assume the chain to be in a sufficiently poor solvent (or in the bulk state) that the long-range expansion from the excluded volume may be disregarded (*unperturbed chain*). The ideal dissolution temperature T_0 will specify our reference chain state, Z_0 being the chain partition function. As soon as the temperature T is taken below T_0, "stable" bundles may form; their overall statistical weight is $\sum[\Omega_i \exp(-E_i/k_B T)]$, Ω_i and E_i respectively being the multiplicity factor and the attractive energy involved in the i-th conformation. Considering that *all the chain conformations contributing to this term are also permitted at* T_0, the partition function $Z(T)$ may be expressed as

$$Z(T) = Z_0 + \sum_i \Omega_i \left[\exp\left(-E_i/k_B T\right) - \exp\left(-E_i/k_B T_0\right)\right] \qquad (1)$$
$$\cong Z_0 - \sum_i \left(\Omega_i E_i/k_B T_0^2\right)\left(T_0 - T\right),$$

and $Z(T) > Z_0$ because $E_i < 0$. The approximate equality derives from the assumption that $(T_0 - T)$ is relatively small; note that, under this assumption, the characteristic ratio C_∞—controlling the chain statistics, in particular the multiplicity factors Ω_i—will not change significantly. The bundle excess free energy $G(T) - G(T_0)$ is, in the same approximation:

$$\Delta G(T) = G(T) - G(T_0) = -k_B T\left[\ln(Z(T)) - \ln(Z_0)\right] \qquad (2)$$
$$\cong \sum_i \frac{\Omega_i E_i}{T_0 Z_0}\left(T_0 - T\right)(< 0).$$

We see that, for small relative undercoolings $(T_0 - T)/T_0$, the thermodynamic drive to bundle formation is proportional to the undercooling itself. This suggests an interesting, although qualitative interpretation. Namely, the probability of bundle formation is also proportional to the undercooling, consistent with the experimental finding with polyethylene that the lamellar fold length tends to be inversely proportional to $\Delta T = (T_0 - T)$ [8, 9]. The interested reader may find more information on the mathematical approach in the Appendix.

We remark that *the formation of bundles implies contraction of the otherwise unperturbed chain.* Apart from the bulk polymer where any chain contraction is strongly contrasted by entanglements, this fact is clearly suggested by Fig. 1. Neglecting the distance between the first and the last atoms in the bundle, we see that the "bundled" chain may be geometrically modeled as a sequence of uncorrelated bridges, whereas the unperturbed chain has bridges plus unbound bundles. Labeling as β^2 the mean-square chain contraction and n_{bundle}, n_{stem}, n_{fold}, n_{bridge} respectively the numbers of chain atoms in the bundle, in a crystal-like stem, in a chain fold, in a bridge and with p the number of stems in the bundle, we have

$$\beta^2 = \frac{\langle n_{\text{bridge}} \rangle}{\langle n_{\text{bridge}} \rangle + \langle n_{\text{bundle}} \rangle} \cong \frac{\langle n_{\text{bridge}} \rangle}{\langle n_{\text{bridge}} \rangle + \left(\langle p \rangle - 1\right) \langle n_{\text{stem}} + n_{\text{fold}} \rangle}, \quad (3)$$

neglecting the correlation between $(n_{\text{stem}} + n_{\text{fold}})$ and p and taking $n_{\text{stem}} \ll n_{\text{fold}}$. We remark that, unlike the usual polymer collapse due to polymer/solvent segregation [15], β^2 is independent from the molecular weight, provided it is large.

As far as crystallization in the bulk is concerned, we point out that here the formation of very small intra-molecular aggregates with very few (i.e., ~ 2–4) stems is quite possible, given the natural tendency of polymer chains to self-pack by back folding (see Fig. 2). In this case, unlike in solution crystallization, the stem arrangement is stabilized by co-packing with topologically distant chain portions. The statistical properties of hairpins in the melt, basically dictating the lamellar thickness, are essentially analogous to those existing in solution.

2.3 Application of the Bundle Model to Selected Experimental Data

2.3.1 Polyethylene Chain-Folded Crystallization

The bundle model can be applied to account for lamellar thickness dependence on supercooling in the case of chain-folded crystallization (see Appendix, Eq. 16). We will discuss in some detail the case of polyethylene. As shown in Fig. 1, bundles as well as bundle aggregates are separated by bridges. From our previous calculations the average length of the loops ($\sim$ 50–200 chain atoms) far exceeds the short-stem length ($\sim$ 2–20 atoms), indicating that the crystal-packed polymer fraction within the bundles is small; conversely, the average length of the bridges is substantially larger than the loops' [8, 9].

In our previous work [7–9] the bundle structure was regarded as sufficiently stable—i.e., enduring in time—to determine the initial lamellar thickness L (or initial fold length [16]) as equal to the average chain length contributed by the short packed stems and by the adjoining loops, see Fig. 4. The bridges would then be accommodated into the structure dictated by the pre-formed bundles (see Fig. 4, old model A [8, 9]). From Fig. 5 [16] the experimental results for PE may be cast in either of the following empirical forms:

$$L = a\Delta T^{-1} + b\,, \tag{4}$$

$$L = \alpha\Delta T^{-\gamma}\,, \tag{5}$$

where the best-fitting parameters from polyethylene data are $a = 240\,\mathrm{nm}\,^\circ\mathrm{C}$, $b = 5\,\mathrm{nm}$, $\gamma \cong 0.70$. In contrast with the last figure, the best-fitting exponent from the calculations based on our previous model is $\gamma \cong 0.45$.

In view of this disagreement, as well as of evidence from polymer mesophases and MD simulations, we also propose an alternative model, based on the concept that the attractive interactions are so short-lived as to be effectively delocalized. As a consequence, *bridges separating consecutive bundles are also taken into account* in the evaluation of the average stem length of the growing crystal, in addition to the crystalline stems and to the loops

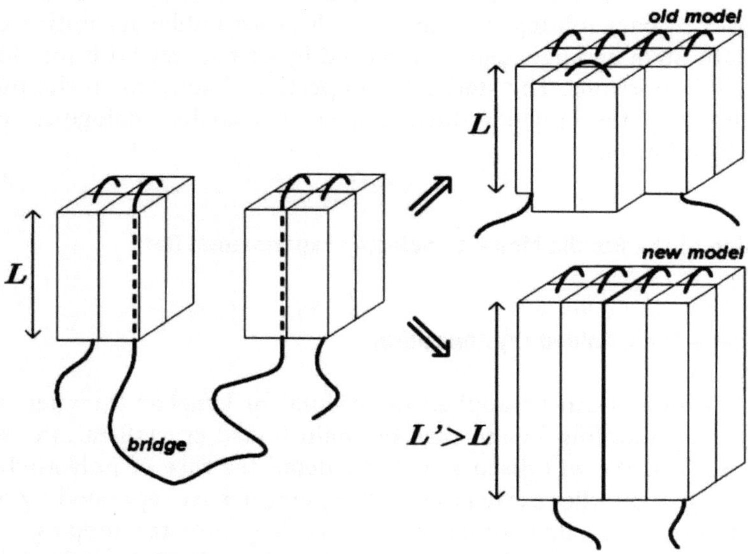

Fig. 4 *Left*: Two bundles with the same lamellar thickness L, separated by a bridge. *Right*: After aggregation, in principle the bundles may either retain the previous thickness L while the bridge produces new crystalline stems (old model), or the bridge is "reeled in" with an increase of L to L' (new model)

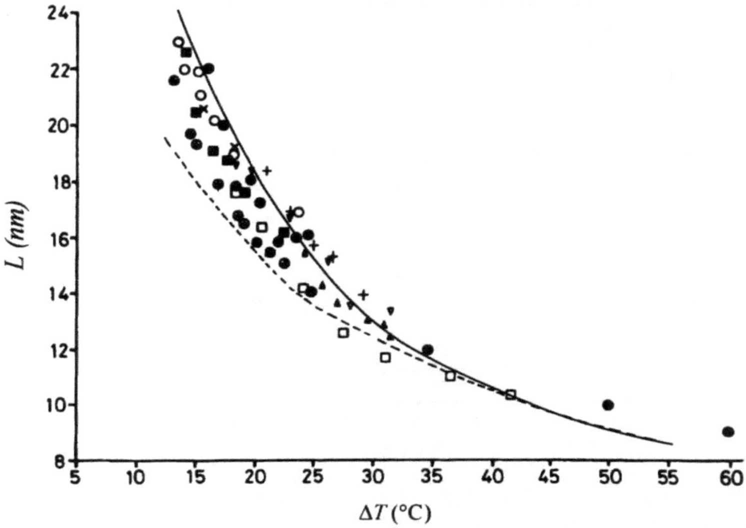

Fig. 5 Initial fold length L against undercooling $\Delta T = T_0 - T$ for both melt- and solution-crystallized polyethylene, from different solvents [16]. *Dashed line* gives previous calculations (Model A [9], old model in Fig. 4), *solid line* shows the results from the new model in Fig. 4 after re-adjusting the energy of fusion per $-CH_2-$ unit from $E = 1.07$ to $E = 1.42$ kcal/mol

constituting the bundles. In mathematical terms the difference between our previous results—only based on the average length of chain strands *internal* to the bundles—and the new ones comprising the bridges, may be seen by comparing the following equations:

$$\bar{n}(old) = \langle n_{\text{stem}} + n_{\text{loop}} \rangle ; \qquad (6)$$

$$\bar{n}(new) = \frac{\langle n_{\text{stem}} + n_{\text{loop}} \rangle (\langle p \rangle - 1) + \langle n_{\text{stem}} + n_{\text{bridge}} \rangle}{\langle p \rangle},$$

where n_{stem}, n_{loop}, n_{bridge}, and p respectively stand for the numbers of chain atoms in a short stem, in a loop, in a bridge and the number of stems per bundle (see Fig. 1), and their average values vs. ΔT are reported in Fig. 6. The lamellar thickness is given by $L = \bar{n}l$, l (= 0.127 nm) being the average advancement per chain bond along the axis. The energy of fusion per ($-CH_2-$) group is taken as equal to $E = 1.07$ kcal/mol (old model), to be compared with an experimental value of 0.98 kcal/mol [17]; we notice that the temperature scale depends on E, other parameters being constant. Upon a change of E, to a good approximation we have $E\Delta T =$ constant.

The new calculations based on the reeling-in of the bridges, i.e., on $\bar{n}(new)$, see Eq. 6, lead to the best-fitting plot of L vs. ΔT reported in Fig. 5 (Model A, solid line). The old plot (dashed line) and the experimental points [16] are

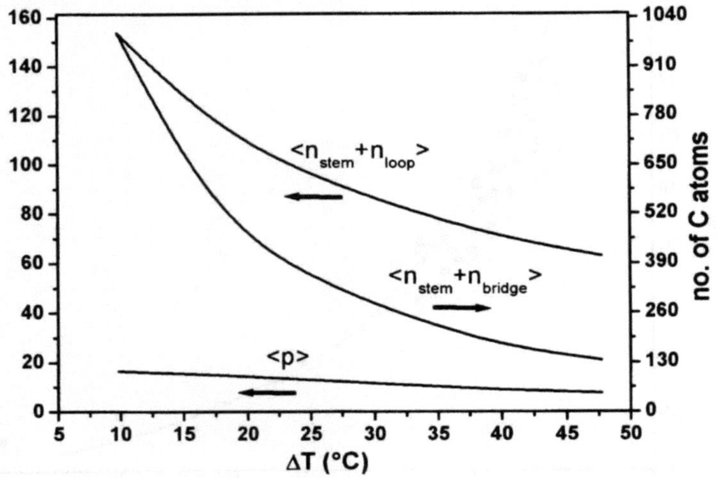

Fig. 6 PE: average values of the length of crystal-packed stems (n_{stem}), of loops (n_{loop}), and of bridges (n_{bridge}), given as no. of $-CH_2-$ units. Also reported is the average number of stems per bundle $\langle p \rangle$, see Figs. 1 and 3 for definition of symbols. Consistent with Fig. 5, calculations were carried out as for previous Model A [9] after changing E from 1.07 to 1.42 kcal/mol

also reported. For the new plot we adopt an energy of fusion per $(-CH_2-)$ group equal to $E = 1.42$ kcal/mol, larger than, although still roughly comparable with, the experimental value 0.98 kcal/mol. We point out that both $\langle p \rangle$ and $\langle n_{\text{stem}} \rangle$ (the latter is not shown separately in Fig. 6) tend to a finite limit for $\Delta T \to 0$.

We see that the bridge contribution also changes the power-law dependence ($L \propto \Delta T^{-\gamma}$), making it closer to experimental observation. In fact the resulting value of the theoretical exponent γ is 0.69, to be compared with the experimental value 0.70; we point out that n_{bridge} becomes the main contribution to $\overline{n}(new)$ for small values of ΔT.

In conclusion, both the value and the temperature dependence of the lamellar thickness $\overline{n}(new)$ is significantly affected by the "reeling in" of the bridges inside the lamella under formation. We notice that, to the extent that such a mechanism will not be complete, the disagreement between our best-fitting value and the experimental figure of the energy of fusion could be partly explained. In fact the two treatments represent limit models and both appear to have something to say; in other words the previous interpretation (see Fig. 4, old model) with $E = 1.07$ kcal/(mol of $-CH_2-$ units) [8, 9] should not be completely discarded, as the reeling-in is contrasted by defects, entanglements etc. Although the issue should be explored in more detail, we point out that some amount of reeling-in should also take place in melt crystallization.

2.3.2 Metastable Mesophases and Cold Crystallization

A number of systems which in polymer literature are normally referred to as mesophases are obtained under kinetic control. Examples are the "smectic" phase of isotactic polypropylene [18, 19], mesomorphic syndiotactic polypropylene [20–22], mesomorphic PET [23, 24], and other instances where intermediate degrees of order result after quenching polymers from the melt to temperatures often close to T_g. In these cases disorder is plausibly more static than in bundles close to T_0 and these phases usually crystallize upon heating to an appropriate temperature in the stable crystal phases.

The processes that lead to these mesophases are usually based on a very rapid quenching over large temperature intervals. Care should therefore be taken correlating the structural and morphological features of these phases with the structure of the undercooled melts at well-defined temperatures and indeed often conflicting data are present in the literature. Furthermore, because of the very high supercooling, the "bundle approach" can be applied only qualitatively. It is unlikely that the reeling-in of bridges (as defined in Sects. 2.2 and 2.3) participates substantially to the formation of domains of this kind of mesophases, although on the basis of the available experimental data it is difficult to confirm this statement quantitatively. On the other hand, consistent with our bundle model in the bulk state, it appears that all these mesophases are formed by relatively small and irregular, loosely chain-folded domains.

In all the known cases sizes reported for mesomorphic domains of the kind we are discussing range typically between 3.0 and 20.0 nm in diameter. The lower value places the ordered domains at the lower limit of sizes Bragg diffraction effects can reasonably be expected for polymers: it is also compatible with the dimensions that can be envisaged for bundles (or baby nuclei), typically containing 5–15 stems from our calculations on PE [8, 9]. The larger dimensions may result from a bundle aggregation mechanism. All these mesophases are also characterized by a more or less disordered packing of parallel chains with significant structural and conformational disorder in arrangements that can be considered as pre-crystalline. In two out of the three cases that will be commented in some detail the proposed packing is pseudo-hexagonal, consistent with the general assumptions of our approach.

Pre-crystalline order in PET has been investigated by a number of different groups and in the present volume the issue is reviewed under different perspectives in two other contributions [25, 26]. Nodular structures measuring ca. 7.5 nm and ca 15 nm apart in PET quenched from the melt close to T_g were first described by Geil [27, 28]. Such structures are essentially amorphous, albeit characterized by some degree of orientational order and by a significantly higher density as compared to the fully isotropic polymer. They are indeed qualitatively compatible with the bundle model we propose [29]. Apparently order in pre-crystalline PET can vary considerably and can range from oriented amorphous to nematic hexagonal, to smectic depending upon thermal and mechanical treatments. Different authors favor a variety of different interpretative models but there is a wide agreement that the cold crystallization process develops through a series of stages characterized by increasing levels of order.

Syndiotactic polypropylene (sPP) tends to adopt, upon quenching and crystallization from the melt at 0 °C, a mesomorphic structure characterized by a disordered but essentially trans-planar conformation [20–22] which is probably present to a significant degree in short sequences also in the amorphous phase. A similar structure had been reported to develop a number of years ago upon stretching the highly defective syndiotactic polymer available at the time [22]. The structure that is obtained does not coincide with the crystalline trans-planar form III of this polymer [30]. Indeed in the mesophase sPP adopts a truly pseudo-hexagonal packing, which upon crystallization in form III becomes substantially distorted by specific intermolecular interactions. In the mesophase the chain packing is disordered but there appear to be correlations between neighboring chains as discrete diffraction features show up on the first layer in fibre diffraction patterns. The available evidence from morphological investigations carried out so far suggests very small dimensions for the semi-ordered domains. Assuming a lamellar morphology for the mesomorphic domains their thickness turns out to be initially ca 2.2 nm for samples "crystallized" for long times at 0 °C and, as crystallization at that temperature proceeds, it grows to ca. 2.7 nm [31, 32]. It is interesting to note that a closely similar value of 2.8 nm is obtained for the lamellar thickness of a 3D crystalline sPP polymorph obtained under marginally different crystallization conditions, i.e., quenching to 0 °C for 1 min and then allowing crystallization to occur at 25 °C [31, 32]. The FWHM of the equatorial peak at $2\theta = 17°$ $(Cu - K_\alpha)$, related to lateral packing of chains in the mesophase measures typically 2–3° after subtraction of the amorphous and other components, suggesting small lateral coherence lengths of the order of 2–3 nm, indeed compatible with the bundle-like structure we envisage [33–35]. The lateral dimensions of the mesomorphic domains may evidently be much larger as indicated by electron microscopy evidence.

The case of isotactic polypropylene (iPP) presents some differences with respect to those just discussed. While both sPP and PET adopt in their mesophases disordered, extended, essentially non-helical conformations, iPP is characterized by a unique, relatively well ordered, stable chain structure with three-fold helical symmetry [18, 19, 36]. More accurately we can state that an iPP chain segment can exist in the mesophase either as a left handed or as the enantiomeric right-handed three-fold helix. The two are isoenergetic and will be able to interconvert only through a rather complex, cooperative process. From a morphological point of view Geil has reported that thin films of mesomorphic iPP quenched from the melt to 0 °C consist of nodular structures measuring ca. 12–13 nm in diameter [37,38]. Grubb & Yoon [39] reported somewhat later that the coherent domain size obtained from FWHM of equatorial WAXD maxima of mesomorphic iPP is about 2.6 nm. This value increases to about 20 nm upon annealing, which coincides with that resulting from electron microscopy. SAXS investigations [38,40] indicate for mesomorphic iPP samples a lamellar long period of 8–9 nm. Considering that the amorphous component is ca. 60–70% in quenched iPP samples, stem lengths around 2.5–3.5 nm (12–16 c.r.u.'s) can be inferred to be typically present in such mesomorphic domains. Various tentative structural models have been proposed for mesomorphic iPP: while solid state NMR evidences [41] suggest significant similarity with the packing of β-iPP [42,43], characterized by hexagonal packing of isochiral helices, diffraction results appear to favor co-presence of helices of opposite chirality in mesomorphic iPP domains [44,45]. As also discussed in another contribution to the present volume [26], Fourier calculations carried out assuming a coherence length of 4 nm along and of 3 nm perpendicular to the helix axis, indicate that the structure probably consists of bundles of less than 20 stems. Best semi-quantitative agreement of observed and calculated scattering intensity suggests coexistence in a disordered arrangement of packing features both for the monoclinic α-modification (with helices of both chiralities) and for the chiral, hexagonal β-form, in a structure that however does not correspond to either [44,45]. This picture agrees with the fact that upon annealing the mesophase the racemic α-phase is obtained, plausibly by a combination of sliding diffusion and lateral stem adjustments at a very local scale. The data pertaining to the crystallization from the melt into the mesophase just discussed appear consistent with the idea that in a very limited temperature range, a few degrees above T_g, iPP has enough intramolecular mobility allowing it to adopt the helical conformation to a significant number of chain segments but limiting chirality reversal once the helix segments are formed. On the other hand it implies very hindered diffusion not allowing inter-stem order to develop substantially beyond a parallel arrangement. These ideas are further corroborated by the observation [38] that the "smectic" mesophase does not develop if iPP is quenched to temperatures substantially below T_g. While the mesophase, once formed, may be stable up to 80 °C, quiescent crystallization either from the melt or from the glass at 25 °C yields preferentially, if not only, small crystals of the monoclinic α-phase. Reports that mesomorphic iPP is composed of small distorted α-form crystals [46] indeed describe samples quenched to −78° or −130 °C, which subsequently are brought to room temperature and observed, with a procedure that hardly gives rise to the mesophase.

A remarkable feature common to both iPP and sPP is that the lamellar morphological units of the mesophase obtained at ca. 0 °C evidenced by electron microscopy [38,39], hardly change in lateral size, thickness, or shape upon annealing and recrystallizing into the stable crystal form at about 80 °C.

Lamellae obtained by direct crystallization from the melt into stable crystalline modifications at temperatures of 60–80 °C have thicknesses which differ only marginally from those of the fully mesomorphic samples. Such a finding, if confirmed, is highly relevant because it suggests that the lamellar organization depends only to a very limited extent on the particular crystal form while, to a first approximation, the pre-existing arrangement in the undercooled melt appears to play a key role.

The mechanism of crystallization by annealing monotropic mesophases or glasses, i.e., cold crystallization, is likely to present key differences from standard nucleated crystallization processes from the melt or from solution. In this respect there is increasing experimental evidence that it may well proceed by aggregation and reorganization of bundle-like pre-crystalline entities into larger domains which eventually transform into crystals as molecular mobility increases with increasing temperature. Pertinent results for PET reported by Kaji, Imai et al. [25], are in this respect convincing and a number of different reports in the literature indicate that similar processes may apply with appropriate modifications in the case of sPP [31, 32] and iPP [39, 47].

2.4 Comparison with Molecular Simulation Results

A large body of molecular-dynamics simulation studies on polymer crystallization were carried out in the last few years. Although comparison of the results with the present theoretical approach is made difficult in view of the different model assumptions, in particular concerning the initial polymer state, some results appear to be relevant to the present analysis. Muthukumar and coworkers subjected to a thorough investigation both heterogeneous and homogeneous nucleation [10]. Using a united-atom model and a classical force field with bond stretching, angle bending, torsion, and 6/12 Lennard-Jones contributions, these authors adopt both Langevin dynamics and Monte Carlo techniques [4, 48, 49]. They find that the initial lamellar thickness increases with temperature according to the $1/\Delta T$ experimentally observed temperature dependence, see Eqs. 3 and 4, and that the lamellae tend to produce further thickening in a quantized manner, reminiscent of secondary crystallization. Consistent with our conclusions for extended-chain mesophases (see following Sect. 3), Muthukumar finds that a liquid-crystalline phase may not be a precursor to chain-folded lamellae, at least with flexible polymers like polyethylene [48]. This author also studies the molecular origins of the shish-kebab crystallization morphology under extensional flow, observing in particular the existence of two populations of stretched and coiled chain strands, correlated in turn to the formation of shish and kebabs, respectively [49].

In basic agreement with our model, it is shown convincingly that preformed bundles essentially comprising single chains, may crystallize as folded chain lamellae ("kebabs") upon extended-chain crystals ("shish") acting as templates. The shish is formed under the action of hydrodynamic forces and the same chain may eventually crystallize either in the shish or in the kebab morphology depending on its initial conformation. A plausible suggestion appears to be that the kebab morphology could result by bundle deposition on the surface of the shish.

In a systematic molecular-dynamics investigation of polymer crystallization with an original molecular bead-and-spring model including the classical stretching, bending, torsion, and Lennard-Jones components, Yamamoto reaches interesting conclusions concerning the present study [51–54]. His simpler 2D model of crystallization leads to very neat chain-folded lamellae, the origin of the folding being recognized as local chain clustering, closely reminiscent of our bundle model; reeling-in of "slack" chains gives an important contribution to regularization of the lamellar folding. 3D crystallization of a polymer chain onto the crystalline surface *in vacuo* also leads to chain-folded lamellae, chain entanglements being pushed into the interlamellar region. Crystallization from the melt was also investigated, with evidence of thickening growth of the lamellae in addition to the usual lateral growth. The lamellar folds were found to be rather short, most of them connecting either the nearest or quasi-nearest crystalline stems.

In a recent molecular-dynamics study of supercooled polymer melts, Meyer and Müller-Plathe [55] (MMP) show that a bundle-like structure may be obtained even with no attractive potential between polyethylene-like parallel strands, merely requiring that the *trans* rotational states along the chain backbone have a lower energy than the *gauche* states. We remark that both the energy and the entropy of their bundles is smaller than that of the pure melt, in analogy with our model. In fact, in the melt, i.e., the reference state, the chains comprise a larger amount of higher-energy gauche rotations than in the bundles where they are only localized on the chain folds, while at the same time the bundle structure is much more ordered than the melt's. As observed with our bundle model, the average fold-to-fold distance is inversely proportional to $1/\Delta T$ for small values of ΔT. The MMP model would probably maintain its essential features if complemented with delocalized attractions between atoms on parallel chain strands; conversely, as it stands the model is obviously unsuited to simulate polymer crystallization from a dilute solution. The main feature common to both our bundle model with localized stem-stem attractions and MMP's model with no attractions is that, for temperatures sufficiently close to T_0 the free energy advantage to have chain folds grows linearly with ΔT. Increasing the number of chain folds we have a proportionately smaller fold-to-fold length L; therefore $L \propto \Delta T^{-1}$ for $\Delta T \rightarrow 0$ (see Eq. 3).

Both our bundle model and MMP's may also be useful approaches to represent the first stage of a polymer mesophase formation from an amorphous system, a process which can be considered a specific mode of polymer crystallization. With reference to the mesophase classification discussed by us in a recent paper [11], the MMP model is better suited to handle the development of polymer mesophases belonging to Class 2, where attractive interactions between parallel chains are in essence identical to those in the melt. In contrast, our bundle model may be naturally applied to mesophases (of Class 1, according to our definition), characterized by some degree of specificity of inter-chain interactions, as apparent from X-ray diffraction or thermal evidence. We shall come back to this issue in Sect. 3.1.

3 Thermotropic Polymer Mesophases as a Pre-Crystallization Step

Upon cooling from the melt under different conditions, various flexible polymers may develop stable phases characterized by different degrees of structural disorder. Examples range from polyethylene [56, 57], to polytetrafuoroethylene [58–60], to a number of polydialkylsiloxanes and polyphosphazenes (see Table 1) [11]. Some of these disordered phases are usually classified as thermotropic mesophases, others are characterized in the literature as crystalline modifications. Since such phases are normally able to transform to ordered crystalline modifications, provided that the polymer is structurally regular, they can in principle be viewed as stages along a path of progressive ordering from the random coil to a crystalline organization characterized by 3D translational order. In this sense we can see them as pre-crystalline entities.

These phases very often share, aside from a relatively high disorder, another important feature with the bundle-like pre-crystalline entities that we discussed up to this point. This feature is hexagonal or pseudo-hexagonal packing of the polymer chains [61]. In this respect molecular order at a local level is similar, although in the thermotropic mesophases we are now discussing disorder is even more dynamic than in previous cases. Indeed they are essentially characterized by a high entropy.

However, very significant differences with respect to bundle-like pre-crystalline aggregates are also experimentally apparent. Thermodynamically stable mesophases of flexible polymers are quite generally characterized by a high propensity towards chain extension, i.e., towards the suppression of chain-folding and hairpins in the polymer. Furthermore, as expected also for liquid crystalline systems, there is a tendency to develop domains which are very large also in lateral dimensions. Last but not least formation of these high entropy phases, as opposed to crystallization, proceeds normally relatively rapidly, even at very modest undercoolings.

A preliminary analysis of the features of thermotropic mesophases of flexible polymers we just described, leads us to envisage a path of polymer crystallization different from chain-folded, fold-preserving crystallization in-

Table 1 Thermal and geometrical data of selected semiflexible polymers giving rise to thermotropic hexagonal mesophases or main-chain disordered crystalline phases (adapted from [11])

	Polymer (→ Class I mesophases)	Ref.	T_{CM} (°C)	ΔH_{CM}[a] (kJ/mol)	T_{ML} (°C)	ΔH_{ML}[a] (kJ/mol)	$D^{\S}$ (Å)	$P^{\S}$ (Å)
1	Polyethylene	[17, 56, 57, 108–110]	240[b]	3.5[b]	250[b]	0.7	4.9[a]	6.5
2	Poly(*trans*-1,4)butadiene	[17, 111–114]	83	1.9	164	0.9	5.0	5.6
3	Polytetrafluoroethylene	[17, 58–60, 98, 115, 116]	19, 30	0.8	332	4.1	5.7	6
4	Poly-*cis*-isoprene[c]	[17, 64, 65, 117, 118]			28	1.1	4.45	5.4
5	Poly(*cis*-1,4)butadiene[c]	[17, 66, 67, 118, 119]			12	2.3	4.34	5.0

[a] Transition enthalpies are given on a per-main-chain-bond basis.
[b] Data at ~ 5 kbar
[c] Main-chain-disordered, high-temperature crystalline phases, rather than columnar mesophases; thermal data refer to the melting of the high temperature crystal phase into the liquid.
§ D are average chain interaxial distances derived from diffraction data. Values of the persistence length P are generally calculated using the expression $P = C_\infty l_o^2/2l$ from C_∞ data, while for l_o the following bond lengths were adopted: C – C (single) 1.54 Å, C – C (double) 1.34 Å, Si – O 1.60 Å, Si – C 1.89 Å, Si – Si 2.35 Å, P – N (phosphazenic) 1.59 Å. P values in parentheses were calculated from the expression $P = \gamma D^2$ using the best fit value 0.16 of γ.

Table 1 (continued)

	Polymer (→ Class II mesophases)	Ref.	T_{CM} (°C)	ΔH_{CM}[a] (kJ/mol)	T_{ML} (°C)	ΔH_{ML}[a] (kJ/mol)	D[§] (Å)	P[§] (Å)
6	Polydihexylsilylene	[68, 69, 120]	41	16.6	> 250	dec.	15.5	31(29)
7	Polydimethylsilylene	[68, 70]	162	0.6	226	0.2	7.8	15 *
8	Polydiethylsiloxane	[17, 121–127]	10	0.9	53	0.2	9.6	8.1
9	Polydipropylsiloxane	[121, 123, 127–132]	70	1.1	206	0.5	11.2	15
10	Polydibutylsiloxane	[121, 123, 127–131]	−19	0.4	310	< 0.1	12.3	(24)
11	Polydipentylsiloxane	[123, 127–131]	−22	1.0	330	< 0.1	13.4	(29)
12	Polydihexylsiloxane	[121, 123, 127–131]	23	2.6	330	< 0.1	14.6	(34)
13	Poly(*bis*-trifluoroethoxy)phosphazene	[133–140]	70	4.4	240	0.4	11.9	(23)
14	Poly(*bis*-phenoxy)phosphazene	[137–142]	160	4.1	390	< 0.1	13.2	24
15	Poly(*bis*-4-chloro-phenoxy)phosphazene	[135–143]	180	4.1	360	< 0.1	14.2	40
16	Poly(*bis*-3-chloro-phenoxy)phosphazene	[136–143]	75	4.0	370	< 0.1	14.2	39 *
17	Poly(*bis*-4-methyl-phenoxy)phosphazene	[137–139, 144, 145]	160	3.9	420	< 0.1	13.5	40 *
18	Poly(*bis*-4-isopropyl-phenoxy)phosphazene	[146–148]	141	4.1	> 330	< 0.1	16.7	40 *
19	Poly(*bis*-propyl)phosphazene	[71]	248	3.4	310	0.4	11.3	(20)

* values adopted from experimental data for polymers which are structurally closely related (e.g., for 17 and 18 the value of poly(*bis*-4-ethylphenoxy)phosphazene is used)

volving bundles and similar structures that we have discussed in the previous section. This alternative or complementary mechanism will be explored in some detail in the subsequent parts of this section. Here we just anticipate what we consider its basic features for the sake of clarity. Because of their dynamically disordered structures, chains giving rise to such mesophases adopt average cylindrical shapes whose intermolecular interactions are weak and very similar to those found in the melt at similar temperatures. Folds and hairpins in these mesophases are energetically (enthalpically) disfavored because they are not "frozen in" by adequate compensatory inter-stem interactions. Note that, especially in cases where the backbone is regularly substituted by a high density of side-chains and thus it is surrounded by them (e.g. in polysiloxanes and polyphosphazenes), folds also substantially reduce the degrees of freedom accessible to the side groups, hence the entropy of the chain as a whole. These disordered chains with an average cylindrical "self-compacted" structure will therefore tend to remove a large fraction of chain folds and hairpins present in the melt. They will pack hexagonally in large domains.

3.1 Polymer Mesophases: the Self-Compacting Model

In a recent paper [11] we propose that a flexible polymer with regularly placed substituent groups may give rise to a stable mesophase even though its glass transition temperature T_g is relatively low ($\leq 0\,°C$). Two classes are distinguished (see Table 1). Class I comprises relatively thin polymers (chain diameter $D < 6$ Å, equal to the distance between axes of contacting chains); their mesophases are characterized by some amount of specific, or crystal-like interactions, as indicated by both X-ray diffraction and transition enthalpy data to the liquid, isotropic state, the associated enthalpy being larger than 1 kJ/(mol of chain bonds). Conversely, Class II comprises thicker chains ($D > 8$ Å), their backbone is less extended than in the crystalline state by about 10%, and their hexagonal chain packing in the mesophase does not show any evidence of specific interactions as the transition enthalpy to the liquid state is about 0.1 kJ/(mol of chain bonds). Class II mesophases are essentially stabilized by the large entropy arising from the conformational disorder of both the main and the side chains, tightly packed around the chain backbone. Unlike in the liquid state, the parallel arrangement of chains with an average straight axis and a hexagonal arrangement optimizes the conformational freedom of the mesophase. In fact, each side group experiences the lowest possible disturbance from both the same-chain and the adjacent-chain neighbors, whereas the entropy loss due to the chain straightening is very modest due to the relatively large value of the chain persistence length. For these polymers the mesophase stability increases with the aspect ratio P/D, where P is the per-

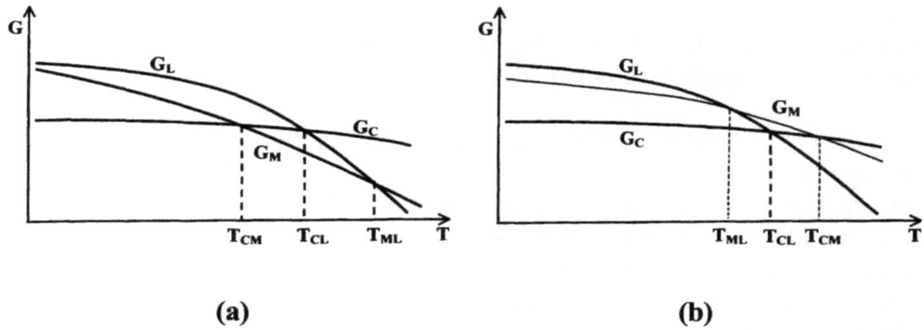

Fig. 8 Free energy per chain bond of the liquid (L), of the bulk mesophase (M), and of the crystalline (C) polymer. T_{XY} is the transition temperature from X to Y ($X, Y = $ L, M, C). **a** In this case the mesophase is stable between T_{CM} and T_{ML} while T_{CL} is virtual. (From [11]); **b** The mesophase is virtual, and so are T_{CM} and T_{ML}; melting of the crystal into the liquid is only observable transition occurring at T_{CL}

with our model at low temperatures the mesophase is stabilized over the liquid essentially because of a higher entropy (i.e., a larger negative slope of G_M than of G_L vs. T). The enthalpy is assumed to be the same in both phases, in analogy with Onsager's original athermal theory for elongated rigid molecules [12]. At higher Ts, on approaching the isotropization temperature T_{ML} the liquid-phase chains increase their bending, which is inherently forbidden to the mesophase, with an increase of both entropy and energy. To provide a quantitative measure of this parallel increase, from $G = H - TS$ and $S = -\left(\frac{\partial G}{\partial T}\right)_P$ we derive the following equation ($P = \text{const}$)

$$H(T) - H(T^*) = T_0\left[S(T) - S(T^*)\right] + \int_{T^*}^{T} [S(T) - S(\theta)]\,\mathrm{d}\theta\,. \tag{8}$$

Since the entropy is bound to increase with temperature, the right-hand side is the sum of two positive contributions, showing directly that the enthalpy is also bound to increase with temperature with a corresponding rate. Obviously enough, in Fig. 8a the increase of both S_L and H_L with increasing T is reflected in the simultaneous increase of the *slope* of G_L and of the *intercept* (on the G axis) of the tangent straight line. We may translate into: an increase of chain bending entails an increase of intramolecular strain and repulsive interactions in the polymer melt. Also, an increase of the chain persistence length—or of the chain diameter as an equivalent statement—implies a larger thermal energy to produce a given chain bending (and a downward curvature of G_L vs. T, see Fig. 8a). This is in qualitative agreement with the larger transition temperature T_{ML} for more rigid polymers, see Table 1 and [11].

3.2
Experimental Data from Polymer Thermotropic Mesophases and Conformationally Disordered Crystals

The data to be briefly discussed pertain to flexible polymers displaying either hexagonal thermotropic mesophases [61], or main-chain conformationally-disordered *crystalline* structures, consistent with Wunderlich's definition [62, 63], or both. Only polymers for which quantitative flexibility measurements are available, or closely related molecules, are listed in Table 1 [11]. From the reported thermal data we can see that the first five entries show comparable, relatively large values of ΔH_{ML} and ΔH_{CM}. The first three polymers give columnar mesophases whereas poly(*cis*-1,4)isoprene [64, 65] and poly(*cis*-1,4)butadiene [66, 67] present main-chain conformationally-disordered crystalline structures deviating from hexagonal coordination. The five polymers just mentioned are a subset of Class I: indeed most thermodynamically stable polymer mesophases arising from the rigidity and anisotropy of mesogenic groups are expected to belong to this class since: (i) they present more or less extensive conformational and positional disorder; (ii) they have comparable values of ΔH_{ML} and ΔH_{CM}; (iii) they adopt modes of packing characterized by directional intermolecular interactions. The formation of thermotropic mesophases and the features of such a transition process will depend on the specifical chemical system. Such mesophases may be in general smectic or nematic depending upon the features of developing order.

The members of Class II in Table 1 present very small enthalpies of the mesophase-liquid transition [$\Delta H_{ML} \leq 0.5$ kJ/(mol of chain bonds)], suggesting that their mesophase is hardly stabilized by specific interatomic interactions. By contrast, we point out that in all cases the crystal-mesophase transition has a significant enthalpy value, mostly $\Delta H_{CM} > 1$ kJ/(mol of chain bonds). Consistent with their relatively flexible character, the polymers listed in the Tables have their glass transition below ambient temperature.

Class I and Class II mesophases differ not only with respect to transition enthalpies, but also for degrees of order evidenced by diffraction measurements. Class I mesophases are normally characterized by a significant intramolecular order and some degree of correlation between neighboring chains in the axial direction, resulting in at least some off-meridional layer-line reflections. Since conformational disorder characterizes both the main and the side chains, for rigorously Class II mesophases off-meridional layer-line reflections are expected to be absent. On a large-domain scale both Class I and Class II present, in the axial projection, an ordered inter-chain packing which is hexagonal for Class 2, as in this case the polymer chains are effectively modeled by continuous cylinders. Class II mesophases are indeed self-compacted, conformationally disordered columnar phases. Conversely, because of possible directional interactions, the packing of Class II mesophases may deviate from the hexagonal arrangement, although it is adopted by chains for which such a lattice is efficient. Thus, a hexagonal packing is a necessary but insufficient feature to identify Class II mesophases.

Geometrical and flexibility data pertaining to the same polymers are also given in Table 1, namely the persistence length and the average chain-to-chain interaxial distance D. The first five polymers in Table 1 have D values smaller than 6 Å, unlike all the following polymers (i.e., no. 6 to 19 in Table 1, Class II). This is a consequence of the relatively bulky substituents carried by Class II polymer chains. For some of the polymers in Table 1 the C_∞ and P literature values are widely scattered or unavailable. In those cases lower-limit values of P from experimentally determined geometrical parameters, are predicted from our model by suitable interpolation and reported within parentheses.

For polymers giving rise to Class II mesophases the chain backbone is at least as flexible as for sp^3 carbon-chain polymers, and carries two identical side groups on every second chain atom, except in the case of polysilylenes [68–70] which are even more crowded; increasing the size of the side groups for a given main chain structure yields both an increasing chain diameter D and a larger persistence length P, see Eq. 7. The experimental distance D between neighboring chain axes is large enough as to ensure that *interdigitation between their side groups is extremely unlikely* and intermolecular interactions occur only at the lateral surface of such cylinders. It is reasonable to expect that, in view of the regular and symmetrical chemical arrangement of the side groups along the chain, the statistically cylindrical structure tends to be maintained also in the melt. In fact for molecules giving rise to Class II mesophases sharp bending of the chain axis is contrasted by non-bonded repulsive interactions between the side groups, filling the volume around the chain axis according to cylindrical symmetry. As a result a relatively large value of the persistence length P is found in spite of the extensive chain conformational disorder. From the data in Table 1 the thermodynamic stability of Class II mesophases as estimated from the melting temperature T_{ML} clearly correlates with the aspect ratio P/D, or more simply with D since $P \propto D^2$.

3.3 Bundles and Mesophases

Two simple thermodynamic considerations are suggested upon examination of Fig. 8. The first is that *at temperatures below T_{ML} the free energy of the bulk mesophase G_M is in general bound to be lower than G_L, the free energy of the amorphous.* In the limit of Class II mesophases, since $\Delta H_{ML} = 0$, we will have $G_M = G_L$ at $T = 0$ K while $G_M < G_L$ at temperatures $0 < T < T_{ML}$ since it is $S_M > S_L$ at temperatures low enough as compared to T_{ML} (Sect. 3.1). In the case of Class I mesophases $\Delta H_{ML} > 0$, i.e., mesophases are enthalpically stabilized with respect to the liquid state, while $S_M < S_L$, so it will be $G_M < G_L$ at temperatures T, with $0 \leq T < T_{ML}$. Note that the above consideration will in general apply also if the mesophase is only metastable (Fig. 8b), i.e., for systems like iPP and sPP etc., and we may logically extend it to cases where this phase is virtual, which we intend as never observed experimentally.

The second consideration, anticipated in Sect. 3.1, is that whereas S_M and S_C increase little with temperature (see Fig. 8) so that in the literature the temperature dependence of G_C and G_M is often approximated by straight lines, S_L *varies moderately up to the glass transition temperature* T_g *but will increase substantially between* T_g *and the isotropization temperature.* The behavior of S_L is due to the fact that, as the temperature increases, polymer chains are progressively more likely to bend sharply in the melt, whereas they are forced to remain straight in the mesophase and in crystals. The downward curvature of G_L in Fig. 8 shifts to lower temperatures with inherently more flexible chains. With very flexible polymers T_{ML} becomes therefore smaller than the crystal-melting temperature T_{CL} and a stable mesophase cannot form.

Below T_{CL} and above T_{ML} (Fig. 8b) crystallization of flexible polymer systems can take place, in our perspective, only through the bundle mechanism. But also below T_{ML} this is likely to be the major route, as we have assumed throughout this contribution, simply because the crystal becomes the stable phase, and rather than growing into large mesomorphic domains, bundles and bundle aggregates crystallize. The fact that for crystallization temperatures typically a few degrees above T_g, irregularly chain-folded, metastable mesophases develop in a number of systems confirms the general picture. In these cases molecular mobility is simply insufficient to allow the metastable mesophase intermediate to evolve into the stable crystal, which appears to be the normal occurrence at somewhat higher temperatures. The bundle structure will indeed evolve according to specific paths, depending on structural features and physical variables like pressure, applied fields, temperature, and concentration. In a slightly different perspective we can roughly state that the more flexible the systems are (i.e., the smaller their persistence length), the higher will be the free-energy barrier to the formation of a true mesophase from the melt, as both attractive energy interactions and entanglements are bound to increase with chain flexibility. As a consequence, under appropriate conditions, flexible chains will tend to form small bundles rather than large mesophase domains, therefore giving rise to lamellar crystallization. The mentioned observation that below T_{ML} the free energy of the bulk mesophase G_M is in general lower than G_L, can be taken in this context as a qualitative indication of the propensity of amorphous chains to form straight stems which organize in parallel chain aggregates closely related to bundles.

Turning to polymers giving thermodynamically stable mesophases we must assume that, since we have described bundles as an inherent structural feature of undercooled polymer melts, such structures should occur, at least in principle, also in such systems, to the extent that attractive interchain interactions which account for bundle formation play a significant role. On the other hand, rigorously speaking Class II mesophases are entropy-stabilized and inter-chain interactions are identical to those found in the melt so that the $M \rightarrow L$ transition occurs virtually with $\Delta H_{ML} = 0$ (see Table 1). Therefore, in a rigorously Class II mesophase model we cannot expect bundles to play any role. What happens in the case of real systems which correspond only approximately to the Class II idealization is largely an open question. The issue is inherently connected with the detailed mechanism by which stable thermotropic mesophases form from the melt and even with the structure of the melt itself in such systems. It should be investigated both experimentally and with molecular simulation approaches. Our present view is that under such circumstances bundle-like precursors may form which evolve rapidly to large mesomorphic domains. If this is correct we could foresee that rapid enough quenching of Class I or approximately Class II polymer melts to temperatures at which the 3D ordered crystal is the stable phase, will result in defective lamellar morphologies originated from bundle-like precursor structures. On the other hand it is clear that bundles, chain-folding and chain-folded lamellar crystallization are ruled out for all polymers with a very high persistence length.

For the isotropization transition of polymers giving rise to rigorously Class II mesophases, $\Delta H_{ML} = 0$ (see Table 1), it is tempting to suggest that it may approach second order, since the absence of localized attractions allows an increasing structural disorder to set in, whereby the chains gradually lose their parallel orientation (order-disorder transition). With the above assumptions the G_L and G_M lines in Fig. 8a should merge with a single slope at the transition temperature T_{ML}, which therefore is expected to be poorly defined experimentally. This is indeed what is found with many Class II polymers for which also X-ray diffraction evidence suggests that local order dissipation in the melt may be difficult.

It is interesting to note that some polymers we classified approximately in Class II (e.g., poly(dipropylphosphazene)) [71] are very poorly soluble in any solvent, consistent with their poor flexibility. On the other hand some Class I polymers giving columnar thermotropic mesophases, and even some Class II systems like polyaryloxyphosphazenes, etc. (see Table 1) are soluble and give rise from solution to chain-folded crystals. This behavior can be understood considering that the solvent may (or may not) influence the side-group behavior depending upon its ability to give specific interactions with the side chains, thus increasing the flexibility of the main chain by decreasing its compactness. We are led to state that in the presence of appropriate solvents polymers that give thermotropic mesophases in bulk may, if they are flexible enough, afford bundle-like structures and therefore lamellar crystals. In summary for polymers giving rise to thermotropic mesophases, depending upon their rigidity, upon pressure, and upon the specific solvent, it is reasonable to expect a variety of crystallization behaviors ranging from bundle-mediated lamellar to chain-extended crystallization: if specific inter-stem interactions are stable enough the tendency to eliminate folds and hairpins will be substantially reduced.

The crystallization of 3D-ordered crystalline phases from thermotropic mesophases, envisaged as stable pre-crystalline partially ordered intermediates, is an additional interesting issue which should be considered with care experimentally, theoretically, and with appropriate simulation approaches. Depending upon the nature of the mesophase it can be seen as a crystal-crystal transition or, for conformationally disordered, columnar mesophases, it approaches a "true" crystallization process. It is quite clear that the pre-existing order will play a major role: for example if the mesophase is chain-extended, bundle equilibria and chain-folding should not play any role. Indeed available experimental evidence supports this idea. Mechanistic and kinetic features should in general differ widely from the standard chain-folded crystallization processes yielding thin lamellar structures. In a number of cases (polyphosphazenes, polysiloxanes, see below) the crystalline polymorphs obtained from the chain-extended precursor differ from those obtained from solution.

4
Polymorphism, Pre-Crystalline Order, and Chiral Crystallization

The development in achiral polymers of chiral crystal structures, which are often metastable polymorphs, may be looked at under different perspectives. Some authors consider crystallization on highly structured, chiral crystalline interfaces or growth fronts as the only relevant mechanistic step. Such crystallization paths may well be applicable in a number of cases and are discussed in different contributions to the present volume. An alternative possibility is that the molecular structure and the aggregation behavior in the melt or in solution play a significant role in chiral crystallization of helical polymers, at least in some instances. In previous work we suggested that a plausible reason for the frequent occurrence of chiral crystallization of helical polymers is related to the persistence in the crystal of a pseudo-hexagonal packing of isomorphous chains already existing in the pre-crystalline state [13]. Such a pre-crystalline state may be broadly identified with the different levels of organization ranging from bundles to hexagonal mesophases which we are discussing in the present contribution.

The suggestion that pseudo-hexagonal pre-crystalline packing determines a bias favoring chiral crystallization rests on two simple ideas: (i) for helical chains with limited bulges and holes, hence with limited interpenetration, pseudo-hexagonal, isochiral chain arrangements yield the largest vibrational entropy as well as the best chain-to-chain contacts; (ii) in the case of small bundles of chains, isochiral aggregates may undergo twist deformations with the largest statistical probability.

Concerning point (i), it is justified by the twofold consideration that the hexagonal packing allows chains to be symmetrically equivalent thus maximizing their vibrational freedom, as suggested by Kitaigorodsky [72], and it also generally allows a good filling of space, unlike the one that may be achieved with non-isomorphous chains, see Fig. 9. Point (ii) takes into consideration twist deformation, or cooperatively spiralized deformation of all chains within a bundle, which represents perhaps the most important departure from straight-chain packing in that it preserves a compact filling of space; double and triple helices are to be regarded as particular cases of twist deformation. With reference to Fig. 10, it may be shown that the partition function of a set of n chains in a bundle undergoing twist deformation ($n = 3$ in the figure) is largest when all the chains have the same chirality. We indicate with n_1 and n_2 the number of chains with opposite chiralities c_1 and c_2 in the bundle ($n_1 + n_2 = n$) and assume that two equal and opposite twist deformations, say $(+\Delta)$ and $(-\Delta)$, are imparted to the bundle. Letting ε_+ and ε_- be the deformation energies of a c_1-chain under the two twist deformations, in view of the chain chirality we have $\varepsilon_+ \neq \varepsilon_-$; for a c_2-chain the energies will be exchanged, and we have ε_- and ε_+ respectively. Labeling with $w_\pm = \exp[-\varepsilon_\pm/k_B T]$ the Boltzmann weights for the two deformations of a single chain, the sum of the Boltzmann weights for the whole bundle is

$$S(n_1, n_2) = w_+^{n_1} w_-^{n_2} + w_+^{n_2} w_-^{n_1}\,, \qquad (9)$$
$$(w_+, w_- > 0\,, \quad w_+ \neq w_-\,, \quad n_1 + n_2 = n)\,,$$

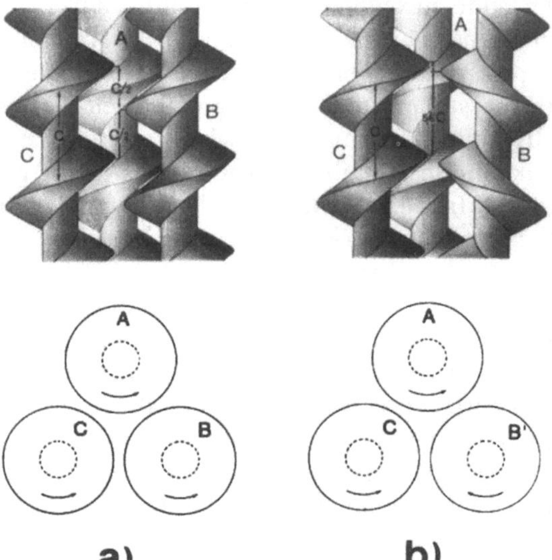

Fig. 9 Two arrangements of helical chains, represented as continuous screws, packing in a hexagonal or a pseudo-hexagonal mode. In **a** the three screws are isomorphous; in **b** the screw B' is enantiomorphous with respect to both A and C and the overall packing is less satisfactory. Reproduced with permission from [13]. Copyright 1998 Am Chem Soc

Fig. 10 Co-operative twist deformation of a bundle with three chains. Reproduced with permission from [13]. Copyright 1998 Am Chem Soc

and we see that S is largest and equal to $(w_+^n + w_-^n)$ if all the chains have the same chirality (i.e., either $n_1 = n$, $n_2 = 0$ or vice-versa) within the bundle.

We notice that the above argument neither implies necessarily that close-to-pseudo-hexagonal polymer crystal structures are chiral nor that polymorphs crystallizing from pseudo-hexagonal pre-crystalline states are necessarily chiral. If a racemic crystal structure is substantially more stable it will be very likely to form, irrespective of the pre-crystalline state. However, literature surveys show that if the crystal packing is hexagonal or close to hexagonal, in the case of non-trivially helical polymers for which enantiomorphic helices can be identified, the likelihood is high that the helices in the crystal are isochiral (see Tables 2 and 3). Indeed the hexagonality index H, defined as the ratio of the largest to the smallest interaxial distance, between the reference helix and its six nearest neighbors, appears to be a reliable indicator of the presence in a given crystal of helices with a single handedness or, respectively, of both. The value of H is 1.00 for a hexagonal or pseudo-hexagonal packing of helices and grows to 1.73 for a trigonal environment. Data pertaining to polymers presenting both chiral and achiral crystalline modifications in which chains adopt helical conformations are in this respect quite illuminating. As shown in Table 2, for any given polymer the polymorph with isochiral helices has a smaller value of the hexagonality parameter H than the racemic polymorph of the same polymer. Some important new cases, the most relevant being sPP [73–76] and polylactide [77, 78], have been added to those already reported in the literature [13] where this approach was developed.

The mechanism we propose for chiral crystallization of helical polymers appears to be also supported by the analysis of the circumstances under which some chiral polymorphs are obtained [13].

Table 2 Selected data of helical polymers (or of enantiomeric polymer pairs) displaying both chiral and achiral modifications (adapted from [13])

Polymer	Phase	ρ_{calc} (g/cc)	Helix	Chir. [a]	$T_{m(exp)}$ (°C)	H [b]	Ref.
it-Polypropylene	α	0.94	3_1	n	175	1.55	[36]
	β	0.92	3_1	y	155	1.00	[42, 43]
it-Poly(1-butene)	I	0.95	3_1	n	136	1.73	[90]
	II	0.92	11_3	n	102	1.41	[92, 93]
	III	0.90	4_1	y	96	1.17	[94, 95]
Polypivalolactone	α	1.23	2_1	n	236	1.57	[85–87]
	γ	1.19	2_1	y	219	1.21	[87, 89]
Poly(-*t*-butylene) sulfide	o.i. *	1.10	3_1	n	210	1.32	[149]
	o.a. *	1.08	3_1	y	162	1.00	[149]
Poly(L-lactide)/ Poly(D-lactide)	o.i. *	1.36	$3_2/3_1$	n	220	1.73	[77], [150–152]
Poly(L-lactide)	α-o.a. *	1.29 (1.35)	10_7	y	185	1.00	[77, 78], [152]
Poly(L-lactide)	β-o.a. *	1.31	3_2	y	175	1.00	[77, 78], [152, 153]
Poly(L-lactide)	γ-o.a. *	1.31	3_2	y	n.a.	1.06	[77, 78]
it-1,4-*cis*-poly (2-methyl-pentadiene)	α	1.00	2_1	n	175	1.57	[154–156]
	β	0.96	2_1	y	165	1.28	[154–156]
st-Polypropylene	I	0.93	2_1	n	163	1.64	[73, 80], [157–160]
	II	0.93	2_1	y	140	1.39	[35, 74–76], [80, 83]
Polydiethylphosphazene	I	1.14	2_1	n	220	1.43	[161, 162]
	II	1.17	2_1	y	220	1.13	[162]

[a] n = achiral structure, y = chiral structure
[b] hexagonality parameter (see text): ratio between the longest and shortest interaxial distance between the reference chain and the six nearest neighbors ($H = 1.00$ for hexagonal structures, i.e 1.41 for tetragonal and 1.73 for tri-coordinated structures)
* o.a. = optically active and respectively, o.i. = optically inactive, i.e., racemic structures; note that similar considerations apply also to 1,4-*cis*-poly(2-methyl-pentadiene)

We examine briefly some specific instances starting with syndiotactic polypropylene (sPP). Aside from the already discussed hexagonal mesophase which can be obtained both drawing fibers and under quiescent conditions, this polymer presents four crystalline forms: phases I [73] and II [74–76] where chains adopt the $(T_2G_2)_n$ helical conformation, forms III [30] and

Table 3 Data relative to selected crystalline polymers for which only chiral crystalline phases, characterized by chiral helical conformations, are known

Polymer	Phase	Helix	H [a]	Ref.
Polytetrafluoroethylene	triclinic	13_6	1.00	[58–60]
	"trigonal"	15_7	1.00	[58–60]
it-Poly (S)-4-methylhexene *	pseudotetragonal (tricl.)	7_2	1.41	[163]
Polyisobutylene	$P2_12_12_1$ (orth.)	$\sim 8_3$	1.00	[164]
Polyoxymethylene	trigonal (pseudohex.)	9_5	1.00	[165–167]
	orthorhombic	2_1	1.05	[168]
Polythiomethylene	triclinic	17_9	1.00	[169]
Polyselenomethylene	pseudo-hexagonal	21_{11}	1.00	[170]
Polyselenomethylene	orthorhombic	2_1	1.00	[171]
it-Poly-(5-methyl-1-hexene)	pseudohexagonal	3_1	1.00	[172, 173]
it-Poly (S)-(5-methyl-1-heptene) *	pseudohexagonal	3_1	1.00	[172, 173]
it-Poly-*t*-butylacrilate	pseudohexagonal	3_1	1.00	[174]
it-Poly(isopropylethylene oxide) *	orthorhombic	2_1	1.01	[175]
Poly(β-ethylpropiolactone) *	orthorhombic	2_1	1.36	[176]
Poly(β-methylpropiolactone) *	orthorhombic	2_1	1.25	[177–179]
Polydiketene	orthorhombic	2_1	1.04	[180]
Poly-*p*-benzamide	orthorhombic	2_1	1.01	[181]
Polyethyleneoxybenzoate	orthorhombic	2_1	1.21	[182]
it-Poly-*cis*-(1,3-pentadiene) *	orthorhombic	2_1	1.08	[183]
it-Poly(2-vinylpyridine)	trigonal (pseudohex.)	3_1	1.00	[184, 185]
it-Poly(styrene-*alt*-carbonmonoxide) *	monoclinic	2_1	1.29	[186]

[a] hexagonality parameter (see text) * optically active chemical repeat

IV [79] which present respectively chains in *trans*-planar and in a non-helical $(T_6G_2T_2G_2)_n$ conformation. In form I (see Table 2), i.e., the stable form of sPP obtained under most crystallization conditions, the chains are packed with an alternation of right-handed and left-handed helices. On the contrary, in the somewhat less stable form II [80], all the helical chains share the same chirality.

The most relevant observation for our analysis is the development of form II by annealing or solvent treatment of non-oriented samples, containing, aside from the amorphous phase, only *pseudo*-hexagonal, *trans*-planar mesomorphic domains [81, 82]. It is quite clear that in this process no ordered crystalline surfaces, which might in principle nucleate the chiral form II, can be present. In the case of sPP, chains with all-*trans* conformation show a hexagonal packing in the mesophase and a very close-to-hexagonal packing in form III. Therefore, *pseudo*-hexagonal bundle arrangements, which we also expect in amorphous sPP, are likely to favor the *trans* conformation. The slow formation of the *trans*-planar, *pseudo*-hexagonal mesophase from the amorphous quenched a few degrees above the glass transition, can be understood in this framework as a *quasi*-homogeneous process by which bundles perfect and aggregate giving rise to mesomorphic domains. With appropriate annealing or solvent treatments of unoriented samples, favoring the mobility of the chains, the *trans* chains adopt the more stable helical conformation. If this occurs at low enough temperatures and rapidly enough to maintain the local memory of the pseudo-hexagonal chain packing, then the chiral form II is favored, as its packing is closer to hexagonal. Since in sPP samples containing the trans-planar mesophase the amorphous fraction may reach values of 0.8 or higher [21, 34], it is not surprising that annealing or solvent treatments can also induce crystallization into the thermodynamically stable, racemic form I. The helical forms I and II are plausibly obtained from the amorphous and from the *trans*-planar mesophase respectively, in proportions that depend on the annealing temperature, as observed in many reported cases. In recrystallization processes of oriented samples at high temperature, where relatively large mesophase domains are unlikely to be stable even transiently, nucleation of form II may plausibly occur from small oriented amorphous domains where chains are likely to pack locally in ways that closely resemble the hexagonal trans-planar mesophase.

The above discussion is consistent with the role played by the pseudo-hexagonal mesophase in the nucleation and growth of the chiral form II; on the other hand the conditions under which form II had been previously obtained, were never examined in this perspective. The chiral modification of sPP had been obtained stretching at room temperature compression molded specimens of low stereoregularity [74–76]. Our present information on sPP polymorphism is consistent with the idea that in such samples the *trans*-planar mesophase, although originally not recognized, is likely to be present. We suggest that when form II develops, some *trans*-planar mesophase, or related oriented amorphous domains, are indeed present and play a key role in the crystallization of the chiral form of sPP. Exceptions occur by (i) self-nucleation i.e., if some of form II is already present; (ii) by epitaxial crystallization of form II, for example as a thin layer on 2-quinoxalinol [83] whose ordered, plausibly chiral crystal surfaces guide the nucleation process; and probably (iii) at very-high pressure where form II is suggested to become the more stable crystalline phase of sPP [84].

The case of isotactic polypropylene is somewhat different: issues related to the crystallization of the chiral β-form [42, 43] of this polymer have been discussed by various authors and probably the model presented by Lotz is the most articulated and detailed [14]. It is known that iPP, unlike sPP, adopts in all its crystalline modifications its stable conformation, namely the 3_1 helix with a 6.50 Å periodicity [36], which may be right-handed or left-handed with identical energy and probability. No regular achiral conformation comparable to the trans-planar form of sPP is accessible to iPP and interconversion of iPP helices of opposite chirality is feasible only in a cooperative process with a significant activation energy. Lotz suggests that the metastable β form [42, 43] is nucleated by chiral surfaces present in the stable α-modification of iPP. Crystallization of the β form is a consequence of its larger growth rate as compared with that of the α-phase, which is proposed to depend on the fact that β-iPP "provides for a favorable, self-repeating nucleating site at the growth front" [14]. The frustrated nature of the β-iPP polymorph is also thought to play a key role on the basis of a rather rough idealization of the crystal structures of α- and β-iPP [14]. According to this approach, qualitatively analyzed details of the hypothetical heterogeneous secondary nucleation site (no molecular modeling is attempted) determine the mode of crystallization of iPP. On the other hand the quoted paper contains a somewhat conflicting general observation, namely that "the less-stable phase form tends to grow faster than the more-stable one". In principle we believe this statement to be reasonable, but in this context its *generality* is quite puzzling. The questions are: why should less stable, lower density chiral polymorphs of achiral helical polymers present *so often* better nucleating sites than the achiral forms? Is the fact that these modifications are very often closer to a hexagonal packing than the racemic polymorphs significant or accidental? The alternative, or at least complementary, explanation that we suggest is that the less stable, often lower-density, close-to-hexagonal chiral polymorphs (see Table 2) tend to present a higher entropy and are structurally closer to the pre-crystalline aggregates which precur crystallization. This will result in a lower activation energy (and consequently a faster growth rate) of the chiral crystal modification, which is structurally closer to the pre-crystalline organization (bundles, hexagonal mesophase etc.) provided that: (i) it is at least meta-stable at the crystallization temperature; and (ii) local chain reorganization or transport allowing the selection of stems (or bundles) of the required helical chirality can occur. These two conditions in the case of iPP and in other similar instances are met at intermediate temperatures. At high temperature—under conditions close to thermodynamic control—the chiral β-form does not crystallize because it is plausibly too unstable as compared to α-iPP. At intermediate temperatures the above-described selection process can occur leading to the pseudo-hexagonal β-polymorph. At low temperatures imperfect α-form crystals are obtained because even very small bundles (of both chiralities) are relatively stable in the undercooled melt, reversal of helical chirality in individual stems and the diffusion processes are inefficient and consequently an intimate mixture of stems of both helical chiralities results, as already discussed for the metastable iPP mesophase.

It is quite interesting that in the case of polypivalolactone PVL (see Table 2) the polymorphic behavior—involving in this case chains with the same 2_1

helical conformation—follows a similar re-entrant pattern. The stable, achiral α-modification [85, 86] is obtained at both high and low crystallization temperatures, whereas within an intermediate temperature window, the less stable chiral γ-PVL form [88, 89], with a lower H value, is also obtained in variable amounts. Looking at the crystal structure of the two modifications, both refined to a considerable detail and highly ordered, it is very difficult to envisage features that may qualitatively favor preferential nucleation of the γ-form, which can hardly be described as a frustrated structure.

It seems also meaningful to recall that, for both PVL and iPP, the metastable chiral modification is not obtained from solution. This fact is hard to rationalize if polymorphic discrimination occurs on the basis of the secondary nucleation site which should exist also in the presence of the solution: it rather points to diffusion and to transport problems in the melt, or thermodynamic control in solution.

The third system we will be briefly review is isotactic polybutene-1 (iPB1), specifically as crystallized from solution. A significant difference of iPB1 with respect to iPP is that its polymorphism does not involve helices all with the same 3_1 conformation. In fact the 3_1 helical $(GTGTGT)_n$ conformation, with main chain torsion angles $\sigma_1 = -60°$ and $\sigma_2 = 180°$, is found only in the achiral trigonal form I [90] and in the untwinned form I' [91] with $H = 1.73$ which is normally referred to as the hexagonal form because of the symmetry of its electron diffraction pattern. The tetragonal form II [92, 93] is also achiral and adopts an 11_3 conformation (main chain torsion angles $\sigma_1 = -77°$ and $\sigma_2 = 163°$) with tetragonal packing ($H = 1.41$), whereas the orthorhombic form III [91, 94] is bound by its space group $P2_12_12_1$ to be chiral ($H = 1.17$) and has been refined by two research groups [94, 95] to yield an approximate 4_1 helix structure with $\sigma_1 = -83°$ and $\sigma_2 = 159°$. The values of the main chain torsion angles suggest that the three conformers, although not necessarily pertaining to the same minimum, are separated by very low barriers: indeed this is the result of conformational calculations which examine the conformational map of iPB1 with special attention to the 3_1 and the 11_3 helices [93]. Interconversion of the three conformations in isolated iPB1 stems should therefore occur readily.

The main result of three papers respectively by Holland and Miller [91] and by Chau and Geil [96, 97] is that the polymorph which results in crystallization at or close to r.t. of solutions of iPB1 depends on the maximum temperature reached by the solutions (T_{MS}), or, more precisely, on the temperature at which the solution has been equilibrated last, *irrespective of the polymorph of the initial crystals.* For example if samples of any of the three crystal forms are dissolved at concentrations of 0.01–0.03% in amyl acetate (clearing point $\sim 65\,°C$) and brought to T_{MS} of 90 °C, kept there for 3 to 48 hours (t_S) and then crystallized at temperatures (T_X) between 24 and 50 °C, then 100% form I crystals are obtained. With the same procedure but using a T_{MS} of 120 °C, 100% of the chiral form III crystallizes. For shorter values of t_S (one hour) more complex situations arise, plausibly because equilibrium is not attained: (i) starting with the tetragonal form II crystals, both with $T_{MS} = 90\,°C$ and with $T_{MS} = 120\,°C$, 0–30% of the tetragonal form II crystals are re-obtained along with 70–100% of form I ($T_{MS} = 90\,°C$) or form III ($T_{MS} = 120\,°C$); (ii) starting with form I crystals 100% form I is obtained with $T_{MS} = 90\,°C$, while ca. 20% form I recrystallizes along with 80% form III if $T_{MS} = 120\,°C$. If a solution is kept at $T_{MS} = 120\,°C$ for 24 hours, cooled to 90 °C and kept there again for 24 hours, when crystallization is allowed to occur at r.t., ca. 70% form I along with 30% form III are obtained implying some reversibility.

The mentioned experiments, which are in essence reproducible in the better solvent o-xylene (clearing temperature 55 °C), suggest that equilibrium features of the solutions are attained in times of the order of hours and determine the crystallization behavior of iPB1. It has been proposed [97] that the temperature-dependent equilibria in solution relate to the helical conformations (3_1 vs. 11_3 and 4_1). Considering the low barriers between the different accessible helical conformations [93], equilibrium in this respect should, however, be reached quickly in isolated stems. Long equilibration times appear to point to cooperative processes involving association of stems in bundle-like, pre-crystalline aggregates which, given the low concentrations of the solutions, should be intramolecular. The fact that memory of the crystal structure of the dissolved crystals is maintained 30 to 50 °C above the clearing point, for times of about one hour and then is lost, is also consistent with the presence of surviving stem aggregates, i.e., subcritical nuclei. We may speculate that such structures are likely to be high-entropy pseudo-hexagonal aggregates at higher temperature (120 °C) which upon crystallization favor the chiral form III. Different, more crystal-like "baby nuclei" with three-fold coordination crystallizing into form I crystals, result at lower temperature. All the mentioned pre-crystalline aggregates should have relatively low concentration and dimensions substantially smaller than visible light wave-lengths. Therefore, their persistence well above the optical clearing point appears to be likely.

Crystallization of helical polymers from their thermotropic hexagonal mesophases appears to be another, particularly interesting, testing ground for the present approach. The idea is that if crystallizing from their mesophases polymers have non-trivial helical conformations, then their packing should tend to be close to hexagonal (i.e., H close to 1) in chiral space groups. A somewhat controversial example is PTFE, for whose low temperature form, aside from the necessarily chiral one-chain cell [58–60], also a two chain unit cell has been proposed [98]. This alternative model, which is non-centrosymmetric, metrically monoclinic, triclinic with enantiomorphous, non-symmetry related helices, is not completely convincing. If confirmed it would represent an exception with respect to our scheme which, as repeatedly stated, should not be considered as mandatory.

Among polymers crystallizing from thermotropic hexagonal mesophases (see Table 1) we have many polysiloxanes and polyphosphazenes. For some of them there is clear evidence for the development of different crystalline polymorphs upon crystallization from the thermotropic mesophase as opposed to crystallization from solution. However, for many of these systems, as all the polysiloxanes, poly(*bis*-trifluoroethoxy)phosphazene etc., the crystal packing has not been described so far and we can only hope that it will be in future work. For other systems, like polydipropylphosphazene, only a single crystalline polymorph has been identified [71]. In this specific case the chain adopts a twofold helical conformation but the crystal structure is non-chiral (space group $P2_1/c$) with a close to pseudo-tetragonal packing ($H = 1.42$) suggesting that this polymorph is strongly favored by thermodynamic effects.

We can conclude this section noting that understanding chiral crystallization of helical polymers is still hardly satisfactory, but quite certainly different mechanisms are likely to apply for different polymers and different crystallization conditions. A lot of experimental work still needs to be carried out if we want to progress beyond speculation and generalization of results that may in fact apply only to very specific instances.

5 Concluding Remarks

In spite of its enduring empirical success, the classical Lauritzen–Hoffman theory [2, 99] is today recognized as inadequate by many scientists, both because it disagrees with the molecular modeling results and because it predicts an unphysical divergence of the lamellar thickness L above some ΔT limit. In addition to offering an alternative to that theory, the present contribution represents an effort to recognize fundamental aspects of polymer crystallization and to evidence structural features which will cause homopolymer systems to crystallize according to different mechanisms. We describe three basic modes of polymer crystallization, i.e., (see Scheme 1): (i) usual lamellar crystallization upon cooling from the reference state (melt or solution at T_0) to a lower temperature T, (for bulk samples $T > T_g$, see Sects. 2.3.1–2.3.2); (ii) cold crystallization quenching the melt to below T_g^- or close to T_g, and subsequent heating above T_g ($T \ll T_0$, see Sect. 2.3.2); (iii) crystallization from the melt through thermotropic mesophases (see Sects. 3.1–3.3). In all three cases we propose that structure transformation from the amorphous to the crystalline organization proceeds via the development of high entropy pre-crystalline aggregates, which in a number of cases influence basic features of the crystalline organization.

Pre-crystalline structures characteristic of *modes* (i) and (ii) can be identified with bundles [7–9], i.e., energy-driven hexagonal associations among chain segments, which are essentially consecutive in solution. Within the

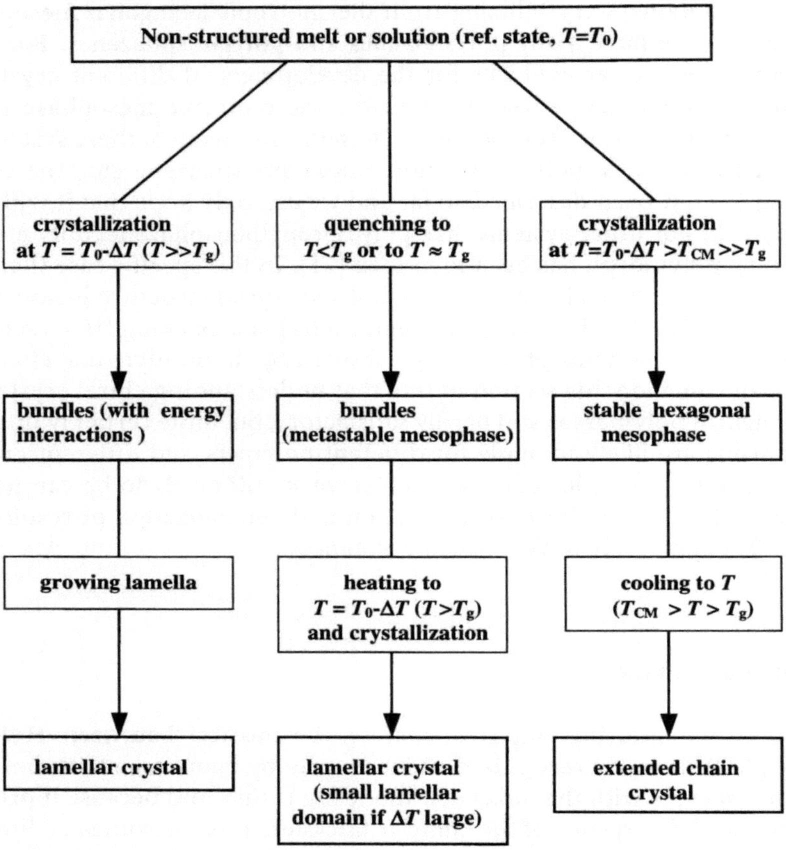

Scheme 1

bundles short chain stems are packed in a crystal-like mode, see Fig. 3. From calculations on PE in solution [8, 9] both the number of chain atoms per stem and the number of stems per bundle is around 10, see Fig. 6. Crystallization is produced if, at temperatures lower than the ideal limiting temperature T_0, aggregation of bundles originates critical crystalline nuclei growing spontaneously (*homogeneous nucleation*). This mechanism is of practical relevance only at very high supercoolings, i.e., close to T_g, where the bundle approach qualitatively applies but quantitative treatments become problematic. On the other hand crystallization will proceed at relative small ΔT values by "standard" *heterogeneous nucleation* processes if bundles encounter a suitable surface on which they can adsorb and crystallize [9]. In the latter case the nucleating surface will have a perturbing influence on bundle equilibrium, which may loose some of its significance with respect to the structure of the growing polymer crystal. In the process of bundle adsorption we make the twofold assumption that both the number of crystalline stems and their adjacency properties are preserved.

Bundle equilibrium is to be regarded as dynamic in character: it is a mean-field description of the metastable equilibrium of an infinite amorphous polymer chain below T_0. As discussed in Sect. 2.1, the bundle picture is especially suited for polymer solutions, as each bundle basically comprises consecutive chain strands. Conversely, in a polymer melt, especially at large ΔT a bundle may also include topologically distant strands and/or strands belonging to different chains, mainly as a result of entanglement effects hindering equilibrium. Obviously enough, if the temperature reaches either the glass transition limit T_g or the limiting dissolution temperature T_0, no bundles can form. The free-energy difference ΔG between the reference state at T_0 and the actual condition at the crystallization temperature $T < T_0$ is proportional to $\Delta T = (T - T_0)$, for sufficiently small values of ΔT; this result, although proven for solution crystallization [9], may be assumed to hold for bulk crystallization as well. As the probability of bundle formation is proportional to ΔG, hence to ΔT, the initial fold length L of the crystalline lamellae is inversely proportional to ΔT ($\rightarrow 0$), see Eq. 4, in agreement with experimental observations for polyethylene [16]. These results show about the same lamellar thickness for both solution and bulk crystallization for a given ΔT, suggesting that the bundle fold length is about the same in both cases and basically depends on the statistical properties of a few ($\sim$2–4) consecutive stems. It is encouraging that different simulation approaches show an analogous dependence of L vs. T: see for example the work by Muthukumar et al. using Langevin dynamics and Monte Carlo techniques [10, 48–50] and the study by Meyer et al. with molecular dynamics [55]. Modeling results at different resolutions and time-scales will clearly be of key importance to guide future theoretical developments. In particular, it would be relevant to compare average predictions derived from the bundle approach [8, 9]—such as the loop, the bridge, and the crystalized stem length at moderate undercooling degrees, see Fig. 6—with molecular modeling results obtained with appropriate conformational and energy parameters. We note incidentally that no surface energy parameter is required by the bundle approach, the fold energy in particular being effectively replaced by the combined definition of the inter-stem attractive energy E and of the characteristic ratio C_∞.

In the perspective discussed in the present contribution, bundle formation occurs within the amorphous phase and in undercooled polymer solutions. It does not imply necessarily a phase separation process, which, however, may occur by bundle aggregation, typically at large undercoolings [*mode* (ii)]. In this case kinetic parameters relating to chain entanglements and to the viscous drag assume a paramount importance. Here again, molecular dynamics simulations can be expected to provide important parameters for theoretical developments; in turn these could orient new simulations in a fruitful mutual interaction.

Depending upon the specific polymer structure and physical variables like pressure etc., a completely different path may be followed, which is essentially entropy driven and corresponds to *mode* (iii), i.e., the development of a hexagonal thermotropic mesophase. This mechanism, in its "pure" form applies to mesophases we define as belonging to Class II [11] (see the discussion in Sect. 3), and leads to the disappearance of folds, hairpins, and also bundles: hence in principle crystallization with fully extended chains will result. In our treatment of these systems we have so far adopted simple single-chain approaches, neglecting intermolecular (or inter-stem) effects. These approximations are less adequate for thermotropic mesophases where enthalpic factors maintain a significant role (Class I mesophases [11]). The latter cases are more difficult to analyze and represent a substantial challenge for future theoretical as well as molecular simulation developments. At present both qualitative considerations and experimental evidence suggest that bundle-like aggregates and chain-folded crystallization play a role in such hybrid instances.

Substantial evidence in a number of existing experimental studies can be easily reconciled with the models discussed in the present contribution. For example *segregation of short chains* reported during crystal growth [1] may be thought to arise with chains which are too short to form bundles and are thus unable to provide a sufficient amount of simultaneous attractive interactions with the crystal to yield stable adsorption. We recall in this respect that one of us obtained the correct trend of the minimum chain length of PE for crystal inclusion vs. the crystallization temperature, using the bundle approach [8].

Neutron diffraction results from partially deuterated PE by Stamm, Fischer and co-workers [100] show that about 20 stems belong to the same chain cluster in sheets within the lamellar crystal, a result compatible with pre-existing building blocks in the crystallization process (from our calculations there should be about 9 stems per bundle at the same ΔT). Similar results were obtained by Sadler [101] who showed that the rms value of the radius of gyration of a long PE chain crystallized from the melt with $\Delta T \cong 40\,°\mathrm{C}$ ($T = 70\,°\mathrm{C}$) is given by $(0.46\sqrt{M_\mathrm{w}})$ Å. If we assume that the chain crystallized from the melt pervades the same volume as in the ideal state, this result corresponds to a reasonable average of the characteristic ratio ($C_\infty = 8.5$). Conversely, a solution crystallized chain with $M_\mathrm{w} = 160\,000$ yields experimentally an average contraction ratio $\beta \cong 0.4$. If we adopt a two-dimensional random walk of bridges connecting different stems as a model for the chain in lamellar crystals, this value is in a reasonable agreement with Eq. 3, predicting $\beta = 0.49$ from the data reported in Fig. 6 ($\langle n_\mathrm{bridge}\rangle \cong \langle n_\mathrm{bridge} + n_\mathrm{stem}\rangle$, see also [8]). Sadler also provides an interesting analysis of the Kratky plots of solution-crystallized polyethylene, showing that scattering data are consistent with a crystal model with some degree of adjacent re-entry which results in some intra-chain stem aggregation along the (110) layers, and also leads to folding between adjacent layers (i.e., superfolding [101]). A general conclusion of both Fischer and co-workers and of Sadler is that in no way the diffraction results may be interpreted in terms of single layers of stems from the same chain; we expect the analysis of these data in terms of the bundle model to yield stimulating results. Analysis of the mean-square size of polymer molecules at appropriate temperatures $T < T_0$ *before the onset of crystallization* would also provide interesting clues to the contraction associated with bundle formation, if we properly account for the polymer contraction due to polymer/solvent demixing. Comparison between atactic (non-crystalline) and isotactic polypropylene could be informative, e.g., as bundles should form only in the latter polymer, whereas the classic polymer/solvent interactions exist in both cases [102].

Additional neutron scattering studies on different polymer systems could prove very important. Strobl [31, 32, 47, 103] provides evidence that, for some polymers, lamellar crystallization is preceded by pre-ordering of the melt followed by formation of planar arrays of blocks. Investigating crystallization from the melt, Kaji and coworkers [25] find pre-ordering phenomena relating to orientational fluctuations of stiff polymer segments which, under appropriate conditions, determine phase separation prior to crystallization.

We have discussed at some length in the present contribution how pre-crystalline aggregates (bundles and hexagonal mesophases) may in principle influence the polymorphic behavior of crystallisable polymers and more specifically the occurrence of chiral crystalline modifications. To analyze this behavior, the relevance of reliable, well-refined crystal structures of different polymorphs for a number of key polymers, can hardly be overestimated. It must be clearly stated that the well-known difficulties associated to predicting polymer crystal structures, are not accidental but result also from the established fact that kinetic effects play a paramount role in polymer crystallization. This is just a different way to express the fact that pre-crystalline states, surfaces, transport phenomena etc., affect the developing crystalline structure and morphology in degrees that remain largely to be quantified. For example in chiral crystallization the effect of specific interactions between a stem and the crystalline surface also needs to be adequately considered. In general the influence of pre-crystalline aggregates on the final crystal structure may turn out to be especially relevant whenever the packing energy surface does not display deep minima.

We note here that all the information presently available on high molecular weight polymer crystal structures is compatible with the bundle model. While very nearly all crystalline polymer polymorphs involve all-parallel chain arrangements, even the only known exception, namely γ-iPP [104, 105], where chains oriented at 80° to each other coexist, is characterized by bilayers of parallel chains with opposite orientation. This structure is thus easily compatible with crystallization mechanisms involving deposition of bundles of 5–10 antiparallel stems on the growing crystal surface. Also the preferred growth direction of γ-iPP crystals [106] normal to the parallel-chain bilayers, is consistent with such a mechanism. On the other hand according to the bundle theory different polymorphs with chains normal to the lamellar surface, if crystallized at the same supercooling and assuming closely similar packing energies, are expected to have identical crystalline lamellar thickness. In the case of γ-iPP, however, the thickness of crystalline lamellae should be smaller than for α-iPP. This is because, according to the bundle approach, it is the stem length and only as a consequence the crystal thickness that are determined by supercooling. Since in the iPP γ-phase chains are tilted by 40° with respect to the normal to the lamellar surface, by crystallization at the same supercooling γ-crystals should be ca. 20% thinner than the α-polymorph and should melt at a lower temperature. This is indeed qualitatively observed for crystals of the two polymorphs obtained at the same crystallization temperature, which correspond to the same supercooling, since the T_0 values of the two polymorphs are very close.

The fascinating issues relating to polymer structures preceding crystallization are still largely open to investigation. More specific and articulated models of such states may provide a better understanding of polymer crystallization, both from the thermodynamic and the kinetic viewpoint. Furthermore, the different mechanisms that lead polymers to crystallize may eventually be understood in a coherent, more unified picture.

Acknowledgements Financial support from "Ministero dell'Istruzione, dell'Università e della Ricerca" (PRIN 2003 and PRIN2004) is greatfully acknowledged.

Appendix

A
The Bundle Model Statistics: A Concise Outline [8, 9]

A.1
Configurational Statistics

Let us consider a polymer chain with $N \to \infty$ identical skeletal atoms, either in solution or in the melt, representing our polymer system. Our reference temperature is T_0, i.e., the temperature above which no bundles may effectively contribute to crystallization. At $T = T_0$ the chain is assumed to be unperturbed and its configurational partition function is $Z_N(T_0) = \lambda^N$ ($N \to \infty$) [107]; for simplicity we use a reduced form $Z_N = Z_N/\lambda^N$ (henceforth simply the partition function) so that $Z_N(T_0) = 1$. Only at $T < T_0$ effective bundles may form, see Fig. 1, and we have $Z_N(T) = 1 + \Delta Z_N(T - T_0)$; note that the unit term corresponds to the bundle-free infinite-chain configuration. Each bundle with n chain atoms ($n \ll N$) will contribute to ΔZ_N

a partition factor ΔZ_n given by $[Z_n(T) - Z_n(T_0)]$, because all conformations appearing at T also exist at T_0 although with a different Boltzmann weight. We deal with the incremental contribution ΔZ_N—dropping henceforth the symbol Δ—which has the meaning of an order parameter measuring the propensity to crystallization; more precisely we shall evaluate the corresponding grand partition function, i.e., the Laplace transform of Z_N with respect to N. We assume the undercooling $\Delta T = T - T_0$ to be small enough compared with T_0 that we may neglect the temperature variation of λ.

The end-to-end vector $\boldsymbol{r}$ of any non-crystallized chain portion comprising n skeletal bonds with a length l will be taken as Gaussian-distributed, i.e.,

$$W(\boldsymbol{r}) = \left(\frac{3}{2\pi l^2 C_\infty n}\right)^{3/2} \exp\left(-\frac{3r^2}{2l^2 C_\infty n}\right), \tag{10}$$

C_∞ being the chain characteristic ratio. Consequently, the statistical weight of a closed loop comprising n chain bonds (i.e., $r = 0$) is given by

$$Z_{n(\text{loop})} = \int W(\boldsymbol{r}) \cdot \mathrm{d}^3\boldsymbol{r} = Qn^{-3/2}, \quad Q = \Delta v \left(\frac{3}{2\pi l^2 C_\infty}\right)^{3/2}, \tag{11}$$

Δv being a small volume wherein the end-to-end loop vector is comprised. As an example, let us consider Fig. 3b showing a bundle comprising 4 self-packing *stems* connected by 3 loops, comprising n_1, n_2, n_3 bonds. Each stem comprises α chain bonds, and we assume the intra-stem conformational freedom to be suppressed by the packing forces. We postulate that each stem after the second one may be placed in two different ways and that an attractive energy $\alpha\eta$ is established between adjacent stems ($3\eta = E = -k_B T_0 \ln \lambda < 0$). Bundles with 2, 3, 4 ... stems have attractive energies $\alpha\eta$, $3\alpha\eta$, $5\alpha\eta$ ... and we assume that the energy stabilization of the 2-stem bundles is poor enough to make their probability negligible. In the example model of Fig. 3b we have a 4-stem bundle with an overall attractive energy $5\alpha\eta$. The total number of chain bonds comprised in the bundle is $n(\text{bun}) = (n_1 + n_2 + n_3 + 4\alpha)$, and the bundle statistical weight is

$$Z_{n(\text{bun})}(\cong \Delta Z_{n(\text{bun})}) = \frac{1}{2}(2Q)^3 (n_1 n_2 n_3)^{-3/2} \times \frac{\exp(5\alpha\eta/k_B T) - \exp(5\alpha\eta/k_B T_0)}{\lambda^{4\alpha}}, \tag{12}$$

the divisor $\lambda^{4\alpha}$ deriving from the conformational freezing of the 4 crystal-packed stems. Following analogous criteria, the statistical weight of less simple bundles (see Fig. 3a, e.g.) and also of bundle aggregates consisting of multiply connected bundles, may be derived. Bundle Model A, from which the numerical results reported in Fig. 6 were obtained, consists of a core of three relatively long stems around which stems of decreasing length are packed, see Fig. 3a. Eventually bundle aggregates were not taken into consideration on account of their modest relevance on the results. After obtaining $Z_{n(\text{bun})}$, the calculations were carried out through the grand partition function as shown in the following.

A.2
The Grand Partition Function of the Chain

The general chain configuration is represented by a sequential arrangement of *Unit Sections* (US); each of them consists of a *bridge*(br) followed by a *bundle* (bun). Equation 12 shows the statistical weight of a representative bundle, whereas any bridge has a statistical weight of unity whatever its length, in our reduced representation. For any Unit Section we have

$$Z_{n(\mathrm{US})} = Z_{n(\mathrm{br})} Z_{n(\mathrm{bun})} = Z_{n(\mathrm{bun})}\,; \quad n(\mathrm{US}) = n(\mathrm{br}) + n(\mathrm{bun})\,, \tag{13}$$

the symbol n always indicating a number of bonds. Considering that the length of any bridge (connecting two bundles), loop (connecting two stems within a bundle), and stem is variable and that the total number N of chain bonds is large ($N \to \infty$), the grand partition function (GPF) formalism is especially suited to tackle the problem. Introducing a new parameter μ, the chain GPF is

$$\Xi(\mu) = \sum_N Z_N(\Delta T) \exp\left[-N\mu\right] = \theta + \theta^2 + \theta^3 + \ldots = \frac{\theta(\Delta T,\mu)}{1-\theta(\Delta T,\mu)}\,, \tag{14}$$

where $\theta = \theta_{\mathrm{br}}\theta_{\mathrm{bun}}$ is the GP factor of the Unit Section. Here

$$\theta_{\mathrm{br}} = 1 + \exp(-\mu) + \exp(-2\mu) + \ldots = \frac{1}{1-\exp(-\mu)}\,,$$

whereas $\theta_{\mathrm{bun}}(\Delta T,\mu) = [\Theta_{\mathrm{bun}}(T,\mu) - \Theta_{\mathrm{bun}}(T_0,\mu)]$ depends on the specific statistical model of the bundle. The requirement of an infinite molecular weight leads to $\theta\Delta T,\mu) = 1$, which defines the value of μ at any $T < T_0$. In turn, this parameter defines the excess free energy $\Delta g = -\mu k_{\mathrm{B}} T$ per chain atom due to bundle formation.

With the knowledge of $\mu(\Delta T)$ we may derive all the bundle statistics. As an example, assuming that $\overline{m}$ is the smallest number of chain atoms in a loop, the average loop length is given by (see Fig. 1, m is a number of chain atoms)

$$\langle m\rangle = m\frac{\partial \ln\theta}{\partial m} = \frac{\sum_{\overline{m}}^{\infty} m^{-1/2}\exp(-\mu m)}{\sum_{\overline{m}}^{\infty} m^{-3/2}\exp(-\mu m)} \cong \frac{1}{2}\sqrt{\frac{\pi\overline{m}}{\mu}}\,\frac{1-2\sqrt{\mu\overline{m}}}{1-\sqrt{\pi\mu\overline{m}}}\,. \tag{15}$$

It is now possible to obtain the average lamellar thickness L. Assuming that the folds are incorporated into the lamellar crystal after a proper conformational straightening, and attributing a 0.127 nm axial advancement to each skeletal atom as in PE, we have

$$L = 0.127\left(\langle m\rangle + \alpha\right)\,\text{nm}\,. \tag{16}$$

In general, a similar procedure based on partial differentiation of $\ln\theta$ with respect to suitable variables (i.e., the stem length, the number of stems/bundle ...) allows us to obtain averages of different sorts.

References

1. Wunderlich B (1976) Macromolecular Physics, vol 1 and 2. Academic Press, New York
2. Hoffman JD, Thomas Davis G, Lauritzen Jr JI (1976) In: Hannay NB (ed) Treatise on Solid State Chemistry, vol 3. Plenum Press, New York
3. Armistead K, Goldbeck-Wood G (1992) Adv Polym Sci 100:219
4. Muthukumar M (2004) Adv Chem Phys 128:1
5. Strobl G (1995) The Physics of Polymers. Springer, Berlin Heidelberg New York
6. Nakatani AI, Dadmun MD (eds) (1995) Flow-induced Structure in Polymers, ACS Symp Ser 597, American Chemical Society
7. Allegra G (1977) J Chem Phys 66:5453
8. Allegra G (1980) Ferroelectrics 30:195
9. Allegra G, Meille SV (1999) Phys Chem Chem Phys 1:5179
10. Liu C, Muthukumar M (1998) J Chem Phys 109:2536
11. Allegra G, Meille SV (2004) Macromolecules 37:3487
12. Onsager L (1949) Proc NY Acad Sci 51:27
13. Meille SV, Allegra G (1995) Macromolecules 28:7764
14. Lotz B (1998) Polymer 39:4561
15. Ganazzoli F, Raos G, Allegra G (1999) Macromol Theor Simul 8:65
16. Barham PJ, Chivers RA, Keller A, Martinez-Salazar J, Organ SJ (1985) J Mater Sci 20:1625
17. Wunderlich B (1990) Thermal Analysis. Academic Press, New York; Gaur U, Wunderlich B (1981) J Phys Chem Ref Data 10:119
18. Natta G, Peraldo M, Corradini P (1959) Rend Fis Accad Lincei 26:14
19. Wyckoff HW (1962) J Polym Sci 62:83
20. Nakaoki T, Ohira Y, Hayashi H, Horii F (1998) Macromolecules 31:2705
21. Vittoria V, Guadagno L, Comotti A, Simonutti R, Auriemma F, De Rosa C (2000) Macromolecules 33:6200
22. Natta G, Peraldo M, Allegra G (1964) Makromol Chem 75:215
23. Bonart R (1966) Kolloid Z 213:1; (1966) Kolloid Z 210:16; (1968) Kolloid Z 231:438
24. Imai M, Kaji K, Kanaya T (1994) Macromolecules 27:7103–7108
25. Kaji K, Nishida K, Kanaya T, Matsuba G, Konishi T, Imai M (2005) Spinodal crystallization of polymers: Crystallization from the unstable melt. Adv Polym Sci (in this volume)
26. Auriemma F, De Rosa C, Corradini P (2005) Adv Polym Sci 181:1
27. Yeh GSY, Geil PH (1967) J Macromol Sci Phys B 1:235
28. Geil PH (2002) In: Fakirov S (ed) Handbook of Thermoplastic Polyesters. Wiley, Weinheim, p 105–224
29. Auriemma F, Corradini P, Guerra G, Vacatello M (1995) Macromol Theo Simul 4:165; Auriemma F, Corradini P, De Rosa C, Guerra G, Petraccone V, Bianchi R, Di Dino G (1992) Macromolecules 25:2490

30. Chatani Y, Maruyama H, Noguchi K, Asanuma T, Shiomura T (1990) J Polym Sci Part C 28:393
31. Al-Hussein M, Strobl G (2001) Eur Phys J E 6:305
32. Grasruck M, Strobl G (2003) Macromolecules 36:86
33. Guadagno L, D'Aniello C, Naddeo C, Vittoria V, Meille SV (2002) Macromolecules 35:3921
34. Nakaoki T, Yamanaka T, Ohira Y, Horii F (2000) Macromolecules 33:2718
35. Guadagno L, D'Aniello C, Naddeo C, Vittoria V, Meille SV (2003) Macromolecules 36:6756
36. Natta G, Corradini P (1960) Nuovo Cimento Suppl 15:40
37. Gezovich DM, Geil PH (1968) Polym Eng Sci 8:202
38. Hsu CC, Geil PH, Miyaji H, Asai K (1986) J Polym Sci B Polym Phys Ed 24:2379
39. Grubb DT, Yoon DY (1986) Polymer Commun 27:84
40. Yan RJ, Li W, Li G, Jiang B (1993) J Macromol Sci Phys B 32:15
41. Gomez MA, Tanaka H, Tonelli AE (1987) Polymer 28:2227
42. Meille SV, Ferro DR, Brückner S, Lovinger AJ, Padden FJ (1994) Macromolecules 27:2615
43. Lotz B, Kopp S, Dorset D (1994) C R Acad Sci Paris 319:187
44. Corradini P, Petraccone V, De Rosa C, Guerra G (1986) Macromolecules 19:2699
45. Corradini P, De Rosa C, Guerra G, Petraccone V (1989) Polymer Commun 30:281
46. Caldas V, Brown GR, Nohr RS, MacDonald JG, Raboin LE (1994) Polymer 35:899
47. Iijima M, Strobl G (2000) Macromolecules 33:5204
48. Muthukumar M (2004) Proceedings of IUPAC World Polymer Congress
49. Dukovski I, Muthukumar M (2003) J Chem Phys 118:6648
50. Muthukumar M, Welch P (2000) Polymer 41:8833
51. Yamamoto T (1997) J Chem Phys 107:2653
52. Yamamoto T (1988) J Chem Phys 89:2356
53. Yamamoto T (2001) J Chem Phys 115:8675
54. Yamamoto T (2003) J Macromol Sci B42:629
55. Meyer H, Müller-Plathe F (2002) Macromolecules 35:1241
56. Bassett DC, Block S, Piermarini GJ (1974) J Appl Phys 45:4146
57. Yamamoto T, Miyaji H, Asai K (1977) Japan J Appl Phys 16:1891
58. Bunn CW, Howells ER (1954) Nature 174:549
59. Sperati CA, Starkweather HW Jr (1961) Adv Polym Sci 2:465
60. Clark ES, Muus LT (1962) Z Kristallogr 117:119
61. Ungar G (1993) Polymer 34:2050
62. Wunderlich B, Gerbowicz J (1984) Adv Polym Sci 60/61:1
63. Wunderlich B, Möller M, Gerbowicz J, Baur H (1988) Adv Polym Sci 87:1
64. Bunn CW (1942) Proc Roy Soc (London) A180:40
65. Nyburg SC (1954) Acta Cryst 7:385
66. Natta G, Corradini P (1956) Angew Chem 68:393
67. Natta G, Corradini P (1960) Nuovo Cimento Suppl 15:111
68. Schilling FC, Bovey FA, Lovinger AJ, Zeigler JM (1990) In: Silicon-Based Polymer Science. A Comprehensive Resource. Zeigler JM, Gordon Fearon FW (eds) Advances in Chemistry Series. Am Chem Soc, Washington, DC 224:1
69. Patnaik SS, Farmer BL (1992) Polymer 33:4443
70. Lovinger AJ, Davis DD, Schilling FC, Padden FJ, Bovey FA, Zeigler JM (1991) Macromolecules 24:132
71. Corradi E, Farina A, Gallazzi MC, Brückner S, Meille SV (1999) Polymer 40:4473

72. Kitaigorodsky A J (1973) Molecular Crystals and Molecules. Academic Press, New York
73. Lotz B, Lovinger AJ, Cais RE (1988) Macromolecules 21:2375
74. De Rosa C, Corradini P (1993) Macromolecules 26:5711
75. De Rosa C, Auriemma F, Vinti V (1998) Macromolecules 31:7430
76. Corradini P, Natta G, Ganis P, Temussi PA (1967) J Polym Sci Part C 16:2477
77. Puiggali J, Ikada Y, Tsuji H, Cartier R, Okihara T, Lotz B (2000) Polymer 41:8921
78. Cartier L, Okihara T, Ikada Y, Tsuji H, Puiggali J, Lotz B (2000) Polymer 41:8909
79. Chatani Y, Maruyama H, Asanuma T, Shiomura T (1991) J Polym Sci Polym Phys Ed 29:1649
80. Palmo K, Krimm S (2002) Macromolecules 35:394
81. Guadagno L, Naddeo C, D'Aniello C, Vittoria V, Meille SV (2004) Macromol Symp 218:125
82. Guadagno L, Naddeo C, Vittoria V, Meille SV (2005) Macromolecules (in press)
83. Zhang J, Yang D, Thierry A, Wittmann JC, Lotz B (2001) Macromolecules 34:6261
84. Rastogi S, van der Burgt F, La Camera D, Terry AE, Cheng SZD (2001) Macromolecules 34:7730
85. Perego G, Melis A, Cesari M (1972) Makromol Chem 157:269
86. Brückner S, Meille SV, Porzio W (1988) Polymer 29:1586
87. Ferro DR, Brückner S, Meille SV, Ragazzi M (1990) Macromolecules 23:1676
88. Meille SV, Konishi T, Geil PH (1984) Polymer 25:773
89. Meille SV, Brückner S, Lando JB (1989) Polymer 30:786
90. Natta G, Corradini P, Bassi IW (1960) Nuovo Cimento Suppl 15:52
91. Holland VF, Miller RL (1964) J Appl Phys 35:3241
92. Turner-Jones A (1963) J Polym Sci Polym Letters Bl:455
93. Petraccone V, Pirozzi B, Frasci A, Corradini P (1976) Eur Polym J 12:323
94. Cojazzi G, Malta V, Celotti G, Zannetti R (1976) Makromol Chem 177:915
95. Dorset DL, McCourt MP, Kopp S, Wittmann JC, Lotz B (1994) Acta Crystallogr B50:201
96. Chau KW, Geil PH (1983) J Macromol Sci Phys B 22:543
97. Chau KW, Geil PH (1984) J Macromol Sci Phys B 23:115
98. Weeks JJ, Clark ES, Eby RK (1981) Polymer 22:1480
99. Lauritzen Jr JI, Hoffman JD (1973) J Appl Phys 44:4340
100. Stamm M, Fischer EW, Dettenmaier M, Convert P (1979) Faraday Disc Chem Soc 68:263
101. Sadler DM (1983) In: Hall IH (ed) Structure of Crystalline Polymers. Elsevier, London
102. Hu W, Frenkel D (2005) Polymer crystallization driven by anisotropic interactions. Adv Polym Sci (in this volume)
103. Strobl G (2000) Eur Phys J E 3:165
104. Brückner S, Meille SV (1989) Nature 340:455
105. Meille SV, Brückner S, Porzio W (1990) Macromolecules 23:4114
106. Lotz B, Wittmann JC, Lovinger AJ (1996) Polymer 37:4979
107. Flory PJ (1969) Statistical mechanics of chain molecules. Wiley, New York
108. Bunn CW (1939) Trans Faraday Soc (London) 35:482
109. Bassett DC (1982) In: Bassett DC (ed) Developments in Crystalline Polymers, Vol. 1. Appl Sci Publ, London, 115 pp
110. Chiang R (1966) J Phys Chem 70:2348
111. Natta G, Corradini P (1960) Nuovo Cimento Suppl 15:9
112. Sueshiro K, Takayanagi M (1970) J Macromol Sci Phys B4:39

113. Rastogi S, Ungar G (1992) Macromolecules 25:1445
114. Mark JE (1967) J Am Chem Soc 89:6829
115. Chu B, Wu C, Zuo J (1987) Macromolecules 20:700
116. Stockmayer WH (1988) In: Nagasawa M (ed) Molecular Conformation and Dynamics of Macromolecules in Condensed Systems. Elsevier, Amsterdam
117. Wagner HL, Flory PJ (1952) J Am Chem Soc 74:195
118. Mark JE (1966) J Am Chem Soc 88:4354
119. Poddubbnyi YaI, Erenberg YeG, Yeremina MA (1968) Vysokomol Soedin Ser A 10:1381
120. Cotts PM, Miller RD, Trefonas PT III, West R, Fickes GN (1987) Macromolecules 20:1046
121. Neuburger N, Bahar I, Mattice WL (1992) Macromolecules 25:2447
122. Godovsky YuK, Papkov VS (1989) Adv Polym Sci 88:129
123. Molenberg A, Möller M, Sautter E (1997) Progr Polym Sci 22:1133
124. Papkov VS, Godovsky YuK, Svistunov VS, Litvinov VM, Zhdanov AA (1984) J Polym Sci Polym Chem Ed 22:3617
125. Beatty CL, Karasz FE (1975) J Polym Sci Polym Phys Ed 13:971
126. Tsvankin DY, Papkov VS, Zhukov VP, Godovsky YK, Svistunov VS, Zhdanov AA (1985) J Polym Sci Polym Chem Ed 23:1043
127. Mark JE, Ciu DS, Su TK (1978) Polymer 19:407
128. Kögler G, Loufakis K, Möller M (1990) Polymer 31:1538
129. Möller M, Siffrin S, Kögler G, Ölfin D (1990) Macromol Chem Macromol Symp 34:171
130. Out GJJ, Turetskii AA, Möller M, Ölfin D (1994) Macromolecules 27:3310
131. Out GJJ, Turetskii AA, Möller M (1995) Macromol Chem Rapid Commun 16:107
132. Lee CL, Emerson FA J (1967) Polym Sci A2 5:829
133. Masuko T, Simeone RL, Magill JH, Plazek DJ (1984) Macromolecules 17:2857
134. Kojima M, Magill JH (1985) Polymer 26:1971
135. Kojima M, Kluge W, Magill JH (1984) Macromolecules 17:1421
136. Schneider NS, Desper CR, Singler RE (1976) J Appl Polym Sci 20:3087
137. Kojima M, Magill JH (1989) Polymer 30:579
138. Sun DC, Magill JH (1987) Polymer 28:1243
139. Tarazona MP (1994) Polymer 35:819
140. Bravo J, Tarazona MP, Saiz E (1991) Macromolecules 24:4089
141. Young SG, Kojima M, Magill JH, Lin FT (1992) Polymer 33:3215
142. Kojima M, Young SG, Magill JH (1992) Polymer 33:4538
143. Desper CR, Schneider NS (1976) Macromolecules 9:424
144. Magill JH, Riekel C (1986) Makromol Chem Rapid Commun 7:287
145. Hozumi H, Kohama C, Yonetake K, Masuko T (1991) J Mater Sci Lett 10:1187
146. Meille SV, Porzio W, Allegra G, Audisio G, Gleria M (1986) Makromol Chem Rapid Commun 7:217
147. Meille SV, Porzio W, Bolognesi A, Gleria M (1987) Makromol Chem Rapid Commun 8:43
148. Nakamura H, Miyata T, Yonetake K, Masuko T, Kojima M (1995) Makromol Chem Rapid Commun 16:189
149. Matsubayashi H, Chatani Y, Tadokoro H, Dumas P, Spassky N, Sigwalt P (1977) Macromolecules 10:996
150. De Santis P, Kovacs A (1968) J Biopolym 6:299
151. Hoogsteen W, Postema AR, Pennings AJ, Ten Brinke G, Zugenmaier P (1990) Macromolecules 23:634

Adv Polym Sci (2005) 191: 137–186
DOI 10.1007/12_010

Published online: 28 October 2005

Topological Mechanism of Polymer Nucleation and Growth – The Role of Chain Sliding Diffusion and Entanglement

Masamichi Hikosaka (✉) · Kaori Watanabe · Kiyoka Okada · Shinichi Yamazaki

Faculty of Integrated Arts and Sciences, Hiroshima University, 1-7-1 Kagamiyama, 739-8521 Higashi Hiroshima, Japan
hikosaka@hiroshima-u.ac.jp

Abstract Direct evidence of nucleation during the induction period of nucleation from the melt is obtained for the first time by means of small angle X-ray scattering (SAXS). This confirmed that the induction period of crystallization from the melt corresponds to the process of nucleation, not to that of spinodal decomposition. This success is due to a significant increase in the scattering intensity (I_x) from the nuclei (10^4 times as large as is normal), which was achieved by adding a nucleating agent (NA) to a "model polymer" of polyethylene (PE). I_x increased soon after quenching to the crystallization temperature (T_c) and saturated after the induction time (τ_i). Lamellae start stacking later than the τ_i.

Power laws of the molecular weight (M_n) dependence of the primary nucleation rate (I) and the growth rate (V) of PE, i.e., I or $V \propto M_n^{-H}$ where H is a constant, were found for both morphologies of folded chain crystals (FCCs) and extended chain crystals (ECCs). As the power law was also confirmed on isotactic polypropylene (iPP), universality of the power law is suggested. It is to be noted that the power H increases significantly with increase of the degree of order of the crystal structure. The power law confirms that the topological nature of polymer chains, such as chain sliding diffusion and the chain entanglement within the interface between the nucleus and the melt or those within a nucleus, adopts a most important role in the nucleation and growth of polymers. This is theoretically explained by improving the "chain sliding diffusion theory" proposed by Hikosaka.

Entanglement dependence of the nucleation rate I is qualitatively obtained for the first time by changing the number density of entanglement (ν_e) within the melt. An experimental formula of I as a function of ν_e was obtained on PE, $I(\nu_e) \propto \exp(-\gamma\nu_e)$ where γ is a constant.

Keywords Crystallization · Degree of supercooling · Entanglement · Extended chain crystal (ECC) · Folded chain crystal (FCC) · Growth · Growth rate · Induction period · Melt relaxation · Molecular weight · Nucleation · Nucleation rate · Nucleus · Optical microscope (OM) · Polyethylene · Polymer · Power law · Sliding diffusion · Small angle X-ray scattering (SAXS) · Topology

1 Introduction

Crystallization can be divided into three processes: the primary nucleation process, the growth process, and the overgrowth process. The growth process is mainly controlled by the secondary nucleation mechanism. The steady (stationary) primary and secondary nucleation mechanisms of atomic or low molecular weight systems have been well studied since the 1930s by applying the "classical nucleation theory (CNT)" presented by Becker and Döring, Zeldovich, Frenkel and Turnbull and Fisher and so on [1–4].

But there are two important unresolved questions regarding the mechanism of crystallization : the first question is "are the primary nuclei actually formed from the earliest stage ("induction period") of crystallization?"; and the second is "what is the role of the "topological nature" of polymers in the polymer crystallization mechanism?"

Purpose

The purpose of this review is to solve these two unresolved problems by confirming the nucleation during the induction period of nucleation and the important role of the topological nature with experimental facts regarding the molecular weight (M)- or number density of the entanglement (ν_e)-dependence of nucleation and growth rates.

The former problem is a general problem not only for polymers but also for any other materials (atomic or low molecular weight systems). Although nucleation is a well-known concept, it has never been confirmed by direct observation due to the low number density of the nuclei to be detected with present experimental techniques, such as small angle X-ray scattering (SAXS). Therefore, one of the most important unresolved problems for basic science is to obtain direct evidence to solve the nucleation mechanism of any material.

The latter problem, that is the important role of "topological nature", such as chain sliding diffusion and entanglement, has not been satisfactorily resolved yet. Crystallization and melting are first-order phase transitions. Figure 1 shows a schematic illustration of the crystallization and melting processes of polymers. The ideal crystallization and melting of polymers can be regarded as the transition between a fully entangled Gaussian chain

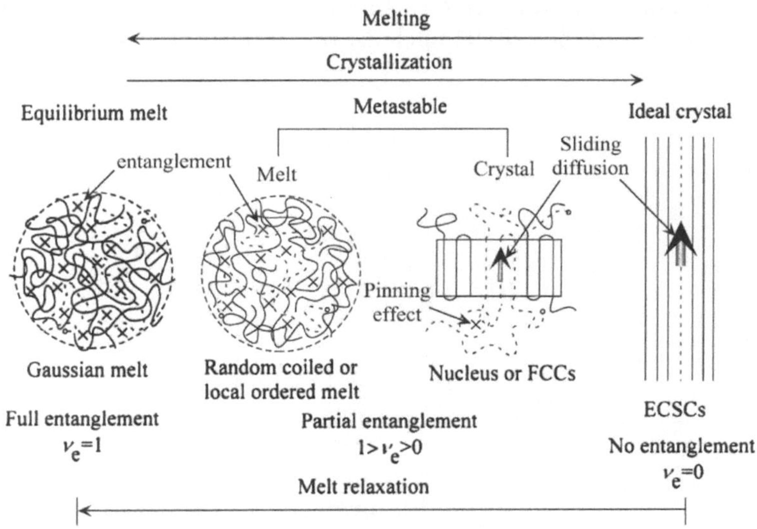

Fig. 1 Schematic illustration of the crystallization and melting processes of polymers. The crystallization process corresponds to processes of disentanglement and chain sliding diffusion. The melting process is the reverse of the crystallization process. Between equilibrium melt and ideal crystal, there exists metastable melt and crystal. *Cross marks* indicate entanglement

melt [5–7] (the equilibrium melt) and fully extended chain crystals [8–13] without entanglements (the ideal crystal).

Hikosaka presented a chain sliding diffusion theory and proposed that the crystallization process is a process where polymer chains are disentangled within the interface between a nucleus and the melt and within a small crystal (nucleus or embryo) and are rearranged into large ideal crystals via chain sliding diffusion [14, 15]. Melting should be the reverse process to crystallization. This means that the topological nature plays an important role in polymer crystallization and melting.

1.1
Nucleation During the Induction Period

The primary nucleation process is divided into two periods in CNT: one is the so called induction period and the other is the steady (or stationary) nucleation period (Fig. 2) [16, 17]. It has been proposed by CNT that small (nanometer scale) nuclei will be formed spontaneously by thermal fluctuation after quenching into the supercooled melt, some of the nuclei could grow into a "critical nucleus", and some of the critical nuclei will finally survive into macroscopic crystals. The induction period is defined as the period where the nucleation rate (I) increases with time t, whereas the steady period is that where I nearly saturates to a constant rate (I_{st}). It should be noted that I is a function of N and t, $I = I(N, t)$. In Fig. 2, N and N^* mean the "size" of a nucleus and that of the critical nucleus, respectively. The size N is defined

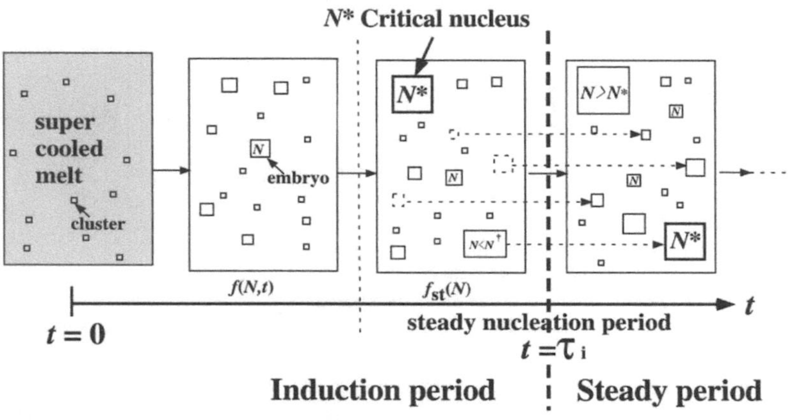

Fig. 2 Illustration of the induction and the steady (stationary) periods during the nucleation process. Small clusters exist in the supercooled melt at $t = 0$. During the induction period ($t < \tau_i$), isolated nuclei of size N, smaller than the critical nuclei (named nanonuclei or embryo), are formed. The nuclei grow larger and larger with increase of time and some of them attain a much larger size than the critical size, N^*

here as the number of atoms or repeating units within the nucleus. In the case of polyethylene (PE), CH_2 corresponds to the repeating unit. t is counted after T reaches a crystallization temperature T_c.

Akupalu and Amis studied the induction period of polyethylene (PE) by means of SAXS and wide-angle X-ray scattering (WAXS) without mixing the nucleating agent (NA) [18]. They observed the scattering intensity (I_x) of the stacked lamellae, but could not observe nuclei due to the very weak signal from the nuclei.

Direct evidence of nucleation during the induction period will also solve a recent argument within the field of polymer science as to whether the mechanism of the induction of polymers is related to the nucleation process or to the phase separation process (including spinodal decomposition). The latter was proposed by Imai et al. based on SAXS observation of so-called "cold crystallization" from the quenched glass (amorphous state) of poly(ethylene terephthalate) (PET) [19]. They supposed that the latter mechanism could be expanded to the usual melt crystallization, but there is no experimental support for the supposition. Our results will confirm that the nucleation mechanism is correct, in the case of melt crystallization.

1.2 Topological Nature in Polymer Crystallization

There is a strong contrast in the crystallization process between atomic or low molecular weight systems and polymer systems due to the topological nature of polymer chains [20]. Figure 3 schematically illustrates this contrast. In the former system the crystallization is a rather simple process where atoms or molecules can be transported and rearranged independently from the isotropic phase into crystalline lattice points (Fig. 3a). In the latter system, the crystallization is a complicated process, because polymers should be transported and rearranged under the strict restriction that molecular chains must not be cut (chemical reaction during the crystallization is excluded here), therefore chains have to slide along the chain axes and disentangle with each other (Fig. 3b). The restriction can be described by introducing a concept of topology that one dimensional topological nature should be invariant during crystallization.

Hikosaka presented a chain sliding diffusion theory and formulated the topological nature in nucleation theory [14, 15]. We will define "chain sliding diffusion" as "self-diffusion of a polymer chain molecule along its chain axis in some anisotropic potential field as seen within a nucleus, a crystal or the interface between the crystalline and the isotropic phases". The terminology of "diffusion" derives from the effect of chain sliding diffusion, which could be successfully formulated as a diffusion coefficient in our kinetic theory.

The theory showed that topological nature assumed an important role in polymer crystallization. The theory succeeded in explaining the ori-

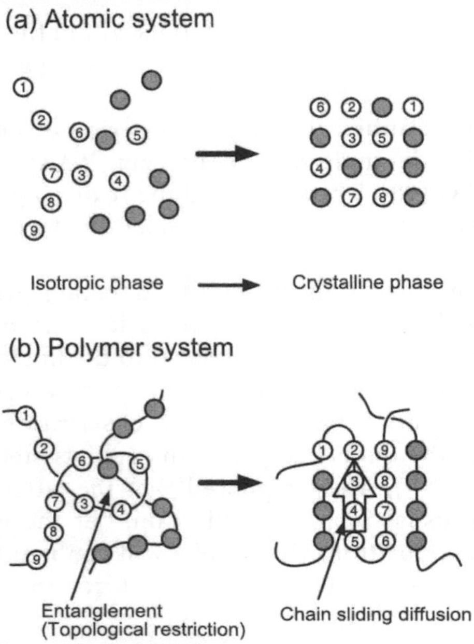

Fig. 3 Difference in crystallization behavior between an atomic or short chain molecular system and a polymer system. **a** Atoms or short chain molecules can be independently rearranged on each lattice point, while **b** the order of the repeating units within a polymer chain is maintained during the rearranging process. Therefore, a chain should slide along its chain axis and disentangle for rearrangement onto the lattice points

gin of folded chain crystals (FCCs) and extended chain crystals (ECCs). It predicted that the FCCs are formed when polymers crystallize into ordered (= immobile) crystals where chain sliding diffusion is difficult, whereas the ECCs are formed when polymers crystallize into a disordered (= mobile) phase where chain sliding diffusion is easy. The prediction has been confirmed experimentally [8–10, 21–23]. Here "order" and "disorder" are defined as the degree of order of chain packing, i.e. high and low, respectively.

Frank and Tosi suggested that "the crystal is likely to change after growth by creeping displacements of the molecular chains tending to even out the segment lengths" [24], which can be regarded as a primitive suggestion of the chain sliding diffusion.

But the topological nature has not been confirmed more directly to date. It is expected that the topological restriction increases with an increase in molecular weight (M) and the number density of entanglement (ν_e). Therefore, the studies of the M or ν_e dependence of crystallization behavior should be important in confirming directly the important role of topological nature in polymer crystallization.

2 Experimental

2.1 Samples

1. In the small angle X-ray scattering and entanglement studies, PE (NIST, SRM1483, $M_n = 32 \times 10^3$, $M_w/M_n = 1.1$) was used. It was named 32 K, where K indicates 10^3. Two kinds of sample were prepared. One was PE mixed with a "nucleating agent (NA)" by 3 weight percent which is named "PE with NA". The NA, NA11-SF, was supplied by Asahi Denka Kogyo K.K. Another was PE not mixed with NA which was named "PE without NA". It should be noted that the 32 K sample accidentally contained a very dilute amount of a different NA which shows equivalent activity to NA11-SF. This means that the nucleation of both samples was heterogeneous nucleation [25].
2. For the study of the M dependence of nucleation rate, seven samples of PE were fractionated from a single sample (called the "mother sample"), named 13 K, 30 K, 50 K, 71 K, 99 K, 139 K, and 256 K, respectively or J-PE1, J-PE2, J-PE4, J-PE5, J-PE6, J-PE7, and J-PE8, respectively. The range of M_n was between 13×10^3 and 256×10^3. The number- and weight-averaged molecular weight (M_n and M_w) and the molecular weight distribution are shown in Table 1.

Table 1 M_n, M_w and molecular weight distribution of polyethylene fractions fractionated from a "mother sample"

Sample Name	$M_n/10^3$	$M_w/10^3$	M_n/M_w	$T_m^0/°C$
J-PE1	13	17	1.28	138.25
J-PE2	30	34	1.15	139.5
J-PE4	50	57	1.14	140.0
J-PE5	71	81	1.14	140.2
J-PE6	99	114	1.14	140.4
J-PE7	139	163	1.17	140.6
Mother sample	47.8	116	2.42	

3. In the study of the M dependence of lateral growth rates, three kinds of fractionated PEs, NIST, SRM1482 ($M_n = 11.4 \times 10^3$, $M_w/M_n = 1.19$), SRM1483 ($M_n = 32 \times 10^3$, $M_w/M_n = 1.11$) and SRM1484 ($M_n = 100.5 \times 10^3$, $M_w/M_n = 1.11$) were used. They were named 11 K, 32 K and 111 K, respectively.

2.2 Crystallization

Isothermal crystallization was carried out at some range of degree of supercooling ($\Delta T = 3.3-14$ K). ΔT was defined by $\Delta T = T_m^0 - T_c$, where T_m^0 is the equilibrium melting temperature and T_c is the crystallization temperature. T_m^0s was estimated by applying the Gibbs–Thomson equation. It was confirmed that the crystals were isolated from each other by means of a polarizing optical microscope (POM).

1. In the SAXS study, samples once melted were kept at the maximum temperature $T_{max} = 160$ °C within an evacuated capillary (1 mm in diameter) for 5 min and were crystallized at $T_c = 129.1$ °C ($\Delta T = 10.4$ K).
2. For the study of the M dependence of I and V of FCCs, film samples (0.1 mm thick) were isothermally crystallized from the melt into the orthorhombic (= ordered immobile) phase at atmospheric pressure. The range of ΔT was 10–15 K. In order to observe isolated single crystals, observation was limited to the earlier stage of crystallization.
3. In the study of the M dependence of I and V of extended chain single crystals (FCSCs), samples were isothermally crystallized from the melt into the hexagonal (= disordered "mobile") phase at high pressure ($P = 0.4$ GPa). The range of ΔT was 3.3–9.4 K.
4. For the entanglement study, FCCs were isothermally crystallized at atmospheric pressure and ECSCs with different l were isothermally crystallized from the melt to the hexagonal (disordered mobile) phase at

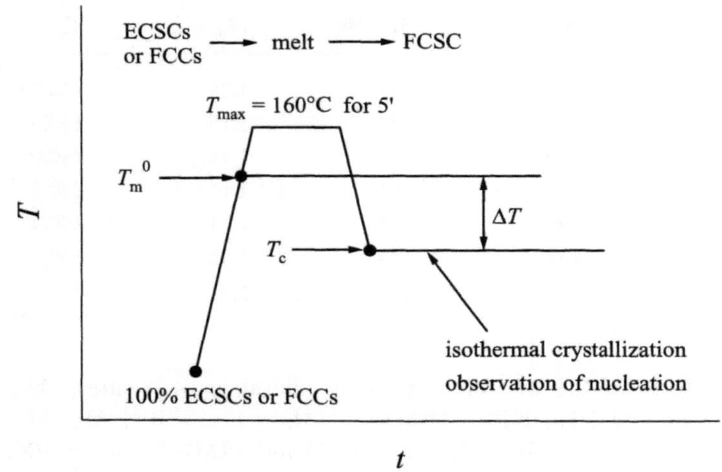

Fig. 4 Crystallization procedure indicated by T against time t. ECSCs or FCCs once melted are kept at T_{max} for 5 min and then isothermally crystallized at a T_c

$\Delta T = 5.5-13$ K under high pressure, $P = 0.4$ GPa. l was estimated by transmission electron microscopy (TEM). As schematically shown in Fig. 4, ECSCs or FCCs once melted were kept at $T_{\max} = 160\,^{\circ}$C for 5 min at atmospheric pressure. After that, samples were isothermally crystallized at various T_{c}s. Hereafter, we abbreviate these processes as ECSCs-melt-FCSC or FCCs-melt-FCSC, respectively, where FCSC means folded chain single crystal. The range of ΔT was about 10–14 K.

2.3 Instrumental

Isothermal crystallization was observed by means of SAXS and a polarizing optical microscope (POM, OLYMPUS, BX or BHS-751-P). The SAXS experiment was carried out using synchrotron radiation on the beam line BL40B2 of SPring8 (SP8) at JASRI in Harima and at the BL-10C small angle installation of the Photon Factory (PF) at KEK in Tsukuba.

In the case of optical observation of FCCs, film samples were put in a hot stage (Linkam, LK-600). Temperature was calibrated using standard materials. In and Sn Nitrogen gas was used at the rate of 50 ml/min. The number of isolated crystals near the center of the sample was counted.

Crystallization of ECSCs was isothermally carried out under high pressure using a piston cylinder high pressure cell with diamond window (PCDW) originally made by us. The formation of isolated ECSCs was confirmed by means of transmission electron microscopy (TEM).

In order to observe nucleation behaviors clearly and evaluate I easily, we have developed a new system combining a digital video camera (Victor JVC KY-F70, image size: 1.3 M pixels) and a hot stage (Linkam LK600PM). The main feature of this combined system is that it is controlled by a computer. In this work, Planetron Co. Ltd., Japan High Tech Co. Ltd. and ourselves have developed new software which is able to control both hot stage and image capturing. Both hot stage and image capturing control software and image analysis software (Media Cybernetics, Image-pro plus) are installed on the computer. The computer is connected to both the digital video camera and hot stage controller. The computer receives time and temperature information from the hot stage controller. The received information is automatically superimposed on the image captured from the digital video camera. Using our system, we can obtain sharp images and easily count the number of isolated crystals.

2.4 M_n Dependence of Equilibrium Melting Temperature

The equilibrium melting temperature T_m^0 was determined on ECSCs using Wunderlich's method [26]. The T_m of ECSCs was estimated from a tem-

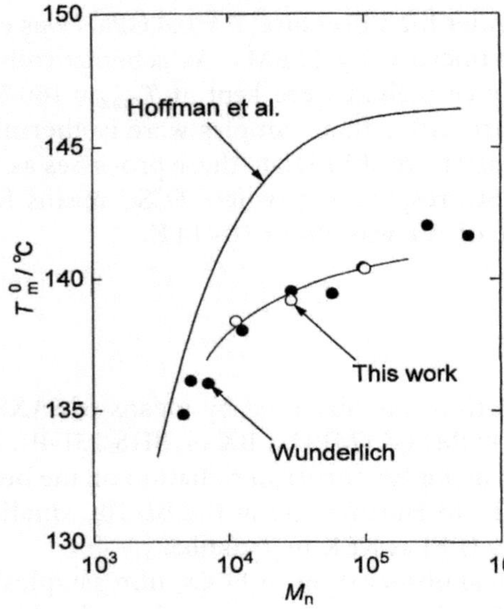

Fig. 5 Equilibrium melting temperature T_m^0 plotted against $\log M_n$. (○) This work, (•) Wunderlich [26], *solid line*: Hoffman et al. [28]

perature where the lateral length (a) of an isolated ECSC started decreasing significantly on heating at a rate of 0.1 K/min. The T_m^0 estimated from T_m (ECSC) was plotted against $\log M_n$ in Fig. 5 [27]. T_m^0 reported by Hoffman et al. [28] and T_ms of orthorhombic extended chain crystals collected by Wunderlich [26] are also shown in Fig. 5. The obtained T_m^0 agreed well with Wunderlich's, while significant deviation (about several K) from the T_m^0 reported by Hoffman et al. [28] was seen.

2.5
Nucleation and Growth Rates

Nucleation rate (I) for N is defined by

$$I(N,t) = \mathrm{d}\nu(N,t)/\mathrm{d}t\,, \tag{1}$$

where $\nu(N,t)$ is the number density of nuclei, which is larger than N at time t defined by

$$\nu(N,t) = \int_N^\infty f(N,t)\mathrm{d}N\,. \tag{2}$$

CNT assumes that I starts increasing after an induction period, increases with increasing time and finally saturates into a steady rate, I_{st}. This is assumed from well-known observations of macroscopic crystals by means of the optical microscope. Figure 6 shows a typical example of this obtained for PE by us. This shows that I increases as a function of t during the induction period and saturates after a period of time which corresponds to the steady period. Here it is assumed that a macroscopic single crystal is generated from a nucleus. Therefore, we will focus on folded chain single crystals (FCSCs) and extended chain single crystals (FCSCs) in our observation of I.

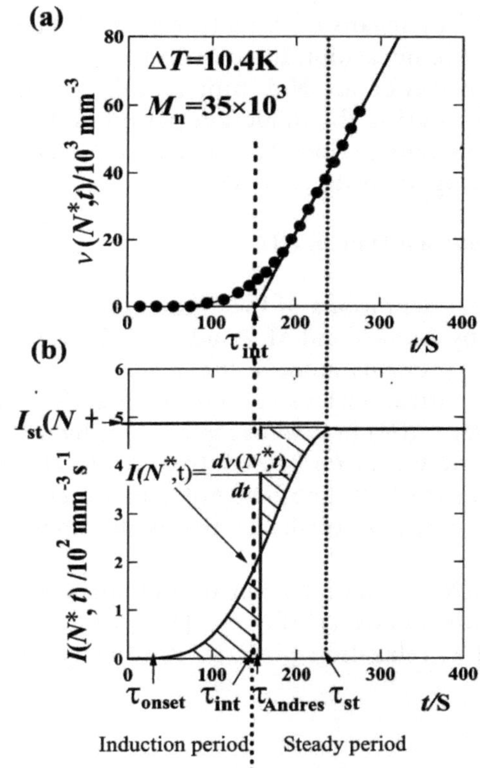

Fig. 6 **a** $\nu(t, N)$ vs. t and **b** $I(t, N)$ vs. t observed by optical microscopy. $T_c = 129.1\,°\mathrm{C}$ and $\Delta T = 10.4\,\mathrm{K}$. Two typical definitions (in CNT) of induction time τ_i are shown

Lateral growth rate V is defined by

$$V \equiv \frac{1}{2}\frac{\mathrm{d}a}{\mathrm{d}t}\,. \tag{3}$$

where a is the lateral size. a usually increases linearly with an increase of t.

3 Direct Evidence of Nucleation During the Induction Period [29]

3.1 Introduction

The purpose of this section is to present direct evidence of nucleation during the induction period by means of synchrotron small angle X-ray scattering (SAXS). In the classical nucleation theory (CNT), the number density distribution function of nuclei of size N at time t, $f(N,t)$, is expected to increase with an increase of t during the induction period and saturates to a steady $f(N,t)$, $f_{st}(N)$ in the steady period. The change of $f(N,t)$ should correspond to that of the scattering intensity of SAXS.

Definitions of the Induction Time in CNT

There are two typical definitions of the "induction time (τ_i)" in CNT given by Frisch [16] and by Andres and M. Boudart [17]. τ_i is an increasing function of N, $\tau_i(N)$. In previous studies, the special case $N = N^*$ was usually focused on. As any critical nucleus can not be directly observed, $\tau_i(N^*)$ has been "estimated" from $\tau_i(N)$ of macroscopic nuclei by optical microscopy by correcting the time necessary for growth from N^* to N. Therefore, $\tau_i(N^*)$ is named $\tau_i(\mathrm{OM})$ in this work. It should be noted that there is no guarantee that the estimated $\tau_i(N^*) = \tau_i(\mathrm{OM})$ is correct, that is also an important unresolved problem.

Frisch defined $\tau_i(N^*)$ as an intercept on the horizontal axis (t) of an extrapolated straight line of $\nu(N^*,t)$ (Fig. 6a) [16]. Andres and Boudart defined $\tau_i(N^*)$ as a kind of a "relaxation time" so as to satisfy the following equation [17],

$$\tau_i(N^*) \equiv \int_0^\infty \{I_{st}(N^*) - I(N^*,t)\}\mathrm{d}t / I_{st}(N^*)\,, \tag{4}$$

which means that the two hatched areas in Fig. 6b should be equal. The two definitions usually give similar $\tau_i(N^*)$, as Fig. 6a and b shows.

Heterogeneous Nucleation

The primary nucleation on the surface of the NA is a kind of "heterogeneous nucleation" [25]. NA is a so called "heterogeneity". In this study, all nucleation is limited to the heterogeneous nucleation. The shape of a heterogeneous nucleus is assumed parallelepiped with length of a stem l and the number of stems m and n. Here l, m, and n are counted by the number of atom or repeating units, therefore they are dimensionless. N is given by

$$N = lmn\,. \tag{5}$$

It is well known that the l, m, and n of a critical nucleus, l^*, m^* and n^*, are given by

$$l^* = 4\sigma_e/\Delta g, \quad m^* = 2\Delta\sigma/\Delta g \quad \text{and} \quad n^* = 4\sigma/\Delta g\,, \tag{6}$$

where σ_e is the end surface free energy, Δg is the free energy of fusion, σ is the side surface free energy, and $\Delta\sigma$ is defined by

$$\Delta\sigma = \sigma + \sigma_{nh} - \sigma_h\,, \tag{7}$$

where σ_{nh} is the surface free energy between the nucleus and the NA and σ_h is that between the NA and the melt.

3.2
Scattering Intensity of the Isolated Nuclei

In the crystallization of polymers, isolated nuclei (= isolated lamellae) are generated first and grow to a bigger isolated nucleus. After a period of time, the isolated nucleus changes to the isolated stacked lamellae through "overgrowth" processes. We will show that first only the scattering intensity of isolated nuclei appear up to the induction time of the nuclei (denoted as τ_i(Nucleus)) and then that of stacked lamellae starts to superimpose after an "onset time", which is denoted as τ_{onset}(L). Therefore, it is important to separate the scattering intensity of isolated nuclei and that of stacked lamellae.

It is well known in SAXS theory that isolated nuclei give a significant scattering intensity at a small scattering vector (q). The range of q between q_1 and q_2 will be simply denoted as q_N, i.e.,

$$q_N = (q_1, q_2)\,. \tag{8}$$

The observed size of the nucleus (N_{obs}) is estimated by applying the Guinier plot [30] at q_N.

Scattering Intensity of the Stacked Lamellae

It is well known that the stacked lamellae of semi-crystalline polymers result in the "long period" which usually gives two (or three) diffuse Bragg reflections. We will focus on the first reflection (named L1) and the secondary reflection (named L2) at two q-ranges, denoted as q_{L1} and q_{L2}, respectively [30, 31].

In the case of PE without NA, only the scattering intensity of stacked lamellae can be observed, so the time evolution of the L1 and L2 reflections can be observed easily. In the case of PE with NA, it will be shown that q_{L1} is accidentally similar to q_N. So the L1 reflection superimposes on the scattering intensity of the nuclei. But it is fortunate that the L2 reflection does not seriously superimpose on the scattering intensity of the nuclei.

Separation of the Scattering Intensity of the Isolated Nuclei After τ_{onset}(L)

It should be noted that there is no difference in structure between isolated stacked lamellae of PE without NA and those of PE with NA in the early stage of the overgrowth process. The difference between them is limited to the difference of the number of isolated stacked lamellae. A significant difference between them should occur when stacked lamellae start colliding with each other in the latter stage of overgrowth. In this work we will not focus on the latter stage.

Therefore, we can regard the relative time evolution between L1 and L2 reflections as being the same between PE without NA and PE with NA. Thus, we can estimate the time evolution of the L1 reflection of PE with NA from that of the L2 reflection using the time evolution obtained on PE without NA.

After obtaining the time evolution of the L1 reflection of PE with NA, we can separate out the scattering intensity of the isolated nuclei at q_N from the observed scattering intensity. Hereafter the range q_{L2} between q_3 and q_4 will be simply denoted as q_L, i.e.,

$$q_{L2} = q_L = (q_3, q_4) . \tag{9}$$

Procedure of SAXS Analysis

a) The observed scattering intensity at q and t ($I_{x\ obs}(q,t)$) was obtained after corrections with respect to incident beam intensity and mass of sample in order to quantitatively compare all data.
b) Information concerning the changes after quenching into the supercooled melt is given by the so called "excess scattering intensity ($I_{x\ d}(q,t)$)" defined by

$$I_{x\ d}(q,t) = I_{x\ obs}(q,t) - I_{x\ obs}(q,0) . \tag{10}$$

It is obvious that information that does not change before and after quenching, such as the effect of NA within the sample, is excluded from $I_{x\ d}(q,t)$.
c) The integrated scattering intensity of $I_{x\ d}(q,t)$ at a range of q_x, denoted as $I_x(q_x,t)$ is defined by

$$I_x(q_x,t) = \int_{qx} I_{x\ d}(q,t)\mathrm{d}q . \tag{11}$$

For example, $I_x(q_x,t)$ at q_N is denoted as $I_x(q_N,t)$ and that at q_L is denoted as $I_x(q_L,t)$.

d) The integrated scattering intensity of PE without NA at q_N and q_L, $I_x(q_N, t)$ and $I_x(q_L, t)$, correspond to those of L1 and L2 reflections, respectively. Therefore, we will simply denote them as $I_x(L1)$ and $I_x(L2)$, i.e.,

$$I_x(q_N, t) = I_x(L1) \quad \text{and} \quad I_x(q_L, t) = I_x(L2) . \tag{12}$$

The relative time evolution of $I_x(L1)$ and $I_x(L2)$ will be described by a function of $\alpha(t)$, i.e.,

$$I_x(L1)/I_x(L2) = \alpha(t) . \tag{13}$$

e) $\alpha(t)$ does not change either for PE without NA or for PE with NA. Therefore, $\alpha(t)$ can be obtained from Eq. 13 of PE without NA and $I_x(L1)$ of PE with NA can be given by

$$I_x(L1) = \alpha(t) I_x(L2) . \tag{14}$$

f) The final goal of the integrated scattering intensity of nuclei denoted by $I_x(\text{Nucleus})$ can be described by

$$I_x(\text{Nucleus}) = \begin{cases} I_x(q_N, t) & \text{for } t < \tau_i \\ I_x(q_N, t) - I_x(L1) & \text{for } t > \tau_i \end{cases} . \tag{15}$$

It is predicted by CNT that $I_x(\text{Nucleus})$ starts increasing with crystallization t and saturates after a much longer time than τ_i. The saturated $I_x(\text{Nucleus})$ is denoted as $I_{x\,st}(\text{Nucleus})$.

Definition of τ_i(SAXS)

$I_x(\text{Nucleus})$ can be regarded to be in proportion of $f(N, t)$. Therefore, the induction time of nuclei with N_{obs} for SAXS, $\tau_i(N_{obs})$ or $\tau_i(\text{SAXS})$, is defined after the definition given by Andres et al. by

$$\begin{aligned} \tau_i(\text{SAXS}) = \tau_i(N_{obs}) &= \int \{f_{st}(N_{obs}) - f(t, N_{obs})\} dt / f_{st}(N_{obs}) \\ &= \int \{I_{x\,st}(\text{Nucleus}) - I_x(\text{Nucleus})\} dt / I_{x\,st}(\text{Nucleus}) . \end{aligned} \tag{16}$$

The Radius of Gyration R_g and the Size of a Nucleus

The Guinier plot will be carried out on $I_{x\,d}(q, t)$ in Eq. 10 as a parameter of t in order to obtain the observed radius of gyration, R_g, and the t dependence of R_g ($R_g(t)$) applying Guinier's law [30],

$$I_{x\,d}(q, t) \propto \exp\{- R_g(t)\}^2 q^2/3\} . \tag{17}$$

PE crystallizes into the orthorhombic form from the melt, which results in folded chain crystals (FCCs) [14, 15]. In this case, the critical nucleus should be a rectangular parallelepiped. It is assumed that the shape of a nucleus is similar to that of the critical nucleus, that means

$$l \propto l^*, \quad m \propto m^* \quad \text{and} \quad n \propto n^* . \tag{18}$$

It should be noted that l, m and n are counted by the number of repeating units (CH_2) and not by the real length. R_g of a nucleus is given by

$$R_g = \{(l^2 + m^2 + n^2)/12\}^{1/2} . \tag{19}$$

The following kinetic parameters for PE are used [29],

$$\begin{aligned} \sigma &= 8.0 \times 10^{-3}\ (\text{J/m}^2) , & \sigma_e &= 87.8 \times 10^{-3}\ (\text{J/m}^2) , \\ \Delta\sigma &= 0.26 \times 10^{-3}\ (\text{J/m}^2) \quad \text{and} & \Delta g &= 0.48 \times 10^{-3}\ (\text{J/m}^3) . \end{aligned} \tag{20}$$

They were evaluated from our analysis of the primary nucleation and lateral growth rates and that of the l dependence to the melting temperature T_m using the Gibbs–Thomson equation. Insertion of the parameters given by Eq. 20 into Eq. 6 shows that the shape of a nucleus is a long thin rectangular parallelepiped with the ratio of

$$l : m : n = 600 : 30 : 1 . \tag{21}$$

Therefore, R_g of the nucleus is approximated by

$$R_g = l/2\sqrt{3} . \tag{22}$$

Thus, observed l (l_{obs}) can be estimated by

$$l_{obs} = 2\sqrt{3}R_g . \tag{23}$$

3.3 Results and Discussions

Observed Scattering Intensity, $I_{x\ obs}(q,t)$

Time evolutions of $I_{x\ obs}(q, t)$ of the mixture of PE with NA and that of PE without NA are shown in Figs. 7 and 8, respectively. They were observed at the same degree of supercooling ΔT as the optical observation shown in Fig. 6. The induction time observed by optical microscopy (τ_i(OM)) was

$$\tau_i(\text{OM}) = 2.5\ (\text{min}) \tag{24}$$

$I_{x\ obs}(q, t)$ at $q_N = (q_1, q_2) = (0.009\ \text{Å}^{-1}, 0.014\ \text{Å}^{-1})$ of the former (mixture of PE with NA) starts increasing significantly soon after quenching, while that of the latter (PE without NA) does not increase at all (Fig. 5) up to τ_i(OM), i.e., during the induction period. The strong contrast between them suggests that the increase of the former should be due to an increase of the number density of nuclei $f(t, q)$.

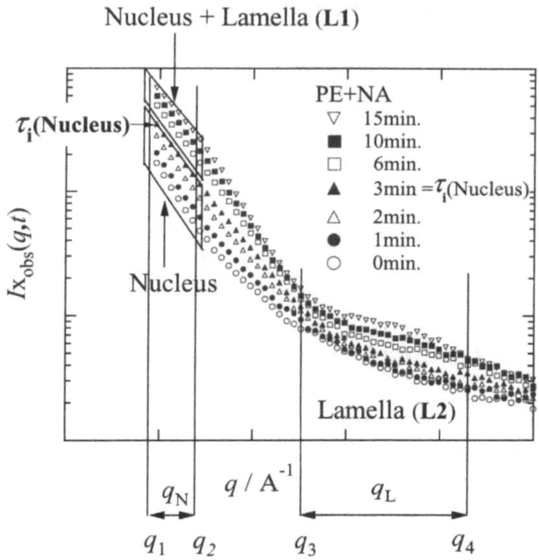

Fig. 7 Observed scattering intensity, $I_{\mathrm{x\,obs}}(q,t)$ vs. q for PE with NA. $T_{\mathrm{c}} = 129.1\,°\mathrm{C}$ and $\Delta T = 10.4\,\mathrm{K}$ are fixed in this work. Up to the induction time, $\tau_i(\mathrm{Nucleus}) = 3\,\mathrm{min}$, $I_{\mathrm{x\,obs}}(q,t)$ at q_N increased significantly, while $I_{\mathrm{x\,obs}}(q,t)$ at q_{L} does not. After $\tau_i(\mathrm{Nucleus})$, the former does not increase significantly, while the latter increases

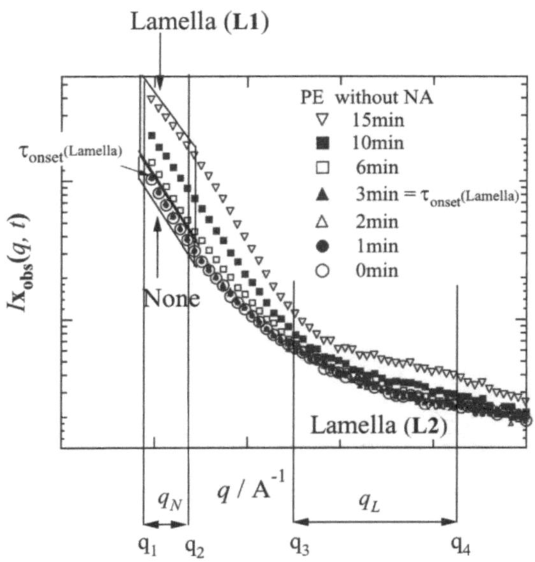

Fig. 8 Observed scattering intensity: $I_{\mathrm{x\,obs}}(q,t)$ vs. q for PE without NA. Strong difference can be seen from that for PE with NA shown in Fig. 4. Up to the induction time, $\tau_i(\mathrm{Nucleus}) = 3\,\mathrm{min}$, both $I_{\mathrm{x\,obs}}(q,t)$ at q_N and $I_{\mathrm{x\,obs}}(q,t)$ at q_{L} did not increase at all

After τ_i, $I_{x\,obs}(q,t)$ at q_N of the former gradually saturates, whereas $I_{x\,obs}(q,t)$ of the latter starts increasing significantly. In the latter part of this paper it will be shown that the saturation of the former is due to the change into steady nucleation and the increase of the latter is due to the effect of onset of stacked lamellae.

$I_{x\,obs}(q,t)$ at $q_L = (q_3, q_4) = (0.025\ \text{Å}^{-1}, 0.046\ \text{Å}^{-1})$ of both the former (mixture of PE with NA) and the latter (PE without NA) do not increase up to τ_i, i.e., during the induction period.

$I_{x\,obs}(q,t)$ at q_L of both the former and the latter starts increasing significantly after τ_i, that should be due to the onset of stacked lamellae.

Integrated Scattering Intensity I_x of the Mixture of PE with NA

The time evolution of the integrated scattering intensity of $I_x(q_N,t)$ and $I_x(q_L,t) = I_x(\text{L2})$ of the mixture of PE with NA is shown in Fig. 9. $I_x(q_N,t)$ starts increasing soon after quenching and increased quickly up to $\tau_i(\text{OM}) = 2.5$ min with the increase of time. The rate of increase slowed after 6 min. This suggests that the formation of isolated nuclei during the induction time has been observed for the first time.

$I_x(q_L,t) = I_x(\text{L2})$ was zero up to $\tau_i(\text{OM})$ and it starts increasing after $\tau_i(\text{OM})$. This is very different to $I_x(q_N,t)$. This suggests that the lamellar stacking is not superimposed during the induction period. The onset time of lamellar stacking $\tau_{onset}(\text{L})$ is estimated as,

$$\tau_{onset}(\text{L}) = 3 \pm 0.5\ \text{min} > \tau_i(\text{OM})\,. \tag{25}$$

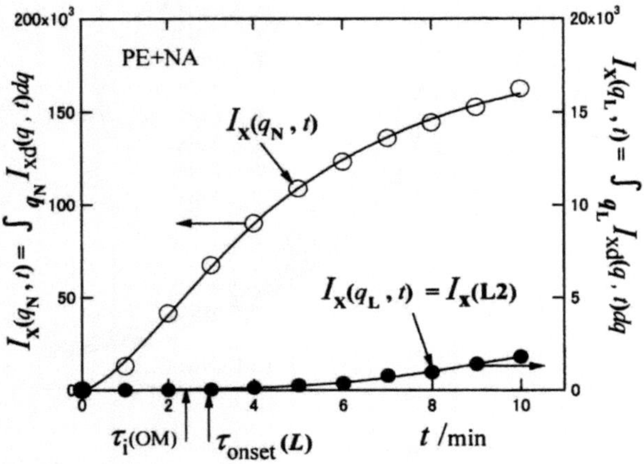

Fig. 9 Integrated scattering intensity of PE with NA, $I_x(q_N,t)$ and $I_x(q_L,t) = I_x(\text{L2})$ vs. t

I_x of the Stacked Lamellae

The time evolution of the integrated scattering intensity from the stacked lamellae of PE without NA, $I_x(q_N, t) = I_x(L1)$ and $I_x(q_L, t) = I_x(L2)$, are shown in Fig. 10a. They were zero up to $\tau_i(OM) = 2.5$ min and after that started to increase. It was confirmed that the time evolution of the lamellar stacking of PE without NA was the same as that of PE with NA.

From observed $I_x(L1)$ and $I_x(L2)$, $\alpha(t)$ was obtained as shown in Fig. 10b. $\alpha(t)$ does not change significantly with time, that is,

$$\alpha(t) = I_x(L1)/I_x(L2) = \text{constant} \quad \text{for PE without NA}. \tag{26}$$

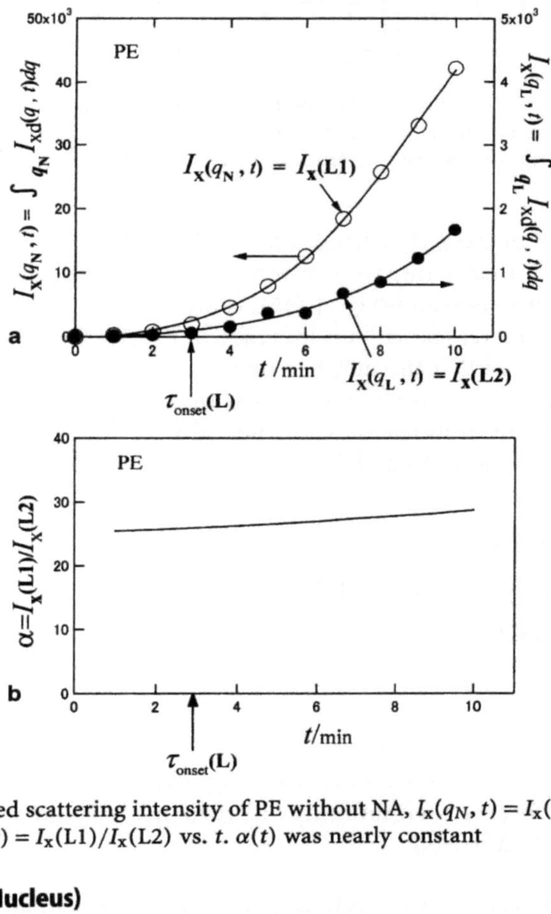

Fig. 10 **a** Integrated scattering intensity of PE without NA, $I_x(q_N, t) = I_x(L1)$ and $I_x(q_L, t) = I_x(L2)$ vs. t. **b** $\alpha(t) = I_x(L1)/I_x(L2)$ vs. t. $\alpha(t)$ was nearly constant

I_x of Nuclei, I_x(Nucleus)

Combination of $I_x(L2)$ of PE with NA and the estimated $\alpha(t)$ gives $I_x(L1)$ of PE with NA, which is shown in Fig. 11. Thus, the final goal of the integrated scat-

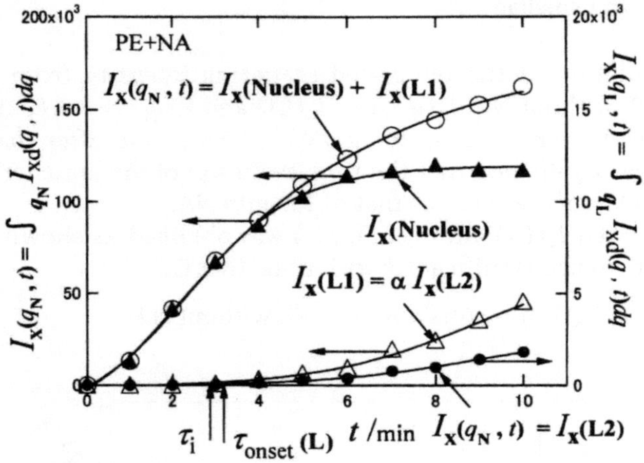

Fig. 11 I_x(L1) vs. t of PE with NA estimated from the relationship, $I_x(\text{L1}) = \alpha(t)I_x(\text{L2})$. I_x(Nucleus) vs. t was obtained for the first time by subtraction of I_x(L1) from $I_x(q_N, t)$, using the relationship, $I_x(\text{Nucleus}) = I_x(q_N, t) - I_x(\text{L1})$. τ_i(Nucleus) = 3 min was obtained

tering intensity of nuclei I_x(Nucleus) was obtained (Fig. 11). This showed that I_x(Nucleus) starts increasing soon after quenching, increased significantly during the induction period up to τ_i(OM), and then saturated after 6 min.

As I_x(Nucleus) increases with an increase of the number density of nuclei, this clearly confirmed that the number density of nuclei increases during the induction period. Thus, it is concluded that the nucleation during the induction period is directly confirmed experimentally for the first time.

The induction time obtained by SAXS τ_i(SAXS) estimated from Fig. 11 using Eq. 16 was

$$\tau_i(\text{SAXS}) = 3 \pm 0.5\,(\text{min}), \tag{27}$$

which was nearly the same as τ_i(OM), that is

$$\tau_i(\text{SAXS}) = \tau_i(\text{OM}). \tag{28}$$

It is also concluded that

$$\tau_{\text{onset}}(\text{L}) = \tau_i(\text{SAXS}) \tag{29}$$

This means that the lamellae start stacking after τ_i(SAXS). Therefore, nucleation can be observed without any “disturbance” of the stacked lamellae during the induction period.

Comparison of I_x(Nucleus) for PE with NA and I_x(L1) for PE without NA

Figure 12 shows a comparison of the time evolution of I_x(Nucleus) for PE with NA and that of I_x(L1) for PE without NA. Nucleation during the induction

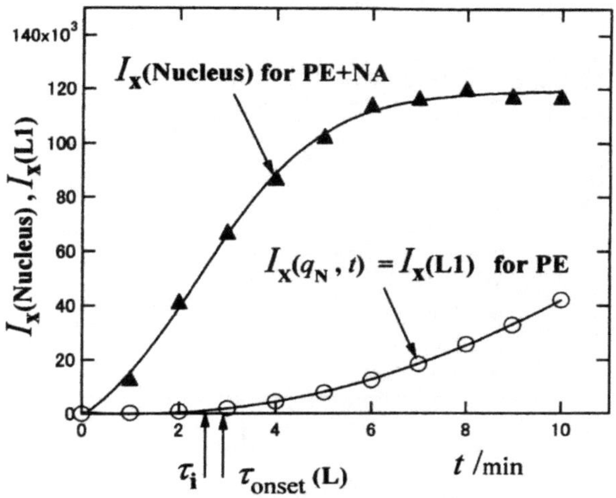

Fig. 12 Comparison of time evolution of I_x(Nucleus) for PE with NA and that of I_x(L1) for PE without NA. Nucleation during induction period was clearly confirmed. Nucleation was observed only when NA was mixed. Without mixing NA, only formation of stacked lamellae was observed

period was clearly confirmed. Nucleation was observed only when NA was added to PE, while only formation of stacked lamellae was observed for PE without NA.

Guinier Plot and R_g and N_{obs}

Figure 13 shows the Guinier plots of $I_{x\ d}(q, t)$ at q_N after quenching up to a little higher t than τ_i(SAXS). Linear lines are obtained and the slope does not change with an increase in time. From the slope, we estimated

$$R_g = 162 \pm 2\ (\text{Å})\,. \tag{30}$$

Substitution of Eq. 30 into Eq. 22 gives l_{obs},

$$l_{obs} = 560\ \text{Å}\,. \tag{31}$$

The size l_{obs} was similar to $l^* = 600$ Å. Therefore, $N_{obs} \cong N^*$ was obtained.

The Mechanism of Induction

The answer to the argument as to whether the mechanism of the induction of polymers is related to the nucleation process (as predicted in CNT [1–4]) or to the phase separation process [19, 32] is that the nucleation process is correct in the case of melt crystallization.

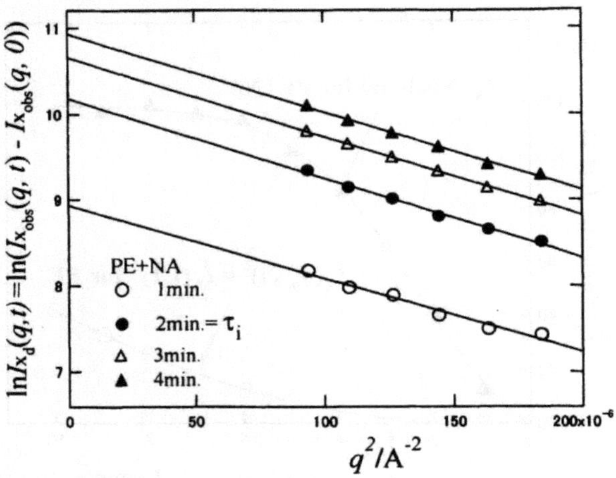

Fig. 13 Guinier plot of $\log I_{x\ d}(q,t)$ vs. q^2 for $t = 1$–4 min. The slopes did not change significantly with t

4 The Power Law of Molecular Weight of the Nucleation Rate of Polyethylene [20, 33, 34]

4.1 Introduction

The molecular weight (M) dependence of the steady (stationary) primary nucleation rate (I) of polymers has been an important unresolved problem. The purpose of this section is to present a power law of molecular weight of I of PE, $I \propto M^{-H}$, where H is a constant which depends on materials and phases [20, 33, 34]. It will be shown that the self-diffusion process of chain molecules controls the M_n dependence of I, while the critical nucleation process does not. It will be concluded that a topological process, such as chain sliding diffusion and entanglement, assumes the most important role in nucleation mechanisms of polymers, as was predicted in the chain sliding diffusion theory of Hikosaka [14, 15].

Topological Nature and Chain Sliding Diffusion in Polymer Nucleation

Classical nucleation theory (CNT) shows that I is a product of the probability of diffusion and that of formation of a critical nucleus [1, 4],

$$I = I_0 \exp(-\Delta G^*/kT) = I_0 \exp(-C/\Delta T^2)\,, \tag{32}$$

where I_0 is a prefactor related to the diffusion constant (D), ΔG^* is the free energy necessary for formation of a critical nucleus, k is the Boltzmann constant, T is temperature, C is a constant, and ΔT is the degree of supercooling. Here a three-dimensional shape of the nucleus is assumed. It is well known that nucleation from the melt is usually heterogeneous nucleation. In the case of heterogeneous nucleation, C is defined by [25]

$$C = 16\sigma \Delta\sigma\sigma_e / kT_m^0 \Delta h\,, \tag{33}$$

where Δh is the enthalpy of fusion. ΔT is defined by $\Delta T \equiv T_m^0 - T_c$, where T_m^0 is the equilibrium melting temperature and T_c is the crystallization temperature. Therefore, it is important to make clear which factor in Eq. 32 (ΔG^* or D) control the M dependence of I. The nucleus is not always 3D but it changes to a two-dimensional nucleus (2D) with an increase of ΔT.

There are three kinds of diffusion: (i) within the isotropic phase; (ii) the interface (between the isotropic and the crystalline phases); and (iii) the crystalline phase. In the case of a polymer system, the topological nature of polymer chains assumes an important role in all three kinds of diffusion, which has been shown in the chain sliding diffusion theory proposed by Hikosaka [14, 15]. It is obvious that any nucleus (a primary nucleus and a two-dimensional nucleus) and a crystal can not grow or thicken without chain sliding diffusion.

4.2 Results

Morphology and log I vs. ΔT^{-2}

Optical morphologies of growing FCSCs did not change with M_n, t, and ΔT. Typical ν vs. t of FCSCs as a parameter of M_n at $\Delta T = 12.5$ K are plotted in Fig. 14. ν increased linearly with an increase of t. I was obtained from the slope of ν vs. t. In the case of heterogeneous nucleation, there is a serious technical problem in that ν vs. t shows significant scatter (see Fig. 14), which adds significant error to the obtained I. This is due to significant scatter in the distribution of heterogeneity accidentally included in the sample. The problem was solved by taking the statistical average of ν vs. t as shown by the thick lines in Fig. 14.

$\log I$ versus ΔT^{-2} of FCSCs and ECSCs are plotted as a parameter of M_n in Figs. 15 and 16, respectively. $\log I$ decreased significantly with an increase of ΔT^{-2}.

The M_n Dependence of I

Lines of $\log I$ vs. ΔT^{-2} in Figs. 15 and 16 are nearly parallel for all M_ns and they shift downward with increasing M_n. Figures 17 and 18 show a plot of C against M_n for FCSCs and ECSCs. This indicates that I_0 decreases significantly

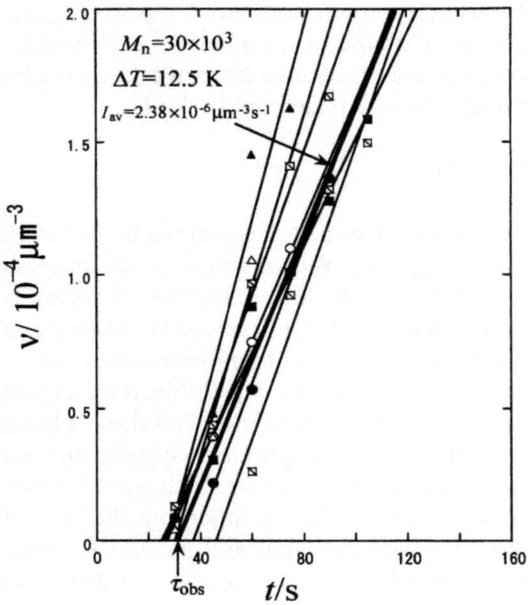

Fig. 14 Plots of ν against t of FCSCs at $\Delta T = 12.5$ K for $M_n = 30 \times 10^3$

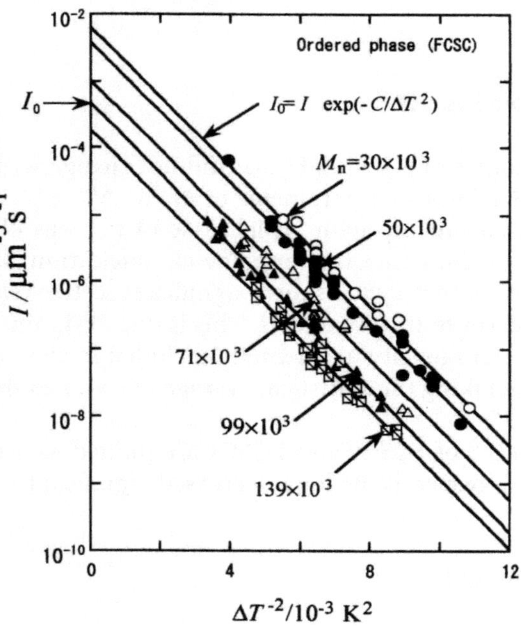

Fig. 15 Plot of $\log I$ versus ΔT^{-2} of FCSCs for $M_n = 30 \times 10^3$, 50×10^3, 71×10^3, 99×10^3, and 139×10^3. The *solid lines* represent the best fit of the plots, which corresponds to the classical nucleation theory. I_0 is the intercept of the vertical axis at $\Delta T^{-2} = 0$

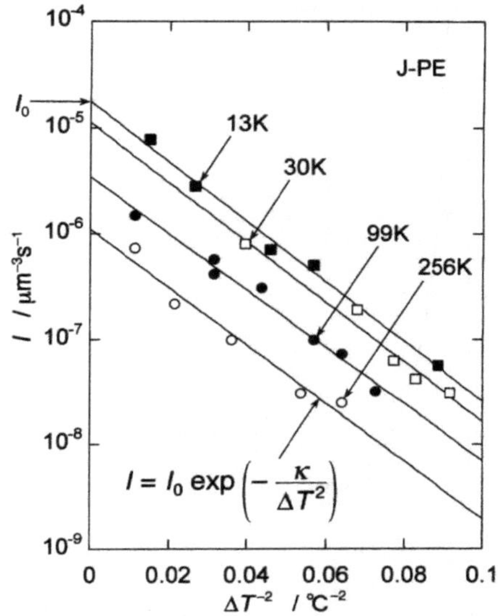

Fig. 16 Plot of log I against ΔT^{-2} of ECSCs in $M_n = 13$ K, 30 K, 99 K, and 256 K. The *lines* show the best fit of the plots, which correspond to the well-known formula of the primary nucleation

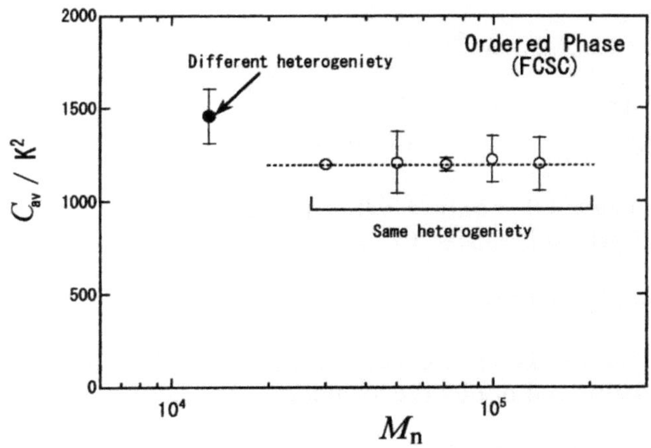

Fig. 17 Plots of C_{av} versus log M_n of FCSCs. The C_{av} is almost the same for all M_n except for $M_n = 13 \times 10^3$, therefore it is concluded that the samples except for $M_n = 13 \times 10^3$ contain the same heterogeneity

with increasing M_n, whereas C does not depend on M_n:

$$I_0 = I_0(M_n) \tag{34}$$

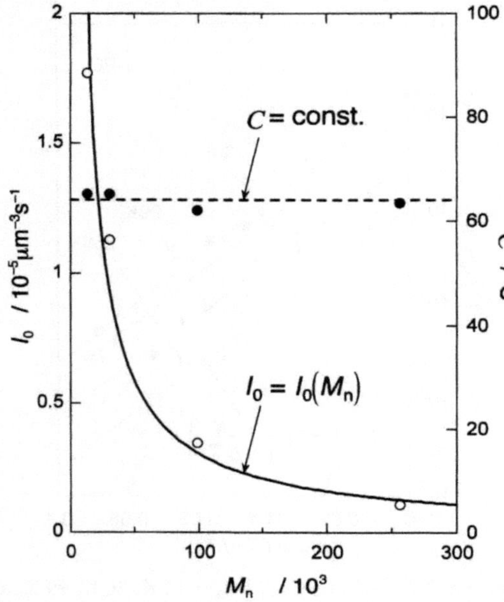

Fig. 18 Plot of I_0 and C against M_n of ECSCs. The *solid line* and *broken line* show the best fit of I_0 and C, respectively. Only C depends on M_n, while I_0 does not

and

$$C \cong \begin{cases} \text{const} \approx 1.2 \times 10^3\ \text{K}^2 & \text{for FCSCs} \\ 63\ \text{K}^2 & \text{for ECSCs} \end{cases} . \quad (35)$$

Thus, it is concluded that only D (which is proportional to I_0) depends on M_n, whereas ΔG^* (which is related to C) does not depend on M_n, i.e.,

$$I(M_n) \propto I_0(M_n) \propto D(M_n) . \quad (36)$$

This means that the M_n dependence of I is controlled by the diffusion process of polymer chains and not by the formation process of a critical nucleus.

The Power Law of Nucleation Rate

$\log I_0$ is plotted against $\log M_n$ for FCSCs and ECSCs in Fig. 19. $\log I_0$ decreases linearly with increasing $\log M_n$. Thus, an experimental power law is obtained,

$$I(M_n) \propto I_0(M_n) \propto \begin{cases} M_n^{-2.4} & \text{for ordered (orth.) phase (FCSC)} \\ M_n^{-1.0} & \text{for disordered (hex.) phase (ECSC)} \end{cases} . \quad (37)$$

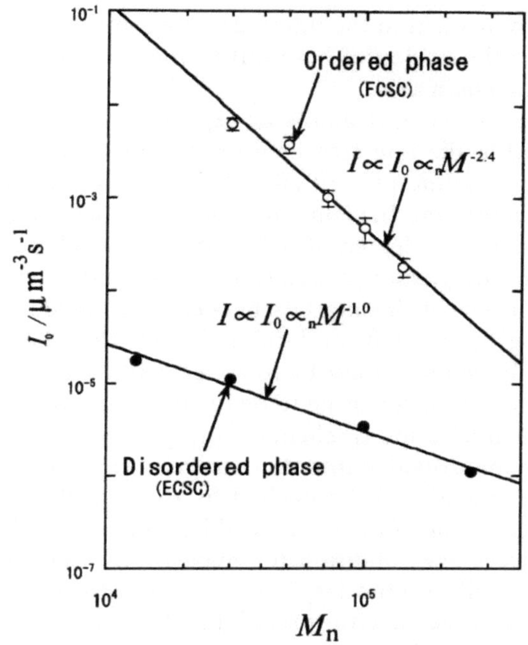

Fig. 19 Plot of log I versus log M_n. The *best fit line* for FCSCs and ECSCs show power laws, $I \propto M_n^{-2.4}$ and $I \propto M_n^{-1}$, respectively

Thus, it is concluded that the power law of I is confirmed for the first time,

$$I(M_n) \propto I_0(M_n) \propto M_n^{-H}, \tag{38}$$

where

$$H = \begin{cases} 2.4 & \text{for ordered (orth.) phase (FCSC)} \\ 1.0 & \text{for disordered (hex.) phase (ECSC)} \end{cases}. \tag{39}$$

4.3 Discussion

Crystalline Phase Dependence of Power H

It is important to consider why H depends on the degree of order of the crystalline phase. Three different types of diffusion process act during the nucleation process. They are diffusion within the melt, within the interface between the melt and a nucleus (or crystal), and within the nucleus. It is obvious that the diffusion of chains within the melt can not be related to the dependence of H on the degree of order of the crystalline phase within the nucleus (or crystal). Therefore, the phase dependence of H should arise from the diffusion of chains within the interface between the melt and nucleus (or crystal) and/or in the nucleus. These diffusions should be the chain sliding diffusion along the chain axis.

The phase dependence of H should be experimental evidence that the sliding diffusion (within the interface between the melt and nucleus or crystal and within the crystalline phase) assumes the most important role in the nucleation and growth mechanisms. It is obvious that the chain sliding diffusion is sensitive to the degree of order of the crystalline phase. Thus, we have a conclusion that primary nucleation is a process of chain sliding diffusion within the nucleus or the interface between the nucleus and the melt that requires disentanglement of the molecular chain within the interface. It is obvious that entanglements can not be included within a nucleus, because the size of an entanglement is too large to be included as a defect in the nucleus.

The topological nucleation is schematically summarized in Fig. 20. The nucleation process can be divided into three stages: (1) small nuclei (= embryos) are frequently generated and diminished by thermal fluctuation within the melt; (2) the size of some of the nuclei will become larger than that of a critical nucleus (= necessary condition for nucleation); and (3) finally a small number of nuclei will become large enough (named the "macroscopic nucleus") to ensure a large survival probability (= sufficient condition for nucleation). In Fig. 20, ΔE^* is the activation free energy necessary for diffusion that includes chain sliding diffusion.

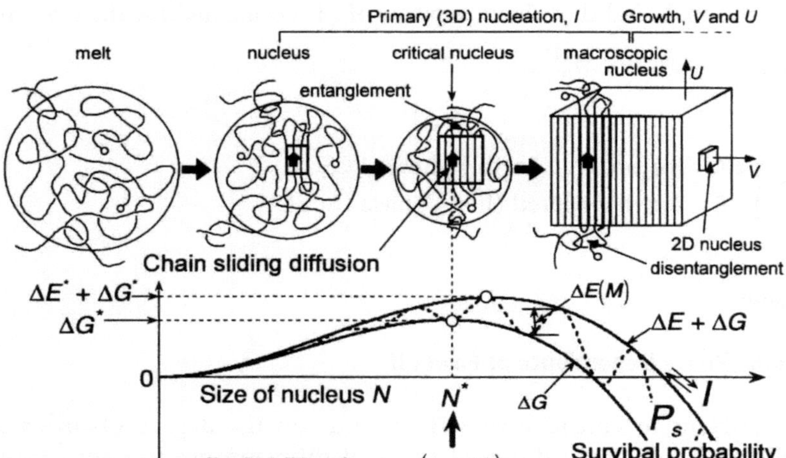

Fig. 20 Chain sliding diffusion model of primary nucleation. Polymer chains are rearranged from Gaussian shape within the melt into a nucleus through chain sliding diffusion within the nucleus and disentanglement within the interface. *Bottom graph* indicates change in free energy of the nucleus against N

The results show that a nucleus can grow into a large nucleus through chain sliding diffusion and "disentanglement" of chains within the nucleus or the interface between the nucleus and the melt. This is due to the topological nature of polymer chains. It is shown that chain sliding diffusion and disentanglement become more difficult with increasing M_n, which is the origin of M_n dependence of I.

Reason Why ΔG^* Does Not Depend on M

The reason why ΔG^* does not depend on M is because only part of the molecular length of one chain is included within a critical nucleus in this study. This means that only a partial length of one chain forms a critical nucleus. In other words, chain ends are not significantly included within a critical nucleus. Therefore, the whole length does not assume an important role in the formation of a critical nucleus. This is quite different from the case of n-paraffin or an oligomer system [35, 36].

5 Power Law of Molecular Weight of the Growth Rate of Polyethylene [21, 27]

5.1 Introduction

Previous Studies

The lateral growth rate (V) of crystals of linear chain polymers strongly depends on molecular weight (M) [37]. Although the M dependence of V of folded chain crystals (FCCs) of polymers has been rather well studied, it is still an important unresolved problem. Magill et al. presented an experimental formula, $V \propto M^{-0.5}$, for poly (tetramethyl-*p*-silpenyline siloxane), poly (ethylene terephthalate), etc [38].

Hoffman et al. [28] and Labaig [39] observed the M dependence of V of a folded chain crystal (FCC) of polyethylene (PE). They found that V decreased with an increase of M. They presented a relation [40, 41],

$$V \propto M^{-\mathrm{H}}, \tag{40}$$

where H is a constant, $\mathrm{H} = 1.0$–1.5, depending on the range of M. Hoffman et al. and Labaig showed similar results for "regime II", but not for "regime I". Hoffman et al. showed that the activated process of formation of a critical nucleus does not depend on M [28] and that the diffusion process depends on M. They assumed that the self-diffusion process within the melt depends on M and proposed a "reeling in" model [40]. Thus, the M dependence of V is not well understood at the moment.

Three Stages in the Lateral Growth Process

The classical nucleation theory (CNT) proposed that a crystal grows via coupling of three stages after primary nucleation [1, 42–44]. The first stage is a self-diffusion process of atoms or molecules from the "environment phase" onto a surface of the crystal (named substrate or growth-surface). The environment phase usually means isotropic phases, such as the melt, solution or gas phases, but it sometimes means anisotropic phases, such as solid or liquid crystal. The second stage is absorption and diffusion processes of atoms or molecules on the growth-surface, interface and within the nucleus. In the case of polymers, the second stage corresponds to the chain sliding diffusion process. The last stage is a nucleation and growth process of the two-dimensional nucleus. The three stages are simply named the first, the second, and the last stages in this chapter. It is important to make clear which stage mainly controls the M dependence of V. V is expressed by

$$V = V_0 \exp(-B/\alpha T_c \Delta T)\,, \tag{41}$$

where V_0 and B are constants and α is a constant ($\alpha = 1$ and $\alpha = 2$ or 3 for single and multi nucleation processes, respectively). In Eq. 41 the following relations are used,

$$V_0 \propto D \quad \text{and} \quad \Delta G^* \propto 1/\Delta T\,. \tag{42}$$

D is usually defined by

$$D = D_0 \exp(-\Delta E/kT)\,, \tag{43}$$

where D_0 is a constant ΔE is an activation energy of self-diffusion of a unit, such as an atom or a repeating unit of a polymer. We have to consider two kinds of D as mentioned above, D within the melt, solution or gas (D_m) and that of chain sliding diffusion (D_s). It is natural to consider that a much slower diffusion process mainly controls D, which is represented by

$$1/D = 1/D_m + 1/D_s\,. \tag{44}$$

Diffusion of Chains Within the Melt

It is well known in the case of self-diffusion of a linear chain polymer within the melt that D_m is in proportion to the power of M,

$$D_m = M^{-H} \exp(-\Delta E_m/kT)\,, \tag{45}$$

where $H = 1$ for $M < M_e$ or $H = 2$ for $M > M_e$, where M_e, is M between entanglements [46] and ΔE_m is ΔE of a repeating unit within the melt. As ΔE_m can not depend on M, Eq. 45 can be given by

$$D_m = M^{-H}\,. \tag{46}$$

It should be noted that D_s has not been formulated so far according to the authors' knowledge, so it is not certain if Eq. 46 can be applied to D_s or not.

Purpose

The purpose of this chapter is to present the power law of the M dependences of V of FCSCs and ECSCs of PE and to make clear what controls the M dependence of V.

Improved Points: Single Crystals

In order to obtain reliable M dependence of V, we need to observe V on a single crystal, because the crystallographic character of the growth-surface of a single crystal is well defined.

In the case of folded chain crystals, Hoffman et al. observed V on polycrystals, such as spherulites or axialites [28, 40, 41, 46]. Labaig observed V on single crystals and axialite [39] and Toda did on single crystals [47]. Toda showed that single crystals could be observed in the early stage of crystallization for the samples with rather low M_n, such as 11 K and 29 K, while they changed easily into polycrystals due to remarkable overgrowth [47] for samples with a rather high M_n, such as 100 K. Therefore, special care was paid in this study so that observation was carried out on single crystals (or at least single crystal-like crystals).

In the case of extended chain crystals, we showed that a single crystal is easily formed even for M_n higher than 10^5. An ECSC shows a cigar-like and tapered shape morphology, observed by polarizing optical microscopy [22] and by transmission electron microscopy [48].

Growth Regime I and II or Type A and B

Hoffman et al. [28], Labaig [39] and Toda [47] showed that the plots of $\log V$ against $1/T_c\Delta T$ showed a breaking at a $1/T_c\Delta T$. Hoffman et al. observed change of the polycrystalline morphology at the breaking point from spherulite to axialite with increase of $1/T_c\Delta T$. They named the two regions, regime I and II, respectively [28]. Toda observed Vs on single crystals and also found change of morphology at the breaking point from lenticular shape to truncated lozenge shape with curved growth $\{200\}$ faces and named them types A and B crystals, respectively [47]. In this paper the term of types A and B will be used, because study will be carried out on single crystals.

5.2 Results

Formation of FCSCs and ECSCs

Formation of isolated FCSCs was confirmed on samples 11 K and 29 K, but single-crystal like crystals were obtained in the case of 100 K. Isolated cigar-like or leaf-like ECSCs were confirmed for all M_n, i.e., even in the case of material with high molecular weight, such as 100 K.

Steady Lateral Growth

Typical observed lateral size a of an ECSC in specimens 11 K, 29 K, and 100 K is plotted against Δt for a $\Delta T = 4.2$ K in Fig. 21. a increased linearly with an increase of Δt for all M_n, that shows that the growth of an ECSC was steady growth for all M_n. The slopes of the lines decreased with increasing M_n where the slope was in proportion to V. Therefore, this indicates that V decreased with an increase of M_n.

ΔT dependence of V

The lateral growth rates (Vs) of FCSCs for samples 11 K, 29 K, and 100 K were plotted against $1/\Delta T$ in Fig. 22. They gave straight lines and breaking points, therefore the well-known experimental formula, $V = V_0 \exp(-B/\Delta T)$

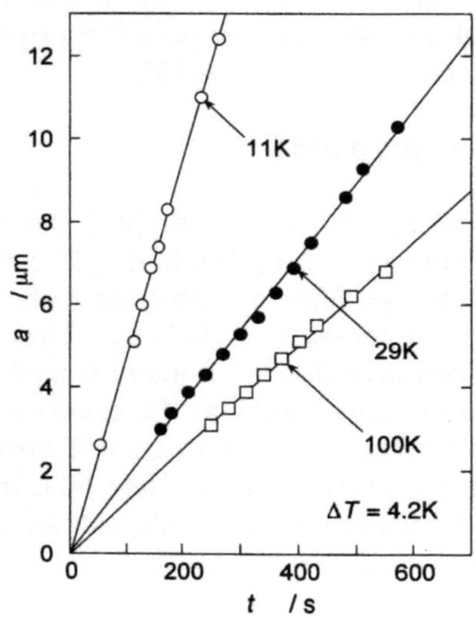

Fig. 21 a as a function of t at $\Delta T = 4.2$ K for $M_n = 11$ K, 29 K, and 100 K. *Lines* show the best fit of the plots. $t = 0$ is defined as the time when the ECSC was generated

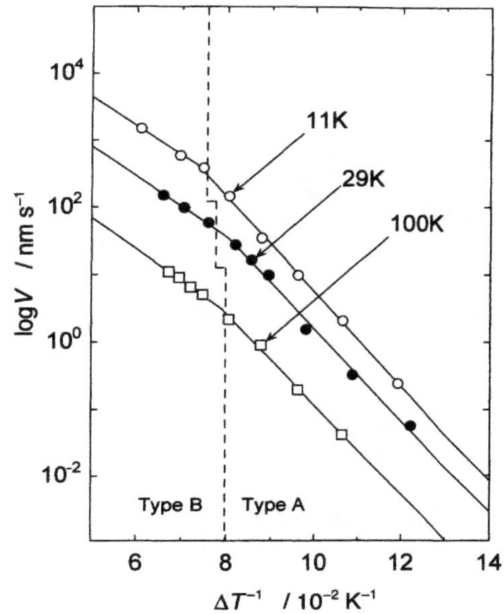

Fig. 22 Lateral growth rate V as a function of $1/\Delta T$ of FCSCs for $M_n = 11$ K, 29 K, and 100 K

was confirmed. The breaking points were confirmed to correspond to the type A to type B transition from morphological evidences observed by optical microscope and TEM after Toda's method [47]. The transition was shown by dotted lines in Fig. 22.

$\log V$ of ECSCs is plotted against $1/\Delta T$ for specimens 11 K, 29 K, and 100 K in Fig. 23. $\log V$ decreased linearly with an increase of $1/\Delta T$ for all M_n. As the $\log V$ vs. $1/\Delta T$ of ECSCs showed linear straight lines, we assume here that the ΔT dependence of the lateral growth is mainly controlled by secondary nucleation on the smooth surface of the hexagonal crystal, which is not well known.

All lines for FCSCs and ECSCs shifted downwards with an increase of M_n. This suggests that V decreases with an increase of M_n. It should be noted that all lines were parallel, which means that the slope B does not depend on M_n. Hence the intercept decreased with an increase of M_n.

The M_n Dependence of V_0 and B

The intercept V_0 and slopes B in $\log V$ against $1/\Delta T$ of FCSCs were plotted against M_n in Fig. 24. This showed that V_0 significantly decreased with an increase of M_n, whereas B did not, as was shown by Hoffman et al. [28] V_0 and B of ECSCs showed similar M_n dependence to those of FCSCs. As V_0 is related to self diffusion of polymer chains and B is related to the activation free en-

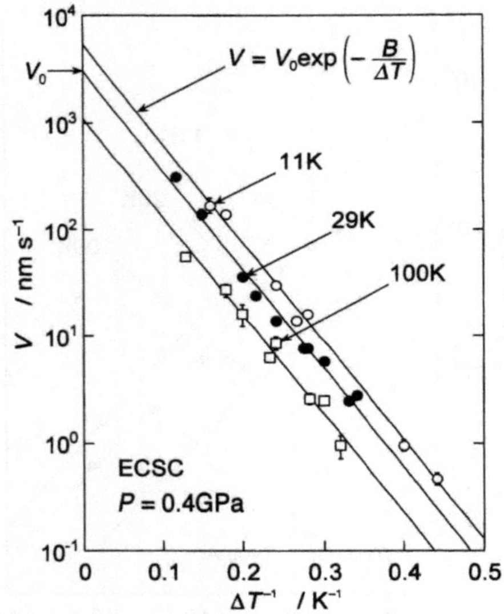

Fig. 23 log V against $1/\Delta T$ of ECSCs for M_n = 11 K, 29 K, and 100 K. *Lines* show the best fit of the plots, which correspond to Eq. 41

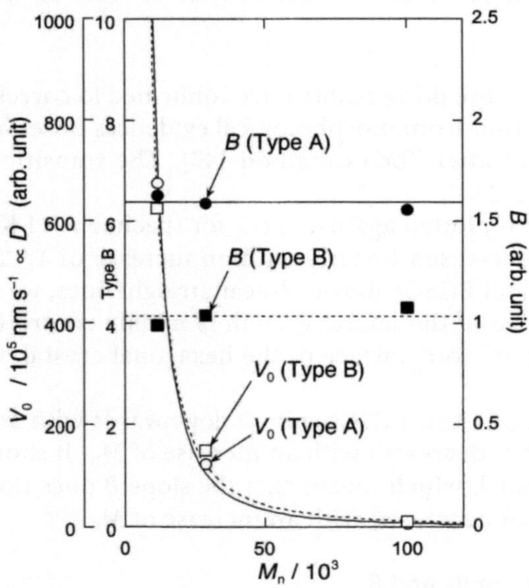

Fig. 24 V_0 and B for type A and B of FCSCs as a function of log M_n

ergy for the formation of a two-dimensional critical nucleus, we arrive at the conclusion that the self-diffusion process of polymer chains mainly controls the M dependence of V, whereas the critical nucleation process does not.

The Power Law of the M_n Dependence of V

log V_0 against log M_n of FCSCs and ECSCs are plotted in Fig. 25. In this figure, log I_0 against log M_n of FCSCs and ECSCs are also plotted. This showed that log V linearly decreased with an increase of log M_n. It should be noted that the slopes for both types A and B of FCCs were nearly the same. Thus, we have the experimental formula of a power law that

$$V(M_n) \propto V_0(M_n) \propto M_n^{-H}, \tag{47}$$

where H is a constant of the power,

$$H = \begin{cases} 1.7 & \text{for both types A and B of FCSCs} \\ 0.7 & \text{for both ECSCs} \end{cases} . \tag{48}$$

In order to compare this with the previous study by Hoffman et al. [28] and Labaig's [39], log V against log M_n at $\Delta T = 10$ K and 15 K for types A and

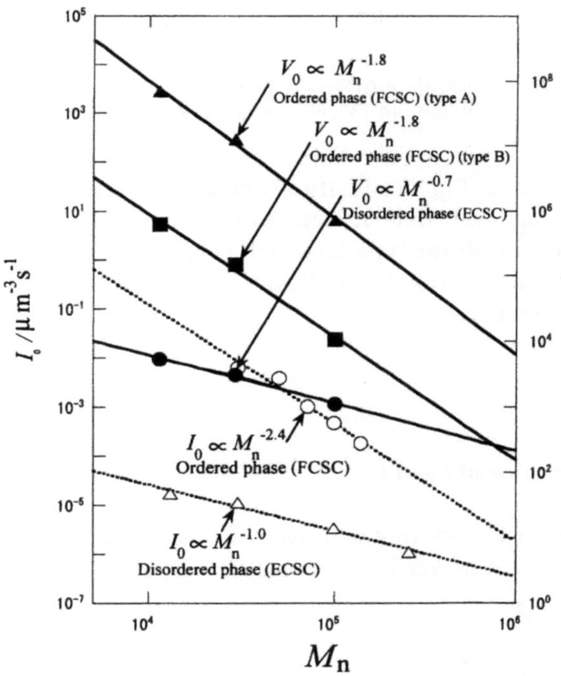

Fig. 25 Plot of log I and log V against log M_n for ordered and disordered phases where FCCs and ECCs are formed, respectively, from which a common power law of I and V for PE, $I, V \propto D(M_n) \propto M_n^{-H}$, is proposed. *Solid* and *broken lines* are the best-fit lines of the experimental power laws. It should be noted that H of the ordered phase is larger than that of the disordered phase

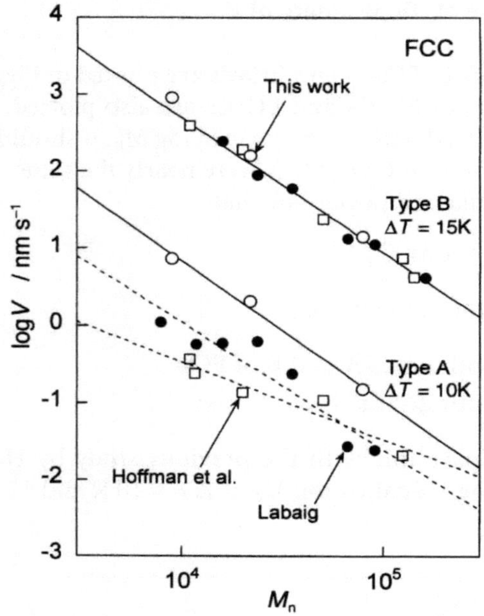

Fig. 26 log V plotted against log M_n for types A and B of FCCs. Hoffman et al. and Labaig plots were re-plotted from the data in [39, 40] using ΔTs obtained in this work

B of FCCs is plotted in Fig. 26. Hoffman et al. [28] and Labaig's [39] data were re-plotted using newly determined ΔTs in this work. They showed good agreement with our result for type B (which corresponds to regime II), while for type A (corresponds to regime I), the data are scattered and it was difficult to estimate the correct H.

5.3 Discussion

The Universal Power Law of *I* and *V*

It is concluded from Fig. 25 that a universal power law of I and V of PE is obtained. The universal power law of I and V is expressed by,

$$I, V \propto D(M_n) \propto M_n^{-H}, \qquad (49)$$

where H is summarized for I,

$$H = \begin{cases} 2.4 & \text{for ordered phase (FCSC)} \\ 1.0 & \text{for disordered phase (ECSC)} \end{cases}, \qquad (50)$$

and for V,

$$H = \begin{cases} 1.7 & \text{for ordered phase (FCSC)} \\ 0.7 & \text{for disordered phase (ECSC)} \end{cases} . \tag{51}$$

It is concluded that H increases with an increase of the degree of order of the crystalline phases,

$$H(\text{ordered}) > H(\text{disordered}) . \tag{52}$$

It is also concluded that H of I is larger than that of V,

$$H(I) > H(V) . \tag{53}$$

Two Kinds of Sliding Diffusion in Growth

The significant difference in H between FCSCs and ECSCs indicates that the M dependence of V can not be controlled by the self-diffusion process within the melt (the first stage) as proposed by Hoffman et al. [40], but it should be controlled by the surface diffusion process (the second stage) as shown in Fig. 27.

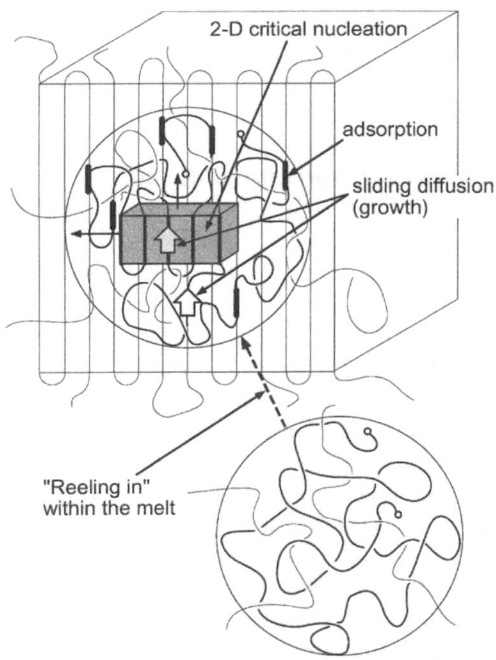

Fig. 27 Schematic illustration of the sliding diffusion model of a secondary nucleation (two-dimensional nucleation) on the surface of a single crystal

One of the authors (MH) showed that the formation of an ECSC or an FCC is related to the order of the crystalline phase [20, 33, 34], that is, an ECSC and an FCC are formed from the melt into a disordered hexagonal and an ordered orthorhombic phase, respectively. It is natural to consider that the surface diffusion process should be controlled by the order of the crystalline phase. This is the reason why H shows a significant difference between ECSC and FCC.

The Reason Why *B* Does Not Depend on *M*

It should be noted that the critical nucleation process does not depend on M. This can be explained by our model of surface diffusion (Fig. 27). In the model a nucleus will be formed from the absorbed chains. We can estimate the number of repeating units within a critical nucleus (N^*) using parameters σ, σ_e, and Δh given in [14]. N^* is the order of 10^2–10^3 for the range of ΔT in our experiment, which is much smaller than the number of repeating units within a molecule (10^3–10^4). This indicates that a critical nucleus should be formed by a part of a molecular chain. Therefore, the nucleation process of the critical nucleus will not depend on M. Thus, it is a natural result that B does not depend on M in this study. This is consistent with the discussion by Hoffman et al. [28] on FCC. They showed that the nucleation process of an FCC does not depend on M_n in the case of $M_n > 10^4$. On the contrary they showed that it depends on M_n for $M_n < 10^4$, because σ_e depends on M_n due to the effect of chain ends on the end surface of the critical nucleus.

6 The Role of Entanglement in Nucleation [49, 50]

6.1 Introduction

Significant changes of number density of entanglement (ν_e) and the chain conformation should occur during the crystallization and melting. In this chapter, ν_e is defined to be unity and zero for the equilibrium melt and for the equilibrium melt crystal of ideal extended chain single crystals (ECSCs), respectively, as shown in Fig. 1. Between two ideal equilibrium states, "metastable" crystalline and molten states exist. Small crystals (nucleus or embryo) or folded chain crystals (FCCs) can be regarded as the metastable crystalline state. Small crystals or FCCs should fully disentangle to grow into ideal ECSCs via chain sliding diffusion in lamellae or interface between a nucleus and the melt [14, 15]. Partially entangled random coiled melt or locally ordered melt can be regarded as the metastable molten state. Therefore, actual crystallization and melting is the transition between the metastable melt and crystals.

Entanglements should assume an important role in the crystallization of polymer crystals, but there has been little direct experimental evidence to date. Therefore, it is important to clearly define the role of entanglements in nucleation and growth of polymer crystals. When a primary or secondary nucleus is formed and grows via chain sliding diffusion, the entanglements within the interface result in the "pinning effect" as shown in Fig. 1. The chain sliding diffusion within the nucleus or interface should be suppressed by the entanglements. Therefore, it is expected that the nucleation rate I decreases with an increase of the number density of entanglement (ν_e). Thus, an important unresolved problem arises: how do ν_e and the chain conformation change during the crystallization or melting? Such changes during melting are sometimes called "melt relaxation".

Psarski et al. reported the effects of the entanglement on the lateral growth of PE [51]. They showed that the lateral growth rate of the spherulite V from the melt of ECSCs is larger than that from the melt of FCCs. They explained this with a model where ν_e of the melt of ECSCs may be smaller than that of the melt of FCCs. However, they did not show the ν_e dependence of V.

The purpose of this section is to obtain the relationship between I and ν_e by changing ν_e within the melt.

6.2 How to Observe $I(\nu_e)$?

How to Change ν_e in the Melt?

The spatial size of entanglement is too large to be included within the crystalline lattice. Therefore, we can assume that the entanglements, which are expressed by cross marks in Fig. 28, can exist only on the surface of a crystal or in the amorphous layers between lamellae. Therefore, thick extended chain single crystals (ECSCs) include little entanglement ($\nu_e \sim 0$), while FCCs composed of stacked thin lamellae include a lot of entanglements (Fig. 28) [49, 50].

Soon after thick ECSCs and thin FCCs are melted, the fresh melt should include small ν_e and rather large ν_e, respectively, i.e., ν_e should decrease with an increase of lamellar thickness (l) of the melted crystals. In this study, we changed ν_e by melting crystals with different l, from 10 nm to several μm.

From the experimental fact that the lamellar thickening growth rate (U) of ECSC of PE is independent of l [8], we can regard ν_e on the end surface of the nucleus as constant and independent of l. For simplicity, we assume that the shape of ECSC is rectangular parallelepiped (Fig. 28). ν_e is given by the ratio of the volume of the crystal to the surface area of the crystal as shown below,

$$\nu_e \propto \frac{\text{surface area}}{\text{volume}} = \frac{4al + 2a^2}{a^2 l}, \tag{54}$$

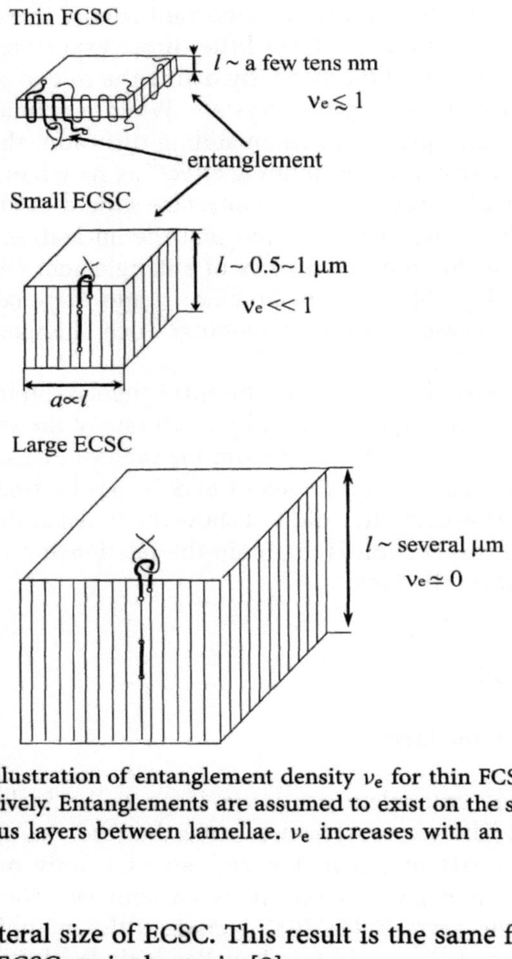

Fig. 28 Schematic illustration of entanglement density ν_e for thin FCSC, small ECSC, and large ECSC, respectively. Entanglements are assumed to exist on the surface of the crystal or in the amorphous layers between lamellae. ν_e increases with an increase of lamellar thickness l

where a is the lateral size of ECSC. This result is the same for the actual "tapered shape" of ECSC, as is shown in [9].

Since we have shown in our previous study that U is proportional to the lateral growth rate V [8], we have the following equation,

$$l \propto a . \tag{55}$$

Here relations $l = l^* + 2Ut$, $a = a^* + 2Vt$, $l \gg l^*$, and $a \gg a^*$ are used, where l^* and a^* are critical lamellar thickness and critical lateral size, respectively. The combination of Eqs. 54 and 55 gives

$$\nu_e(l) \propto \frac{1}{l} . \tag{56}$$

It should be noted here that ν_e will increase with an increase of annealing time (Δt) at a temperature above the melting temperature. Hereafter we will call Δt the "melt annealing time".

How to Obtain $I(\nu_e)$?

We will obtain the experimental formula of I as a function of l

$$I = I(l)\,. \tag{57}$$

With substitution of Eq. 56 into Eq. 57, we can obtain the formula of $I(\nu_e)$,

$$I = I(\nu_e)\,. \tag{58}$$

How to Change l of ECSC?

In the case of PE, we can prepare ECSCs with different l using the established technique [10]. The pressure-temperature ($P - T$) phase diagram consists of liquid, hexagonal, and orthorhombic phases [10–13]. When PE is isothermally crystallized at the relevant crystallization temperature (T_c) under the triple point pressure (P_{tri}), ECSCs are generated in the metastable hexagonal phase [8–13]. In the hexagonal phase, l of ECSC increases linearly with an increase of crystallization time (t) [8, 9]. After a period of time, lamellar thickening growth is stopped when the metastable hexagonal phase transforms into the most stable orthorhombic phase [23]. It was shown that l decreases with a decrease of T_c, i.e., increase of ΔT. From this mechanism, we can prepare ECSCs with different l by controlling $P(< P_{tri})$, ΔT, and t.

6.3
Results

Morphology

The formation of isolated FCSCls was confirmed from the melt of samples with different l (ECSCs-melt-FCSC or FCCs-melt-FCSC). The morphology is the same as the usual one of spherulite or axialite, as reported by Toda [47], irrespective of the morphology before melting.

Nucleation Rate

The number density of crystals (ν) increased linearly with an increase of t for all samples, which indicates a steady nucleation process. Figure 29 shows the plot of $\log I$ against ΔT^{-2} for different l. I obeyed the well-known equation, $I = I_0 \exp(- C/\Delta T^2)$ where I_0 and C are constants. These straight lines were parallel to each other. This indicates that the slope of the straight line C is almost constant irrespective of l. We obtained the average of C ($\langle C \rangle$),

$$\langle C \rangle \cong 955 \pm 30\,(\mathrm{K}^2)\,. \tag{59}$$

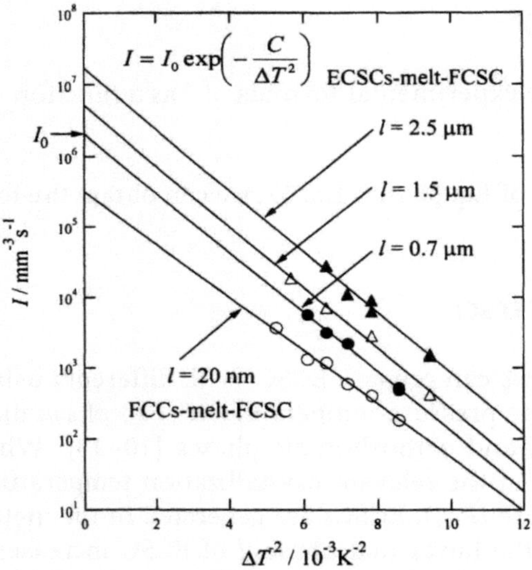

Fig. 29 Plots of $\log I$ against ΔT^{-2} for different l = 20 nm, 0.7 μm, 1.5 μm, 2.5 μm, respectively. The *solid lines* show the best fit of the plots

Moreover, the straight line shifts upward with an increase of l. Thus, it is concluded that $I \propto I_0$ increases with an increase of l for any ΔT.

Since I_0 and C are proportional to the diffusion coefficient (D) and activation free energy for formation of a critical nucleus (ΔG^*), respectively, it is concluded that the l dependence of I is mainly determined by the diffusion process of the polymer chain and not by the formation process of a critical nucleus.

Formulae of $I(l)$, $I(\nu_e)$ and $\nu_e(\Delta t)$

l Dependence of I

Figure 30 shows a plot of $I_0(\propto I)$ against l. It was found that I_0 increases gradually at first and then rapidly with an increase of l. We obtained the following experimental formula,

$$I(l) \propto I_0(l) \propto \exp(-\alpha/l)\,, \tag{60}$$

where

$$\alpha = 3.27 \pm 0.17\ (\mu\text{m})\,. \tag{61}$$

ν_e Dependence of I

An experimental formula of $I(\nu_e)$ was obtained for the first time from a combination of Eqs. 60 and 56,

$$I(\nu_e) \propto I_0(\nu_e) \propto \exp(-\gamma\nu_e)\,, \tag{62}$$

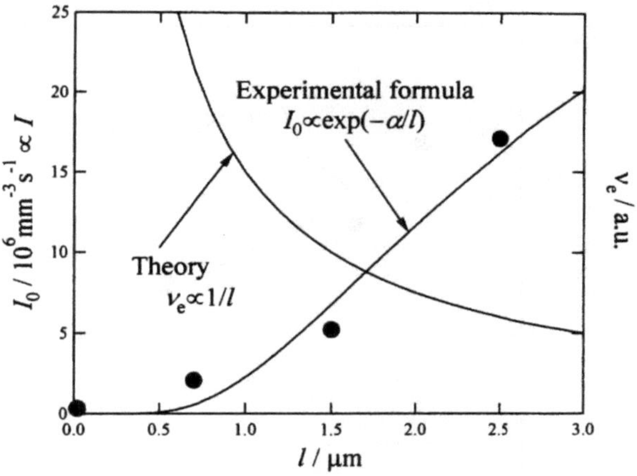

Fig. 30 Plots of $I_0(\propto I)$ against l. The experimental formula was obtained by fitting to experimental data. The other *solid curve* shows the equation $\nu_e \propto 1/l$

where γ is a constant. Figure 31a shows the plots of I_0 against ν_e. The experimental data is fitted with Eq. 62. As $I(\nu_e)$ decreases exponentially with an increase of ν_e (Fig. 31b), it is confirmed that nucleation is suppressed by entanglement.

Effect of Melt Relaxation on Nucleation

We have found recently that I of PE decreases exponentially with an increase of annealing time at a temperature above the melting temperature (Δt) [52]. Hereafter, we will call Δt the melt annealing time. We consider this phenomenon to be a type of melt relaxation. We obtained I as a function of Δt as,

$$I \propto \exp(-\Delta t/\tau_m) + \text{const.}, \quad (63)$$

where τ_m is the melt relaxation time.

We have speculated that when the melt is kept above the melting temperature, ν_e gradually increases with an increase of Δt and approaches the equilibrium ν_e ($\nu_e = 1$). It was shown that the melt relaxation process takes a long time, i.e., it takes several hours or a few days depending on the annealing temperature (T_{max}) and M_n. However, this experimental fact was indirect evidence of the role of entanglement in the nucleation of polymers.

"Melt Memory Effects" on Nucleation

It was found that melt memory effects are significant in polymers due to the topological nature [53]. It is considered that melt memory effects are mainly

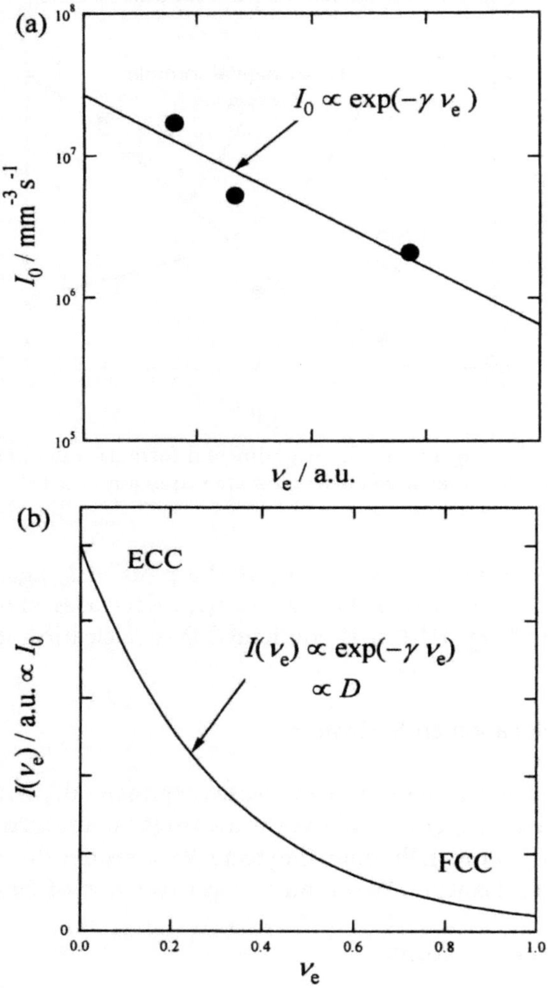

Fig. 31 **a** Plots of I_0 against ν_e. **b** The function of $I(\nu_e)(\propto I_0)$ exponentially decreases with increase of ν_e. ECC and FCC in the figure correspond to morphologies before melting

controlled by the following two factors. Firstly, ν_e changes with Δt, i.e.,

$$\nu_e = \nu_e(\Delta t) \tag{64}$$

as mentioned in the above section. Secondly, a change of chain conformation during the melting. Though it is expected that the chain conformation in the melt significantly affects nucleation, the mechanism has not as yet been resolved. One way to solve this problem is to observe nucleation from the melt of ECSCs, because it is expected that the melt memory effect of this melt is different from that of the melt of FCCs.

Δt Dependence of ν_e

As mentioned above, we have already obtained the experimental formula between I and Δt as Eq. 63. Combination of Eqs. 63 and 62 gives the formula,

$$\nu_e(\Delta t) \propto -\ln\left\{\text{const.} + A\exp(-\Delta t/\tau_m)\right\}, \quad (65)$$

where A is a constant.

This equation can be approximated for limit cases,

$$\nu_e \propto \Delta t \quad \text{for small } \Delta t \quad (66)$$

$$\cong \text{const.} \quad \text{for large } \Delta t\,. \quad (67)$$

Figure 32 shows a plot of ν_e against Δt as derived from Eq. 65. This figure also shows the experimental $I(\Delta t)$. We found that a decrease of $I(\Delta t)$ with an increase of Δt correspond to an increase of $\nu_e(\Delta t)$ with an increase of Δt. Therefore, it is concluded that an increase of ν_e with an increase of Δt is clearly an important mechanism of the melt memory effect.

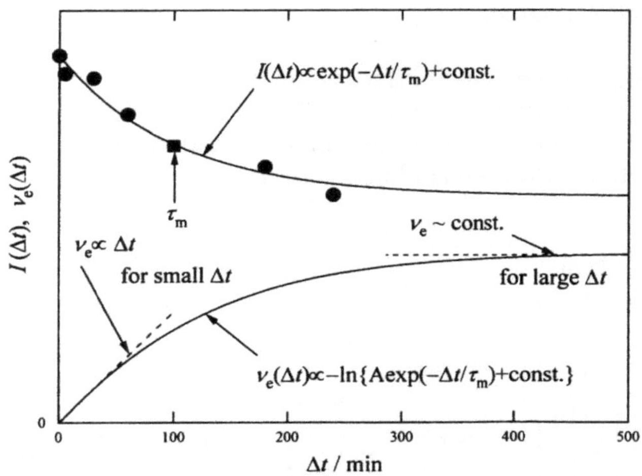

Fig. 32 Plots of I and ν_e against Δt. Decrease of I with an increase of Δt corresponds to an increase of ν_e with an increase of Δt

6.4 Discussion

Effects of Entanglement on Nucleation of Polymers

We have shown that only $I_0(\propto D)$ depends on ν_e, while $C(\propto \Delta G^*)$ does not depend on ν_e. This means that the topological nature of nucleation is reflected only on the kinetic factor (D) and not on the thermodynamic factor (ΔG^*) as shown in our previous papers [20, 33, 34]. This is an important conclusion for the effects of entanglement on the nucleation of polymers.

As shown in Fig. 1, chain sliding diffusion becomes difficult due to pinning effect within the interface between a nucleus and the melt. Since I_0 is proportional to the topological diffusion constant D, D is related to ν_e from Eq. 62,

$$D(\nu_e) \propto \exp(-\gamma\, \nu_e) \quad (68)$$

The Fold Nucleus is Formed From Both the FCC-Melt and ECSC-Melt

It is shown that C does not depend on l. As C is proportional to ΔG^*, ΔG^* is a constant irrespective of l. Therefore, σ_e is constant irrespective of l. In the case of ordinary melt crystallization such as FCCs-melt-FCSC, a fold type nucleus is formed. Therefore, it is concluded that the fold type nucleus should be also formed from ECSCs-melt because the σ_e is constant irrespective of l, i.e., morphology before melting.

Two-Stage Melt Relaxation

When the fold type nucleus is formed from the melt, the chain conformation within the melt should be a random coiled one. Therefore, the chain conformation within the melt of ECSCs should be a random coiled one. The crucial difference between the melts of ECSCs and FCCs at small Δt is only that of ν_e.

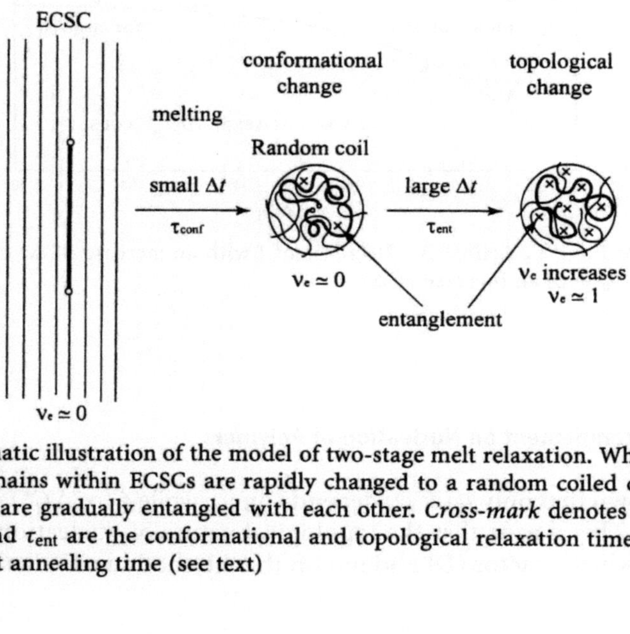

Fig. 33 Schematic illustration of the model of two-stage melt relaxation. When ECSCs are melted, the chains within ECSCs are rapidly changed to a random coiled conformation. Then, chains are gradually entangled with each other. *Cross-mark* denotes the entanglement. τ_{conf} and τ_{ent} are the conformational and topological relaxation time, respectively. Δt is the melt annealing time (see text)

From this we will propose a "two-stage melt-relaxation" for the melting of ECSCs, i.e., consisting of conformational and topological changes (Fig. 33). The chain conformation rapidly changes from the extended one to the random coiled one when ECSCs are melted. Then chains will be gradually entangled with each other with an increase of Δt and ν_e will approach to equilibrium one ($\nu_e = 1$). Therefore, the conformational relaxation time (τ_{conf}) is much smaller than the topological relaxation time necessary for entanglement (τ_{ent}),

$$\tau_{conf} \ll \tau_{ent} \,. \tag{69}$$

7 Conclusions

In this work all experiments were carried out with polyethylene (PE).

1. Nucleation during the induction period from the melt was directly confirmed for the first time by means of small angle X-ray scattering (SAXS), which was enabled by increasing the number density of nuclei as large as 10^4 times the usual case by adding a nucleating agent to the sample. The number density of nuclei starts increasing after quenching into a crystallization temperature and then saturates, this corresponds to the "induction and steady (stationary) states" of the nucleation process.
2. Lamellae start stacking much later than nuclei start developing. The onset time of stacked lamellae was similar to the induction time. Therefore, nucleation during the induction period can be observed without being affected by the stacked lamellae.
3. Power laws of molecular weight of the primary and lateral growth rates,

   $$I \propto M_n^{-H} \quad \text{and} \quad V \propto M_n^{-H} \,,$$

   were confirmed for the first time, where H is a constant. It is to be noted that the power H significantly increases with the increase of degree of order of the crystals. In the case of primary nucleation,

   $$H = \begin{cases} 2.4 & \text{for the orthorhombic "ordered" phase} \\ 1.0 & \text{for the hexagonal "disordered" phase} \end{cases} ,$$

   is obtained. The former corresponds to formation of FCC and the latter ECC, respectively. In the case of the lateral growth,

   $$H = \begin{cases} 1.7 & \text{for the orthorhombic "ordered" phase} \\ 0.7 & \text{for the hexagonal "disordered" phase} \end{cases} ,$$

   is obtained.

4. The free energy necessary for the formation of a critical nucleus ΔG^* in both primary and secondary nucleation processes does not depend on M_n, i.e., $\Delta G^* \approx$ const, while only the diffusion coefficient D depends on M_n, i.e., $I \propto D(M_n)$. Therefore, the M_n dependences of I and V are not controlled by the formation process of a critical nucleus but are mainly controlled by the chain sliding diffusion process.
5. The power laws of the nucleation and lateral growth rates confirmed that the topological nature of polymer chains assumes the most important role in the polymer crystallization. It is concluded that the primary nucleation is a process where chains rearrange from the melt into a crystalline lattice via chain sliding diffusion and disentanglement within the nucleus or the interface between the nucleus and the melt. In the case of lateral growth, chain sliding diffusion on the substrate is also important.
6. When ECCs are melted, the number density of entanglement within the melt ν_e should be small soon after melting, because entanglements exist only on lamellar surfaces or amorphous layers between crystalline lamellae. Applying this fact, we succeeded in changing ν_e by melting crystals with a significantly different lamellar thickness (l), between a few μm (ECCs) and a few tens of nm (FCCs). We showed logically the relationship, $\nu_e \propto 1/l$. The nucleation rate (I) increases with an increase of l, $I \propto \exp(-\alpha/l)$ where α is a constant. With the combination of these relationships, we showed that I decreases with an increase of ν_e, $I(\nu_e) \propto \exp(-\gamma\, \nu_e)$. We propose a two-stage melt relaxation, i.e., a fast conformational one from the extended to random coiled conformation and slow topological relaxations of reconstruction of entanglement.

Acknowledgements The authors are grateful to Prof. Akihiko Toda, Dr. Isao Wataoka, Dr. Swapan K. Ghosh of Hiroshima University, Dr. K. Yamada of SunAllomer Co. Ltd., Dr. Katsuaki Inoue of the Japan Synchrotron Radiation Institute (JASRI) and Dr. Zdenek Kozisek of the Institute of Physics, Academy of Sciences of the Czech Republic for their help with the experiments and discussions. SAXS experiments were carried out at the BL40B2 of SPring8 (SP8) at JASRI (Proposal No. 2001B0187-NDL-np—2004A0224-NL-2b-np) in Harima and at the BL-10C small angle installation of the Photon Factory (PF) at KEK in Tsukuba. The authors also thank Asahi Denka Kogyo K.K. for supplying the nucleating agent. This work was partly supported by the Grant-in-Aid for Scientific Research on Priority Areas B2 (No.12127205) and Scientific Research A2 (No. 12305062). The authors are grateful to the financial support from the International Joint Research grant, NEDO, 1996–1998.

References

1. Becker R, Döring W (1935) Ann Phys 24:719
2. Zeldovich YaB (1943) Acta Physicochim USSR 18:1
3. Frenkel J (1946) Kinetic Theory of Liquids. Oxford University, London
4. Turnbull D, Fisher JC (1949) J Chem Phys 17:71

5. Flory PJ (1953) Principles of Polymer Chemistry. Cornell University, Ithaca, New York
6. de Gennes PG (1979) Scaling Concepts in Polymer Physics. Cornell University, Ithaca, New York
7. Doi M, Edwards SF (1986) The Theory of Polymer Dynamics. Clarendon Press, Oxford
8. Hikosaka M, Amano K, Rastogi S, Keller A (2000) J Materials Sci 35:5157
9. Hikosaka M, Amano K, Rastogi S, Keller A (1997) Macromolecules 30:2067
10. Hikosaka M, Tsukijima K, Rastogi S, Keller A (1992) Polymer 33:2502
11. Bassett DC, Block S, Piermarini GJ (1974) J Appl Phys 45:4146
12. Yasuniwa M, Enoshita R, Takemura T (1976) Jpn J Appl Phys 15:1421
13. Hikosaka M, Minomura S, Seto T (1980) Jpn J Appl Phys 19:1763
14. Hikosaka M (1987) Polymer 28:1257
15. Hikosaka M (1990) Polymer 31:458
16. Frisch HL (1957) J Chem Phys 27:90
17. Andres RP, Boudart M (1965) J Chem Phys 42:2057
18. Akpalu YA, Amis EJ (1999) J Chem Phys 111:8686
19. Imai M, Mori K, Kizukami T, Kaji K, Kanaya T (1992) Polymer 33:4457
20. Nishi M, Hikosaka M, Ghosh SK, Toda A, Yamada K (1999) Polym J 31:749
21. Nishi M, Hikosaka M, Toda A, Takahashi M (1998) Polymer 39:1591
22. Rastogi S, Hikosaka M, Kawabata H, Keller A (1991) Macromolecules 24:6384
23. Hikosaka M, Okada H, Toda A, Rastogi S, Keller A (1995) J Chem Soc Faraday Trans 91:2573
24. Frank FC, Tosi M (1961) Proc Roy Soc A263:323
25. Price F (1969) Nucleation in polymer crystallization. In: Zettlemoyer AC (ed) Nucleation. Marcel Dekker, Inc, New York
26. Wunderlich B (1980) Macromolecular Physics. Academic Press, London
27. Okada M, Nishi M, Takahashi M, Matsuda H, Toda A, Hikosaka M (1998) Polymer 39:4535
28. Hoffman JD, Frolen LJ, Ross GS, Lauritzen JI (1975) J Res NBS 79A:671
29. Hikosaka M, Yamazaki S, Wataoka I, Das NC, Okada K, Toda A, Inoue K (2003) J Macromol Sci B42:847
30. Guinier A (1967) Theory of technique of the radiocrystallograpy, (Japanese ed). Rigaku Denki, Tokyo
31. Roe RJ (2000) Methods of X-ray and neutron scattering in polymer science. Oxford Univ Press, New York
32. Olmsted PD, Poon WCK, McLeish TCB, Terrill NJ, Ryan AJ (1998) Phys Rev Lett 81:373
33. Ghosh SK, Hikosaka M, Toda A (2001) Colloid Polym Sci 279:382
34. Ghosh SK, Hikosaka M, Toda A, Yamazaki S, Yamada K (2002) Macromolecules 18:6985
35. Garti N, Sato K (eds) (2001) Crystallization Process in Fats and Lipid Systems. Marcel Dekker, Inc, New York
36. Nozaki K, Hikosaka M (2000) J Material Sci 35:1239
37. Wunderlich B (1973) Macromolecular Physics, vol 1&2. Academic Press, New York
38. Magill JH, Kojima M, Li HM (1973) the IUPAC Symp Macromol, Aberdeen, UK
39. Labaig JJ (1978) PhD Thesis, Faculty of Science, University of Strasbourg
40. Hoffman JD (1982) Polymer 23:656
41. Hoffman JD, Miller RL (1988) Macromolecules 21:3038
42. Kossel W (1927) Nach Ges Wiss Gottingen 135
43. Volmer M (1939) Kinetik der Phasenbildung
44. Burton WK, Cabrera N, Frank FC (1950-1951) Phil Trans Roy Soc A243:299

45. Watanabe H (1986) Kobunnshi High Polym Jpn 35:111046
46. Hoffman JD (1994) International Polymer Physics Symposium (Honoring Prof Kawai) p 19
47. Toda A (1992) Colloid Polym Sci 270:667
48. Hikosaka M, Rastogi S, Keller A, Kawabata H (1992) J Macromol Sci Phys B31:87
49. Yamazaki S, Hikosaka M, Gu F, Ghosh SK, Arakaki M, Toda A (2001) Polym J 33:906
50. Yamazaki S, Hikosaka M, Toda A, Wataoka I, Gu F (2002) Polymer 43:6585
51. Psarski M, Piorkowska E, Galeski A (2000) Macromolecules 33:916
52. Yamazaki S, Hikosaka M, Toda A, Okada K, Gu F, Watanabe K, submitted to Polymer
53. Alfonso GC, Scardigli P (1997) Macromol Symp 118:323

Adv Polym Sci (2005) 191: 187–240
DOI 10.1007/12_013

Published online: 25 October 2005

Spinodal Crystallization of Polymers: Crystallization from the Unstable Melt

Keisuke Kaji[1,2] (✉) · Koji Nishida[1] · Toshiji Kanaya[1] · Go Matsuba[1] · Takashi Konishi[1] · Masayuki Imai[3]

[1] Institute for Chemical Research, Kyoto University, Gokasho, Uji, 611-0011 Kyoto-fu, Japan
kajiksk@ybb.ne.jp, knishida@scl.kyoto-u.ac.jp, kanaya@scl.kyoto-u.ac.jp, gmatsuba@scl.kyoto-u.ac.jp

[2] 1-1-5-304, Uchihirano-machi, Chuo-ku, 540-0037 Osaka, Japan
kajiksk@ybb.ne.jp

[3] Department of Physics, Faculty of Science, Ochanomizu University, Otsuka, Bunkyo-ku, 112-0012 Tokyo, Japan
imai@phys.ocha.ac.jp

Abstract This paper reviews the authors' investigation into polymer crystallization, especially involving a spinodal decomposition (SD) type phase separation due to the orientation fluctuation of stiff segments prior to crystal nucleation. Evidences for SD obtained from small-angle X-ray and neutron scattering (SAXS and SANS), depolarized light scattering (DPLS), Fourier-transform infrared spectroscopy (FT-IR) are discussed in detail in the case of the glass crystallization of poly(ethylene terephthalate) (PET) just above T_g. SD-like optical micrographs are also shown as a function of crystallization temperature for the melt crystallization of PET; their characteristic wavelengths Λ, which are of the order of μm above 120 °C, follow a van Aartsen equation derived from the Cahn–Hilliard theory for SD. By fitting the equation to the observed characteristic wavelengths the spinodal temperature T_s was determined to be $T_s = 213 \pm 5$ °C for the PET melt, above which the SD pattern suddenly changed to the usual spherulite pattern. On the basis of a theory by Olmsted et al. [4], the general mechanisms of polymer crystallization are also discussed; the crystallization from the metastable melt causes the nucleation and growth (N&G) of dense (nematic) domains while that from the unstable melt causes SD into the dense (nematic) and less dense (isotropic) domains. Furthermore, the secondary phase separation of the SD-type phase separation into smectic and amorphous domains subsequently occurs inside the nematic domain for both these cases.

Keywords Induction period · Melt and glass crystallization · Nucleation and growth · Optical microscopy · Scattering techniques · Spinodal decomposition

Abbreviations

b	excluded volume parameter being nearly equal to the cross-sectional diameter of a stiff segment
CRR	cooperatively rearranging region introduced to explain the α process of the glass transition
c	velocity of light
d	dimension or cross-sectional diameter of a stiff segment
d_{Bragg}	Bragg spacing
D	dense domain size
D_1	dense domain size for the early stage of SD
D_2	dense domain size for the late stage of SD
D_c	crystalline lamellar thickness
DSC	differential scanning calorimetry
DPLS	depolarized light scattering
FT-IR	Fourier-transform infrared spectroscopy
G	gauche
I_m	the maximum intensity of a scattering peak
$I(Q, t)$	scattering intensity as a function of Q and t
iPS	isotactic polystyrene
k_B	Boltzmann's factor
l	average length of stiff segments
l_p	persistence length

l_{rmi}	range of molecular interactions
L	long period
N&G	nucleation and growth
OM	optical microscopy
PE	poly(ethylene)
PEN	poly(ethylene naphthalate)
PET	poly(ethylene terephthalate)
$P(R)$	distance distribution function
q	length of scattering vector of light
Q	length of scattering vector
Q_m	maximum position of a scattering peak
$R(Q)$	growth rate of SAXS intensity or density fluctuations as a function of Q
R_n	growth rates of the DPLS intensity for time range n in the induction period
$R_m(t)$	average diameter of clusters as a function of time
$R_{\perp}(q)$	invariant of the Rayleigh factor for DPLS
SANS	small angle neutron scattering
SAXS	small angle X-ray scattering
SD	spinodal decomposition
$S(x)$	universal scaling function of $x = Q/Q_m$
sPS	syndiotactic polystyrene
t	time
t_0	beginning time of the late stage of SD
T	trans
T	temperature
T_{bx}	boundary crystallization temperature
T_c	mode coupling critical temperature
T_g	thermal (or calorimetric) glass transition temperature
T_m	melting temperature
T_s	spinodal temperature
T_{s2}	spinodal temperature of the secondary phase separation
T_x	crystallization temperature
V_{excl}	excluded volume
WAXD	wide-angle X-ray diffraction
y	$= r\,Q_m$
y_1	minimum value of y
$\gamma(r)$	spatial density correlation function
$\Gamma(y)$	Fourier transform of the universal scaling function $S(x)$
$\langle\delta^2\rangle$	mean-square anisotropy
ΔE_n	the activation energy for time range n in the induction period
$\eta(r)$	local density function at point r
$\langle\eta^2\rangle_{av}$	average density fluctuation
λ	X-ray wavelength
Λ	characteristic wavelength of spinodal decomposition
ν^*	critical concentration of stiff segments
ν(glass)	concentration of stiff segments in the glassy state
ν(melt)	concentration of stiff segments in the molten state
ρ_a	macroscopic density of amorphous glassy sample
θ	a half of the scattering angle
Θ	angle between the neighboring stiff segments
ω	angular frequency of incident radiation

1 Introduction

Spherulites are a typical crystal morphology derived from polymer melts; they are composed of ribbon-like chain-folded lamellar crystals that radiate from their centers (e.g., see [1]). Polymer crystallization kinetics have been examined mainly with use of the theory of Lauritzen and Hoffman [2, 3]. However, initial crystallization processes prior to crystal nucleation are not well understood even in the case of homogeneous crystal nucleation. It has so far been considered that primary crystal nucleation occurs directly from the melt; spontaneous fluctuations acting on embryos of sufficient size in a supercooled liquid create the crystal nuclei. Such a mechanism is valid only for crystallization from the coexistence region in the phase diagram of a polymer melt proposed by Olmsted et al. [4] and that of colloids with short-range attraction by ten Wolde and Frenkel [5], however, the reality is much more complicated. Thus, the supercooled polymer melt has two states, metastable and unstable, other than the liquid-crystal coexistence state, depending on temperature and the stiffness of the polymer molecules. In order to avoid confusion in the terminology used, the word "metastable" without any proviso will be used in the meaning of Olmsted's phase diagram in this paper. This review is mainly based on the authors' former finding of spinodal decomposition (SD) from the unstable state [6–20], which is due to orientation fluctuations of stiff polymer segments. The characteristic wavelength of SD depends on the crystallization temperature but it has been observed that its size drastically changes from tens of nm to μm below and above a critical temperature near the glass transition temperature T_g (see Fig. 28). As will be discussed in Sect. 4.3, however, this phenomenon may only be a problem of observation. In the case of SAXS measurements we needed a sufficiently long induction period because of the sensitivity of the SAXS detector. Below the critical temperature, which probably corresponds to a mode-coupling critical temperature T_c, the induction period was long enough to make the SAXS observations while above T_c it was too short to make time-resolved SAXS measurements. If they could be possible even above T_c, the characteristic wavelength of tens of nm would be observed in addition to that of μm. Thus, it is presumed that there exist two kinds of SD type phase separation with characteristic wavelengths of the orders of μm and tens of nm, which can be assigned to the primary and the secondary phase separation, respectively. This will be discussed in Sect. 5. Such phenomena that occur prior to crystal nucleation were studied by means of small-angle X-ray scattering (SAXS) [6, 7, 9, 14], small-angle neutron scattering (SANS) [10, 11], depolarized light scattering (DPLS) [8, 15], Fourier-transform infrared spectroscopy (FT-IR) [12, 13], and optical microscopy (OM) [16, 17]. We have previously reviewed these investigations [18–20], but in this paper a more comprehensive description of the initial mechanisms involved in polymer crystallization will be given.

Polymer crystallization mechanisms depend not only on crystallization temperature, but also on the initial state of a sample from which crystallization starts. For an example of the latter, van Krevelen [21] showed that many small spherulites are produced when a homopolymer without a nucleating agent is crystallized from glass, while a small number of large spherulites are produced when crystallized from the melt. The number densities of such spherulites are different by six orders of magnitude between these two cases. The former and latter crystallization processes are called "glass crystallization" and "melt crystallization", respectively. In this review these terms will be used to distinguish crystallization conditions. As a surprising phenomena in polymer crystallization, which we were the first to note [6,7, 16, 19], we will show that a spinodal structure appears prior to crystal nucleation at low and intermediate temperatures as described above but a drastic morphological change occurs from the spinodal pattern to the usual spherulitic pattern at some critical temperature which corresponds to the spinodal temperature defined in the equation for the temperature dependence of characteristic wavelength by van Aartsen [22, 23]. The elucidation of these phenomena was also made based on a kinetic theory for the phase transition of polymer liquid crystals by Doi et al., [24, 25] and the theoretical phase diagram of polymer melt as a function of normalized density of the polymer melt and temperature, which was proposed by Olmsted et al. [4]. Such a drastic morphological change suggests that the common spherulites are produced when a polymer is crystallized from the metastable melt but not from the unstable melt.

In recent years several research groups have tried to establish—morphologically and/or kinetically—the mechanisms for structural formation processes in the early stages of polymer crystallization from the melt; from this body of work we will remark upon particularly important papers. Strobl and coworkers [26] have extensively studied the crystallization mechanisms both kinetically and morphologically, proposing a major route for the melt crystallization that the melt transforms to lamellar crystals through a mesomorphic state of a liquid crystal-like structure and a granular crystal state. The present authors essentially agree with their model though the polymers they studied were limited to flexible ones. In most flexible polymers such as polyethylene (PE), polypropylene (PP), and polycaprolactam (PCL) crystallization occurs very commonly from the metastable region of Olmsted's phase diagram [4] (see Fig. 30) at the usual crystallization temperatures because the normalized melt densities of these polymers are considerably higher than that at a critical point of the binodal curve, suggesting that their melts fall into the metastable state by the usual quenching. For example, the normalized melt density of PE is 0.685, which is much higher than the critical normalized density (0.53) [4]. In this case, as will be shown in Sect. 5, it may be considered that through a nucleation and growth (N&G) type of liquid-liquid phase separation droplets with a nematic-like structure are first formed sporadically in the isotropic matrix and grow in size with time. These droplets are considered to be of the order of tens of μm in size. Subsequently, however, another important phase separation occurs inside the droplets when they grow beyond a certain critical size. This secondary phase separation is an SD type where the nematic structure is separated into smectic and amorphous structures. In this process molecular entanglements contained in the nematic structure are excluded from the ordered domains having the smectic structure to the amorphous domains, and after the late stage of SD the spinodal structure turns into many much smaller spherical particles of the order of tens to hundreds of nm in size owing to surface tension, which probably corresponds to the mesomorphic state as defined by Strobl [26]. An evidence for this will be shown in Sect. 5. In the case of crystallization from the metastable melt it is therefore expected that SAXS would first provide a central scattering from the shape factor of the droplet of the pre-ordered denser region caused by N&G, and then it would show a peak due to the interferences among the small particles inside the droplet. Thus, when we observe a SAXS peak in the usual crystallization of flexible polymers, we are seeing secondary SD-type phase separation. This peak would later transform into the so-called long period peak from the lamellar crystals which are produced by fusion of these small particles with decreasing the interparticle spacing. A general scheme of polymer crystallization including such a case will be given in Sect. 6.

Wang, Hsiao and coworkers [27, 28] have criticized the concept of phase separation such as SD occurring in the early stage of polymer crystallization prior to crystal nucleation, and raised two issues of the detection limit of wide-angle X-ray diffraction (WAXD) and crystallization mechanisms. For the first issue, Ryan, Heeley and coworkers [29–31] showed—using a newly developed position sensitive high countrate microstrip gas chamber (MSGC) detector with a high count rate operation at the DUBBLE, ESRF, Grenoble—that the induction period where SAXS appears before the emergence of WAXS can actually be observed (for examples, see Figs. 3 and 6 of [29] and Fig. 4 of [30]) and that it is not a problem of the detection limit of WAXD though the length of the period, of course, depends on the crystallization temperature. Regarding the crystallization mechanism a misunderstanding seems to exist over two points. As already described above, the most important point is that the crystallization of isotactic polypropylene (iPP), which they investigated, usually occurs from the metastable state but not from the unstable state, and hence SD as a primary phase separation would never be expected as far as the sample is not quenched at extremely low temperatures. Even if SD were to occur as a primary phase separation, the characteristic wavelengths would be of an order of μm under the experimental conditions used, so that SAXS would not be able to detect a characteristic peak due to SD because of the lower angle resolution limit of SAXS cameras. The SAXS peak they observed is not due to the primary phase separation of the SD type but to the interference between many densely packed small particles of tens of nm in size, which are produced by the secondary phase separation of SD type inside the droplet. A strong evidence for the secondary phase separation of SD type seems to have been given in terms of a Cahn–Hilliard plot of $R(Q)/Q^2$ versus Q^2 by Ryan et al. [29–31] as will be shown later in Sect. 5. Furthermore, the densely packed clusters in a transmission electron micrograph observed by Wang et al. (Fig. 1 of [28]) seem to correspond to the small particles produced by the secondary phase separation of SD type inside the droplet because they are distributed rather homogeneously but not sporadically and they correctly pointed out that the SAXS peak is due to their average adjacent spacing though they did not recognize at which stage such clusters emerge. From this correspondence we can also understand the reason for their interesting observation as to why the lamellar orientation was uniform over the whole field of view when the quenched sample is annealed; the droplets caused by N&G have a nematic-like structure, which means that each droplet has some preferred molecular orientation such as a tangential molecular orientation. Hence, locally at least polymer chain segments may orient parallel to each other, and this orientation would be kept even among the neighboring small particles (or the clusters) inside each droplet. Anyway it should be noted again that the droplets caused by N&G do not provide a SAXS peak because they are produced sporadically and too large in size (of the order of tens of μm). Therefore, it is natural that the mechanism for crystallization from the metastable melt is different to crystallization from the unstable melt. The other point is that they confuse the crystallization process and the heating process; the latter is not a pure crystallization process but it usually involves partial melting and recrystallization even from a crystalline state. Furthermore, in the case of isothermal crystallization at high temperatures the small particles (or the clusters) are also observed prior to crystal nucleation for PET having no helical conformation in the crystalline state as will be shown in Sect. 5. Though the change of helical hands might practically take place in the case of iPP, that is another problem. Hence, they should not use their DSC data as an evidence for questioning the Strobl's model [26]. Apart from this it should be emphasized that the Avrami theory is a phenomenological theory which predicts well the crystallization processes in terms of macroscopic parameters such as crystallinity, but this does not necessarily mean that it describes true processes microscopically. The good fit of the crystallinity data obtained by SAXS and WAXD with the Avrami equation is a necessary condition but not a satisfactory condition for its validity; many natural phenomena can be described by exponential functions.

As described above, Ryan and coworkers [29–31] have shown very clearly the phenomenon of the so-called SAXS before WAXD, that is, that an induction period exists where larger density fluctuations begin before crystal nucleation. They also used iPP samples and crystallized them at high temperatures between 130 °C and 142 °C, to ensure that the crystallization is from the metastable state, i.e., the first step of the crystallization mechanism must be N&G of the nematic phase, and the second step which subsequently occurs inside the droplets with a spherical shape caused by the N&G is probably an SD-type phase separation. According to this concept, their analysis based on the Cahn–Hilliard theory can be well understood. Thus, a maximum in the Cahn–Hilliard plot $R(Q)/Q^2$ versus Q^2 (Fig. 5 in [30]) where $R(Q)$ is the growth rate of density fluctuations as a function of wavevector Q can be considered to be a result of the overlap of the N&G and SD; the straight line with a negative slope above a critical Q value (designated as q_{ic} by the authors) shows that SD occurs inside the droplets and the part below it may correspond to the N&G. Furthermore, it should be noted that the spinodal temperature obtained by them from a plot of D_{eff} versus $1/T$ corresponds to the secondary phase separation and above this spinodal temperature no SD occurs though it is unknown at the moment what happens above this temperature. This secondary spinodal temperature should of course be above the primary spinodal temperature, which was confirmed by comparing their data [29] and ours [16, 17] as will be shown in Sect. 5.

Muthukumar and coworkers [32, 33] investigated the molecular mechanisms of primordial stages of polymer crystallization using computer simulations and theoretical models. Such simulations were, however, made for crystallization from solutions, which might correspond to the crystallization from the melt in the co-existence region of the Olmsted phase diagram where single crystals are assumed to grow. It is therefore natural that the kinetics are different from that for crystallization from the unstable melt, or SD. He pointed out that in the case of crystallization from solution the growth rate of density fluctuations $R(Q)$ as a function of wave vector Q is proportional to Q^4 for small Q values while a plot of $R(Q)/Q^2$ versus Q^2 is linear with a negative slope for intermediate Q values. With their model such relations would be valid, but this mechanism is not applicable to crystallization from the unstable melt where the SD mechanism works. Of course, when SD occurs homogeneously in the whole system, the plot of $R(Q)/Q^2$ versus Q^2 should be linear with a negative slope even for small Q values, however, when the two-step phase separations occur in the system, e.g., N&G as the primary phase separation and SD as the secondary phase separation, such linearity would be broken as described above.

Very recently Li and Jeu [34, 35] discovered a smectic SAXS peak, indicating that bundles with smectic ordering are produced by a step shear even in a supercooled iPP melt. This finding seems to be very important not only for the understanding of fiber structures such as the shish-kebab structure, but also to understand the structure of polymer melts from which crystallization initiates. The emergence of bundles under shear might be due to the fusion of the small particles standing in line along the shear direction, which without shear would be produced inside the droplet caused by the primary phase separation of the N&G type because we observed by scanning electron microscopy (SEM) that the small particles are aligned parallel to the drawing direction of a PET film though they were observed at low temperatures near T_g [36].

2 Discovery of Spinodal Decomposition (SD) Prior to Crystallization—Glass Crystallization Near T_g

2.1 Motivation and Finding of SD

In 1967 Yeh and Geil [37] reported the novel structure of the so-called nodules, with ball-like particles with an average diameter of 7.5 nm and average interparticle spacing of about 12 nm, which was observed in the melt-quenched amorphous glass of poly(ethylene terephthalate) (PET) by electron microscopy. On the basis of this observation, Yeh [38] proposed a model with ordered domains of loosely folded chains for the amorphous structure of polymers, called the folded-chain fringed micellar grain model. This model caused great controversy at the time [39, 40] because it contradicted Flory's well-known theoretical model [41] predicting that the amorphous structure is orderless and consists of homogeneously interpenetrated Gaussian (or ideal) chains. Through subsequent extensive investigations, mainly made by SANS, this controversy was ended with the final conclusion that Flory's model was correct [42]. Because of this, the nodular structure was regretfully regarded as false at the same time, representing only a surface artifact or ghost resulting from the defocus electron micrograph, though Geil [40] refuted the claim. In recent years Geil [43] has revealed through experimentation that the nodular structure is observable and real.

From the viewpoint of polymer crystallization, Kaji and coworkers [6, 7] looked at the fundamentally important question of what happens during the induction period of polymer crystallization. They presumed that the nodules were precursors of subsequent crystal nucleation because PET is essentially a crystalline polymer and an ultra-quenched PET film did not show the nodules [44]. On the basis of this idea, Imai et al. [6, 7] carried out SAXS studies on the structural changes that take place during the induction period of the crystallization of PET, which led to the surprising finding that SD really occurs during the induction period before primary crystal nucleation. It was noted that the initial characteristic wavelength was about 15 nm, which agrees well with the above-mentioned internodular spacing. It was also confirmed by DPLS that the cause for this was the orientation fluctuations of rigid polymer segments [8]; the orientation domains may be considered to have a liquid crystal-like structure, i.e. nematic. It was then evidenced from Fourier-transform infrared (FT-IR) spectroscopic observations for isotactic polystyrene (iPS) [12] that in the amorphous state prior to glass crystallization polymer chain molecules first begin to partly assume a helical conformation, 3/1 helix for iPS, which is almost equal to the crystalline one. The helical conformation parts correspond to the stiff segments, and their extension with time leads to an increase of excluded volume, thereby causing the driving force for the orientation fluctuations.

Furthermore, the annealing temperature dependence of the integrated intensity of DPLS was investigated for several polymers [15]; a detailed analysis of the results showed that the induction period can be separated into three regimes: the conformational change to the stiff segments, the initiation of orientation of the stiff-segments (the early stage of SD), and the growth of the oriented domains (or clusters) with self-similarity (the late stage of SD). Inspired by our finding of SD, Olmsted and coworkers [4] proposed a theoretical generic phase diagram for a polymer melt as functions of density and temperature, which was calculated based on an idea of conformation-density coupling. This phase diagram predicts three states, an equilibrium melt-crystal co-existence region, a metastable region, and an unstable region depending on the quenching depth or the crystallization temperature. As described in the Introduction, when a polymer is crystallized from the metastable melt at higher temperatures, an N&G-type phase separation into an ordered dense liquid and a disordered less-dense liquid occurs, while when crystallized from the unstable melt at lower temperatures, an SD type liquid-liquid phase separation occurs. In this review the latter phase separation (occurring prior to crystal nucleation) will first be explained in detail, and then the former phase separation will be referred to briefly.

2.2 Details Concerning the Finding of SD

2.2.1 Determination of the Induction Period

In order to investigate the structural change before crystal nucleation the induction period of crystallization should first be determined. To do this the isotherm by DSC, the macroscopic density with a density gradient column, and the distance distribution function by wide-angle X-ray diffraction (WAXD) of a glassy PET sample were measured as a function of annealing time at given temperatures [6, 7]. The thermal (or calorimetric) glass transition temperature T_g of our PET sample was 75 °C and the crystallization temperature T_x was first chosen to be 115 °C. Figure 1 shows the isotherm ϕ and the macroscopic density as a function of annealing time. As seen from Fig. 1, neither exotherm nor endotherm was observed in the first 100 s, but after that a rapid exotherm was detected, suggesting an induction period of

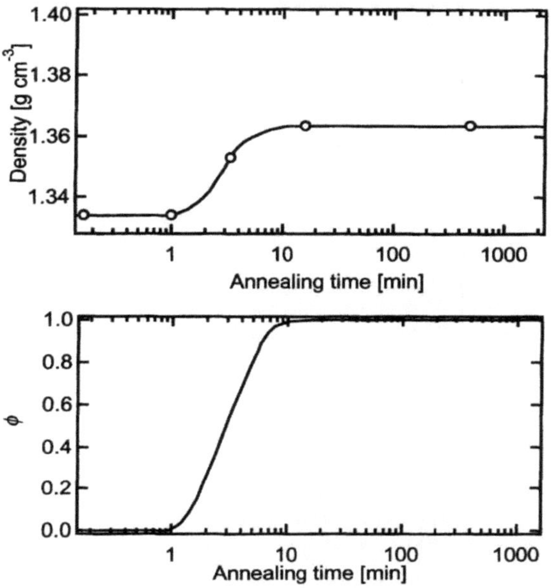

Fig. 1 Annealing time dependences of the crystallization isotherm ϕ (*below*) and the macroscopic density (*above*) of PET: annealed at 115 °C [6]

about 100 s. During this time the macroscopic density did not change either, but after 100 s it suddenly increased. In order to check that no crystallization occurs during this time, WAXD measurements were also carried out. Figure 2a shows the annealing time dependence of the WAXD profiles and Fig. 2b shows the distance distribution functions $P(R)$, which were derived by making the inverted Fourier-transform of Fig. 2a. The latter confirms that up to 100 s only the short-range order assigned to the amorphous structure exists, but thereafter long-range order due to crystallites suddenly appears. Hence, we concluded that the induction period is about 100 s. This incubation time is, however, too short to perform quantitative experiments to study the structural formation processes during the induction period. We therefore employed a lower crystallization temperature $T_x = 80$ °C, or only 5 K above T_g. As seen from the isotherm of DSC in Fig. 3, the incubation time in this case was obtained as 100 ∼ 120 min, which was enough time to perform the quantitative experiments with SAXS etc., as will be shown below.

2.2.2
The Finding of a New Peak in SAXS

It has been shown above that during the induction period no change occurs in DSC, macroscopic density, and WAXD. Does nothing change in this period? In order to answer this question, we made real-time SAXS measurements.

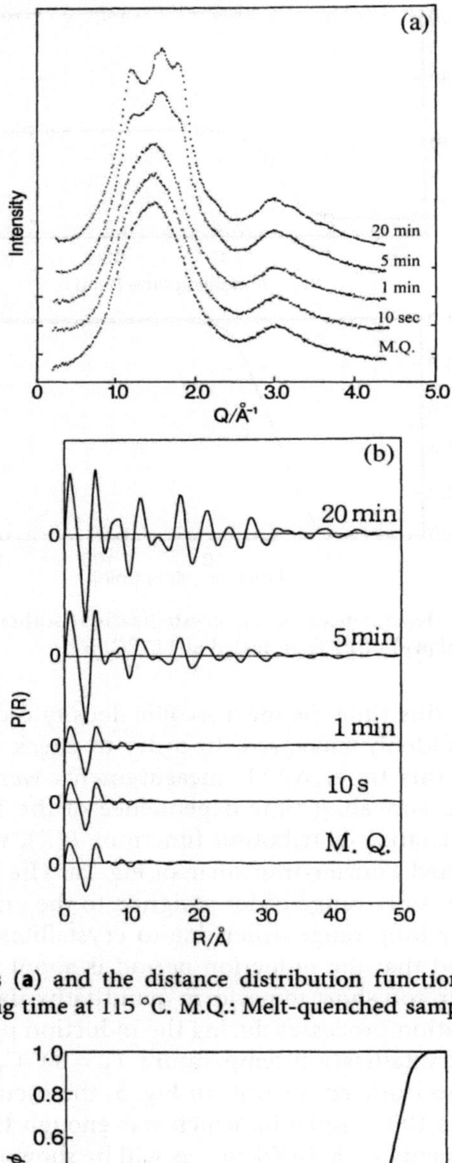

Fig. 2 WAXD profiles (**a**) and the distance distribution functions $P(R)$ (**b**) of PET as a function of annealing time at 115 °C. M.Q.: Melt-quenched sample [6]

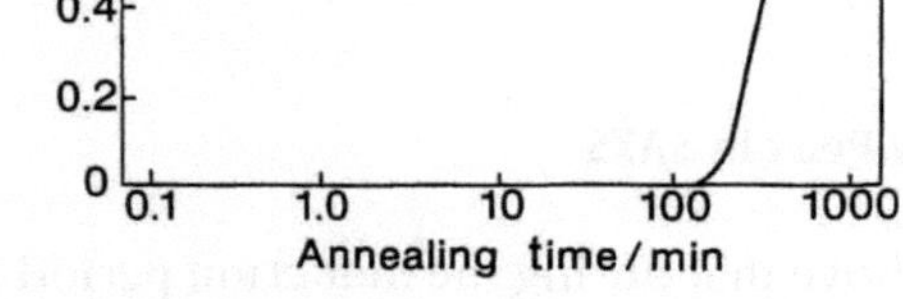

Fig. 3 Annealing isotherm of the amorphous melt-quenched PET sample at 80 °C [7]

Figure 4 shows the time-resolved SAXS intensity $I(Q)$ in log-log scales for the glass crystallization of PET when a glassy sample was annealed at 80 °C. Here, the magnitude of scattering vector Q is defined as $Q=(4\pi \sin\theta)/\lambda$, θ and λ being a half of the scattering angle and X-ray wavelength, respectively. As can be seen from the Figure, the intensity for the melt-quenched glassy sample increases monotonously with decreasing Q. This excess intensity may be considered to correspond to the tail of the light scattering intensity from the usual glass-forming materials in which the so-called Fischer's clusters [45, 46] are believed to exist. The correlation length of such a density fluctuation is in the range of several hundreds of nm and independent of other correlations appearing in the SAXS range, but the origin is not well understood. On the other hand, the SAXS curves of the samples sufficiently annealed for 243 and 313 min indicate an intense broad peak of the well-known long period at around $Q=0.06$ Å^{-1}, which is due to the alternation of crystalline and amorphous layers. However, at the very initial stage of annealing a new peak (different from the long period peak) appears at around $Q=0.04$ Å^{-1} and increases in intensity with time. Though this peak looks weak and broad in the logarithmic expression, it actually exists and can be seen more clearly in the linear expression. In Fig. 5 the difference intensity vs. Q curves are plotted as a function of annealing time in the linear

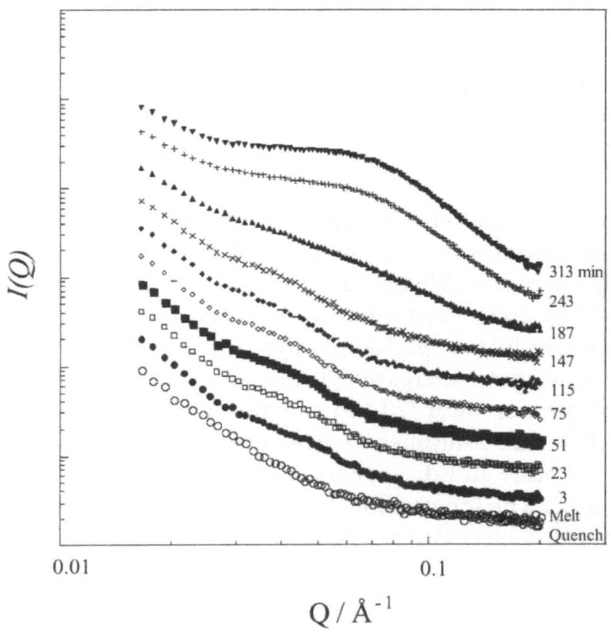

Fig. 4 SAXS curves of PET crystallized from the glassy state at 80 °C as a function of annealing time [7]

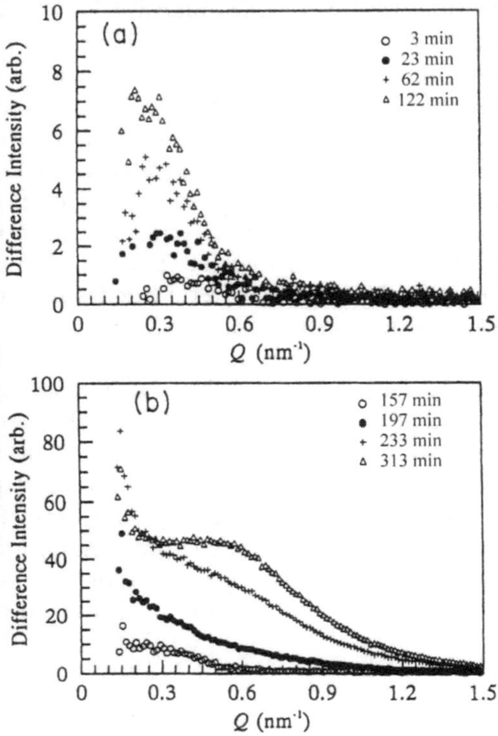

Fig. 5 Difference intensity SAXS curves of PET after subtraction of the intensity of the melt-quench sample: crystallized from the glassy state at 80 °C for 3–122 min (**a**) and 157–313 min (**b**) [7]

expression where the difference intensity means the scattering intensity of the annealed samples from which the intensity of the melt-quenched sample was subtracted. This is reasonable because the latter intensity may be considered independent of the concerned intensity as described before. Figure 5a, corresponding to the induction period of crystallization (< 120 min), shows how the new peak develops; it increases in intensity with annealing time and the position shifts towards lower Q from the initial value $Q = 0.041$ Å^{-1}. As seen from Fig. 5b, this peak continues to grow even after entering the crystallization stage (> 120 min) and disappears outside the resolution window of the SAXS camera used. The long period peak begins to appear near the initiation of crystallization and increases in intensity, but the peak position hardly shifts from $Q = 0.06$ Å^{-1}.

2.2.3
The Finding of SD

As a next step, the annealing time dependence of the new peak was examined quantitatively to understand its meaning. The time evolution of the logarithmic intensity of the new peak at several given Q's, which is not shown here, gave two regions whose boundary is at around 20 min; in the former stage the difference intensity at a fixed Q increased exponentially, and in the latter stage it leveled off. Furthermore, the time dependence of both the maximum position Q_m and the maximum intensity I_m of the new scattering peak have been examined as seen in Fig. 6. Here, it is seen that the induction period can be divided into two stages at 20 min; Q_m remains constant before 20 min but it decreases after 20 min, following a power law of time t: $Q_m \sim t^{-0.25}$, while I_m increases exponentially in the former stage and then obeys another power law, $I_m \sim t^{-0.75}$, in the latter stage.

This behavior reminds us of an SD type of phase separation. Thus, the process of SD may be divided into at least two stages, an early stage and a late stage. The early stage is described theoretically by the linearized SD theory by Cahn and Hilliard [47], and it is the early process where the amplitude of density fluctuations grows with time, keeping constant the period of density fluctuations of the so-called characteristic wavelength λ. In other words, the peak position does not change while the peak intensity increases exponentially with time. These features agree well with the observations de-

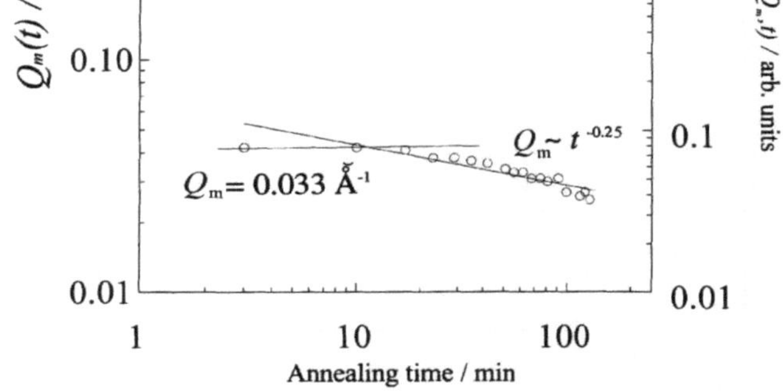

Fig. 6 Time evolutions of the maximum position Q_m and the maximum intensity I_m of a lower angle SAXS peak of PET, crystallized from the glassy state at 80 °C [7]

scribed above. On the other hand, the late stage of SD has been studied by many authors; among them Binder's group [48] who gave the following relations depending on the ratio of crystallization temperature T_x and spinodal temperature T_s from a computer simulation for 1 : 1 blends:

$$Q_m(t) \sim t^{-0.21} \qquad (T_x/T_s \cong 0.6) \tag{1}$$
$$Q_m(t) \sim t^{-0.25} \qquad (T_x/T_s \cong 0.8, 0.9)\,.$$

In the present experiment $T_x = 80\,°C$ was chosen and $T_s = 213\,°C$ was determined from the optical microscopic observation which will be explained in Sect. 4.2, and so $T_x/T_s \cong 0.73$. Hence, the lower relation of Eq. 1 would be expected, which agrees well with the observed relation.

Furthermore, Furukawa [49] proposed a scaling theory for the late stage of SD, which is the process where the amplitude of the density fluctuation is saturated to an equilibrium value and the characteristic wavelength grows with time. When the structure grows whilst maintaining self similarity, the scattering function $I(Q, t)$ can be scaled by $Q_m(t)$. Furukawa derived such a function for an off-critical concentration mixture where the interface between the different phases is smoothly curved. For a three-dimensional system it assumes the following form because the scattering intensity must be proportional to the mass or the volume of a scatterer and so $Q_m^{-3}(t)$.

$$I(Q, t) = Q_m^{-3}(t) S(x)\,, \tag{2}$$

where $S(x)$ is a time-independent function representing self-similarity, called a universal scaling function, and given by

$$S(x) = 3x^2/(2 + x^6)\,. \tag{3}$$

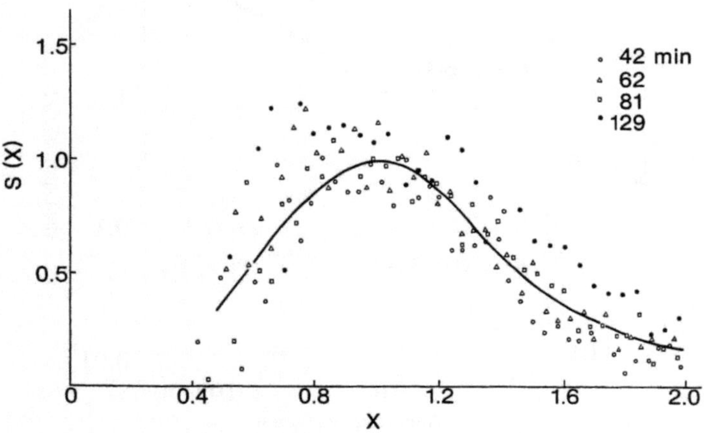

Fig. 7 Universal scaling function $S(x)$ plot for the late stage of SD, calculated from the SAXS data of PET. The *solid curve* indicates the theoretical Eq. 2 [7]

Here $x = Q/Q_m(t)$ and $S(1) = 1$. The peak intensity $I_m(t)$ is obtained by substituting $x = 1$ into Eq. 3:

$$I_m(t) = Q_m^{-3}(t)\,. \tag{4}$$

Therefore, the ratio of exponents of time t dependence in $I_m(t)$ and Q_m should be 3 for the space dimension. This requirement is also fulfilled in the experimental results, which were $Q_m(t) \sim t^{-0.25}$ and $I_m(t) \sim t^{0.75}$ for the late stage. Furthermore, as shown in Fig. 7, the universal scaling function $S(x)$ calculated from Eqs. 2 and 3 is independent of the annealing time within experimental error, indicating that the structure grows while maintaining self-similarity. All of these results support the late stage of SD. The above quantitative examinations therefore confirm that a kind of spinodal decomposition actually occurs during the induction period prior to crystallization.

2.3 Structure Sizes in Crystallization Processes

Several structure sizes caused by microphase separation occurring in the induction period as well as by crystallization were determined as a function of annealing time in order to determine how crystallization proceeds [9, 18]. The characteristic wavelength $\Lambda = 2\pi/Q_m$ was obtained from the peak positions Q_m of SAXS while the average size of the dense domains, probably having a liquid crystalline nematic structure as will be explained later, was estimated as follows. The dense domain size D_1 for the early stage of SD was calculated from the spatial density correlation function $\gamma(r)$ defined by Debye and Buche [50]

$$\gamma(r) = \langle \eta(r_1)\eta(r_2)\rangle / \langle \eta^2 \rangle_{av}\,, \tag{5}$$

where $\eta(r_1)$ and $\eta(r_2)$ are the local density fluctuations at points r_1 and r_2 with $r = |r_1 - r_2|$ from the average value $\langle \eta^2 \rangle_{av}$. The function $\gamma(r)$ is given by the inverse Fourier transform of SAXS intensity $I(Q)$, and D_1 is estimated from the minimum value of r at $\gamma(r) = 0$. The dense domain size D_2 for the late stage of SD is obtained according to a method by Komura et al. [51]. Thus, the Fourier transform $\Gamma(y)$ of the universal function $S(x)$ of Eq. 2 is given by

$$\Gamma(y) = 1/2\pi^2 \int_0^\infty x^2 S(x) \sin(xy)/xy\,dx\,, \tag{6}$$

where $y = rQ_m$, and the minimum value y_1 of y at $\Gamma(y) = 0$ gives D_2. This calculation leads to $y_1 = 2.517$ and hence $D_2 = 2.571/Q_m$. After crystallization, the long period L due to the alternation of crystalline and amorphous layers was determined from the usual SAXS peak appearing at a higher Q. Furthermore, the values D_c of crystalline lamellar thickness were estimated from the one-dimensional density correlation function based on a method by Strobl et al. [52].

The results for the glass crystallization of PET annealed at 80 °C as before are shown in Fig. 8. In the early stage of spinodal decomposition up to 20 min, the characteristic wavelength Λ remains constant at a value of 15 nm, which agrees with the theoretical expectation that only the amplitude of density fluctuations increases whilst keeping a constant characteristic wavelength. In the late stage from 20 to 100 min it increases up to 21 nm just before crystallization. Such a time dependence of Λ in nm can be represented by

$$\Lambda = 15.0(t/t_0)^{0.2}\,, \qquad \text{for } t > t_0\,, \tag{7}$$

where $t_0 = 20$ min. Correspondingly, the average size D of the dense domains remains constant during the early stage, and it increases from 6.0 to 8.5 nm in the late stage; D is almost always about 40% of Λ. The long spacing (8.6 nm) after crystallization is nearly equal to D; the periodicity of the dense domains seems to develop into the long period structure. This suggests that the initially phase-separated less dense (isotropic) domains may also change to the dense (nematic) phase just before the beginning of crystallization. As will be shown in Sect. 4.2, this presumption is reasonable. Thus, in the case of higher temperature crystallization where the characteristic wavelength of the SD pattern becomes much larger so as to be observable in the field of a microscope, we observed that the initial SD pattern disappeared when crystallization started, meaning that the less dense isotropic domains were also later converted to the dense nematic domains. Here, it is considered that the

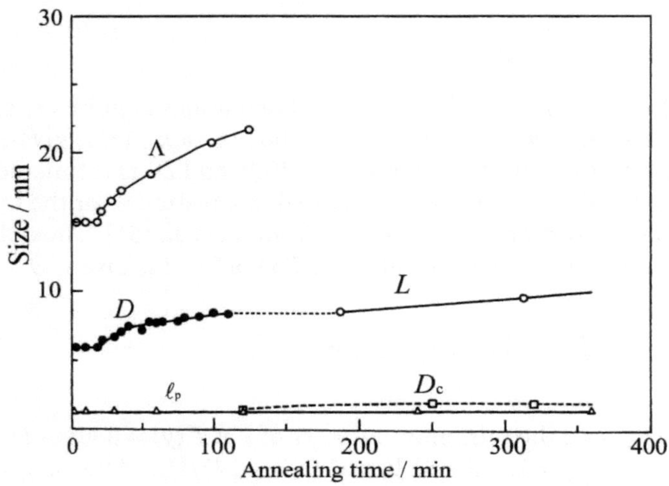

Fig. 8 Various structure parameters appearing in the crystallization process of PET at 80 °C: Λ, characteristic wavelength of SD; D, dense domain size; L, long period; l_p, persistence length; D_c, lamellar stem length [19]

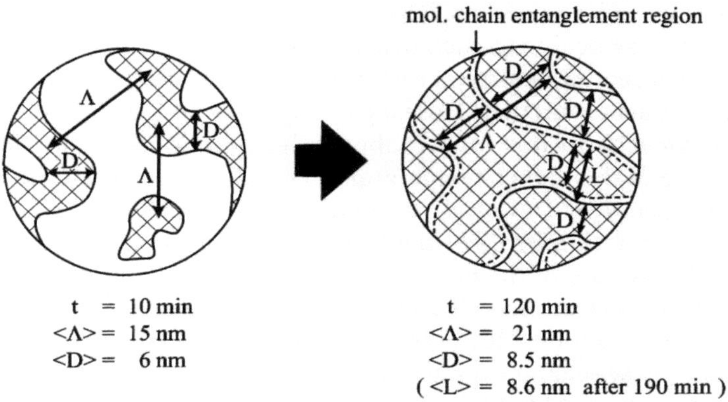

Fig. 9 Spinodal decomposition model for glass crystallization at a low temperature just above T_g. The indicated structure parameters mean: $\langle \Lambda \rangle$, the average characteristic wavelength; $\langle D \rangle$, the average size of the dense domain; $\langle L \rangle$, the average long period after crystallization. The numerical values are for PET crystallized at 80 °C

chain entanglements were excluded into the thin regions between two neighboring dense (nematic) domains. The critical size of the dense domains above which they can crystallize is considered to be 8.5 nm; only above this critical size the stiff segments in the oriented phase can slide or diffuse along the chain axis to attain the closest packing. A kinetic model for such spinodal decomposition is schematically shown in Fig. 9. Here it should be noted that in the final stage of the induction period the spinodal structure may transform to a particle structure because of interfacial tensions though it is not shown in the scheme.

The crystalline lamellar thickness D_c obtained by Strobl's method is initially 1.4 nm and grows to about 2.0 nm, which is roughly equal to the crystallite size in the chain direction of 2.8 nm estimated from the wide-angle X-ray diffraction (WAXD) [7]. Interestingly, the persistence length $l_p = 1.45$ nm just before crystallization measured by SANS (also see Fig. 11) [9, 10] is almost equal to the crystal thickness.

3
Origin and Mechanism of Spinodal Decomposition

The SD is a phase separation process usually occurring in systems consisting of more than two components such as in solutions or blends. However, in the present case the system employed is composed of one component of pure PET. In this case, what triggers such an SD type phase separation? Doi et al. [24, 25] proposed a dynamic theory for the isotropic-nematic phase transition for liquid crystalline polymers in which they showed that the orientation process of rod-like molecules causes an SD-type phase separation. Our experimental finding of SD may be due to such orientation fluctuations. In 1956 Flory [41] proposed a concept of two-step crystallization, a parallel orientation process of the stiff segments and their closest packing process. We believe our finding is a direct evidence for this concept though he did not predict the SD.

In the case of the higher temperature crystallization, as will be shown in Sect. 4, the theory of Doi et al. is applicable without doubt since the primary phase separation involves the transition from the isotropic to nematic phase, but in the case of the glass crystallization near T_g described above its applicability is unclear since the observed data may correspond to the secondary phase separation. However, if the secondary phase separation occurs, the primary phase separation must have proceeded prior to that. In a rapidly quenched glass even if the primary phase separation had already taken place, it would be still incomplete, so that it will re-start by heating.

Below, Doi's theory is first elucidated in some detail, and then our experimental results will be shown to support the partial extension of polymer chains during the induction period of crystallization which triggers the parallel orientation of the stiff segments.

3.1 Doi's Kinetic Theory

Figure 10 shows a schematic diagram to explain Doi's theory [24, 25]. As is well known, polymer molecules in the melt assume a random coil or Gaussian conformation and are entangled with one another. When the sample is quenched at a temperature below T_m where the polymer molecules can crystallize, the polymer chains tend to assume a crystalline conformation, generally a helix, which is the most stable energetically. Here it should be noticed that the stiff segments almost correspond to the crystalline sequences. However, this is a hard process because of two resistant factors; one is the chain entanglements and the other is the excluded volumes of the stiff segments, which increase in length with the growth of the helical sequences. The entanglements greatly delay the process and the extension of the helical sequences make the system unstable because the excluded volume V_{excl} of a stiff segment rapidly increases with the length l (or persistence length l_p) of the stiff segment according to the following equation:

$$V_{excl} = 2bl^2 |\sin\Theta| , \tag{8}$$

where b is nearly equal to its cross-sectional diameter d and Θ is the angle between the neighboring stiff segments. When the length of crystalline sequences exceeds a critical value, they start to orient parallel to one another to reduce the excluded volume or the free energy of the system. The critical concentration of the stiff segments when the orientation begins is given

as [24]

$$\nu^* = 4.19/dl^2 \, . \tag{9}$$

The complete parallel orientation makes the excluded volume zero while the complete perpendicular orientation gives the maximum value. Doi's theory predicts that such parallel orientation does not occur homogeneously in the system, but it involves an SD-type microphase separation into the oriented and unoriented domains. These theoretical predictions actually agree with our observations as described in the previous section.

3.2 Increase of Persistence Length and Parallel Orientation of Stiff Segments (PET)

In order to estimate the critical concentration ν^* from Eq. 9 the values of d and l are necessary. In the case of PET the value d may be assumed to be 0.66 nm from the van der Waals size of a benzene ring, and l may be taken to be the persistence length l_p for flexible polymer molecules. The persistence length can be measured by SANS with a labeling method [42]. Since neutrons distinguish hydrogen and deuteron atoms even though they have almost the same chemical properties, the chain conformation can be measured from a dilute solid solution of deuterated PET. Figure 11 shows the annealing-time dependence of the persistence length of PET measured using such a SANS labeling method [10, 11]. The persistence length of PET in the quenched glass

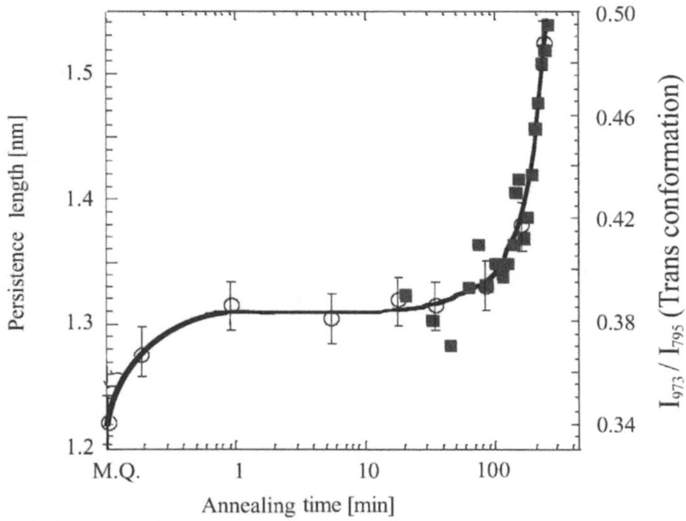

Fig. 11 Persistence length (○) [10, 11] and fraction of trans conformation (■) of PET as a function of annealing time at 88 °C, detected by SANS and FT-IR, respectively [19]

is 1.22 nm and within an annealing time of 1 or 2 min it increases to 1.32 nm. Thereafter, it keeps almost constant until the crystallization begins. After crystallization it starts to increase again.

The next problem is whether such a slight increase of 0.1 nm in persistence length actually causes the segment orientation. Let us examine the critical concentration of the stiff segments from Eq. 9 using these persistence lengths. The calculated critical concentrations for the melt-quenched glass with $l_p = 1.22$ nm and the annealed sample with $l_p = 1.32$ nm are $\nu^* = 4.5$ and 3.6 segments/nm^3, respectively. If we assume the hypothetical freely jointed chain model proposed by Flory [41], where the chain is considered as consisting of connected stiff segments having the length of the persistence length, the concentration ν(glass) of stiff segments in the melt-quenched glassy sample can be calculated from the macroscopic density $\rho_a = 1.333$ Mg m^{-3} of the sample and the molecular weight of the segment, which is almost equal to the monomer mass of PET ($M_0 = 192$). The calculated result is $\nu(glass) = 3.9$ segments nm^{-3}. Since the macroscopic density of the melt can be considered to be lower than that of the glass (a frozen melt), the stiff-segment concentration of the melt, ν(melt), may be lower than ν(glass). This means that the SD due to orientation fluctuations does not occur in the molten state. The annealing for 1 to 2 min increased the persistence length slightly. This lowers the critical concentration to 3.6 segments nm^{-3}, which is lower than ν(glass). In this state the sample is unstable, so that the orientation fluctuations involving SD start to occur in order to lower the free energy. Here it should be noted that the above discussion does not certify the absolute values of the segment concentrations, but it only shows that the system becomes unstable as the persistence length increases.

The above-mentioned theoretical possibility of the segment orientation has been confirmed using a depolarized light scattering technique. According to Stein et al. [54, 55], the orientation fluctuations can be estimated from the total integrated intensity or the invariant of the Rayleigh factor $R_\perp(q)$ for DPLS from a system with randomly correlated orientation fluctuations. Thus, the invariant is given by

$$I_{\text{Orient}} = \int_0^\infty R_\perp(q) q^2 \mathrm{d}q = (2\pi^2/15)(\omega/c)^4 \langle \delta^2 \rangle \,, \tag{10}$$

where q is the magnitude of the scattering vector of light, ω is the angular frequency of incident radiation, c is the velocity of light, and $\langle \delta^2 \rangle$ is the mean-square anisotropy. However, it is practically difficult or even impossible to obtain the invariant by integrating the observed intensity over the infinite q-range. Instead of this, therefore, we employed the integrated intensity over an observed q-range for the following reason. The observed scattering intensity during the induction period is almost constant independent of q in the observed q-range, which is reasonable because the oriented domain size of

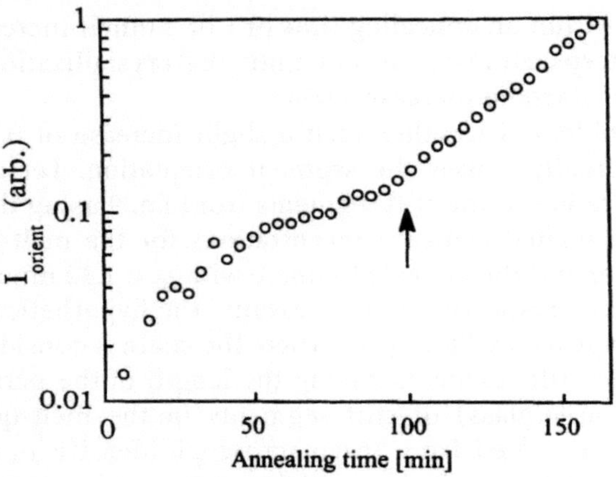

Fig. 12 Normalized invariant I_{orient} of DPLS intensity for PET as a function of annealing time at 80 °C [11]. The invariant giving orientation fluctuations is normalized to that for a sufficiently annealed sample which was taken to be unity. *Arrow* indicates the initiation of crystallization

less than 10 nm (see Fig. 8) is sufficiently small compared with the wavelength of the used light; the q-independent intensity is due to the point scattering. Therefore, we can employ the integrated intensity within a limited q-range instead of the invariant when the orientational change is discussed. Figure 12 shows the invariant of PET as a function of annealing time at 80 °C. In this plot the integrated intensity is normalized to that for a sufficiently annealed sample, which was taken to be unity. As will be shown later, the integrated DPLS intensity vs. annealing time curve in the induction period may be separated into four regions; Region I: constant intensity (0–3 min); Region II: the first exponential intensity increase (3–20 min); Region III: intensity increase following a power law of time t with an exponent of 1/2 (20–55 min); Region IV: the second exponential intensity increase (55–100 min). After the induction period, the rate is suddenly accelerated and the intensity increases exponentially again, which is due to the formation of spherulite texture. The first exponential increase of the intensity up to 20 min agrees with the prediction of Doi's theory, suggesting that the parallel ordering actually occurs during the induction period. Region II and Regions III + IV correspond to the two stages in the induction period of the above SAXS results.

3.3 Conformational Change and Parallel Orientation of Stiff Segments (PSs)

As described earlier, Doi's kinetic theory leads to a prediction that the SD is triggered by extension of unoriented crystalline sequences prior to crystal nucleation. In order to confirm this prediction the conformational change of polymer chains during the induction period was measured by a time-resolved FT-IR spectroscopic technique using two polymers of syndiotactic and isotactic polystyrenes (sPS & iPS). Figure 13 shows the crystalline conformations of these polymers corresponding nearly to their stiff segments and their growth by annealing. sPS assumes a planar zigzag structure with trans sequences $(TT)_2$ and iPS assumes a 3/1 helix with alternating trans and gauche $(TG)_3$.

First, let's examine the case of sPS [12]. Figure 14 shows the FT-IR spectra of this polymer; the lower and upper parts of the figure are the spectra for samples, melt-quenched and annealed at 120 °C for 400 min, respectively. The latter shows that the crystal form is α phase with an all-trans conformation [56, 57]. As seen from this figure, several absorption bands at 538, 935, 1224, and 1335 cm^{-1} vary with annealing or crystallization. Of these, the band at 1224 cm^{-1} was assigned to the crystalline packing, and some bands around 538 cm^{-1} were related to the conformational changes by Kobayashi et al. [56].

When the melt-quenched sample of sPS was crystallized at 120 °C, 25 K above the glass transition temperature $T_g = 95$ °C, the induction period was determined to be ca. 30 min from the crystallization isotherm measurements. FT-IR measurements under the same conditions supported this induction period as shown in Fig. 15. The band at 1224 cm^{-1} does not appear until after 30 min, but after 30 min it starts to emerge increasing in intensity with time. The strong peak at around 540 cm^{-1} in Fig. 14, which is related to the chain conformations, may be separated into four components: 511 cm^{-1} and 572 cm^{-1} for T$\underline{\text{TG}}$G, 548 cm^{-1} for G$\underline{\text{TT}}$G and 538 cm^{-1} for G$\underline{\text{TT}}$G/T$\underline{\text{TT}}$T, T and G being trans and gauche, respectively, where the

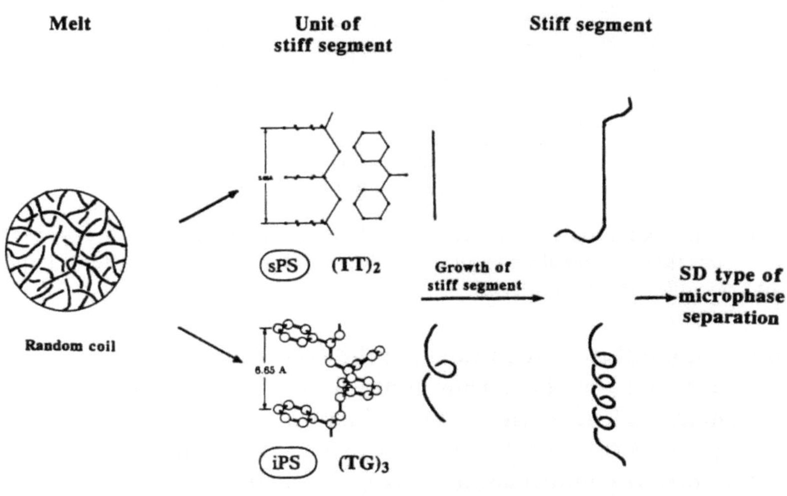

Fig. 13 Schematic diagram showing the change of chain conformation, i.e., coil to helix, during the induction period of crystallization for sPS and iPS

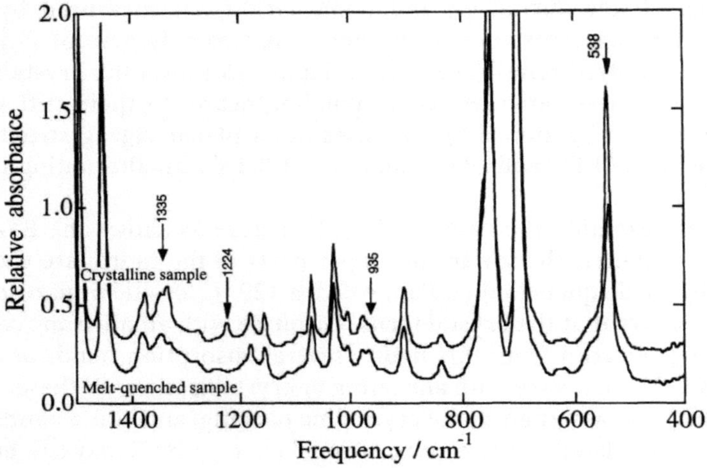

Fig. 14 FT-IR spectra of crystalline (*upper*) and amorphous (*lower*) sPS samples in the region 400–1500 cm^{-1} [12]. The bands marked by *arrows* appear after crystallization. The crystalline sample was annealed at 120 °C for 400 min

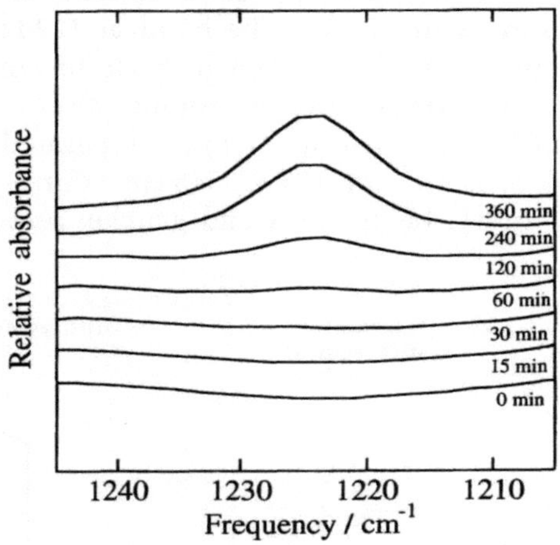

Fig. 15 1224 cm^{-1} band assigned to the crystalline chain packing of the α-crystal phase of sPS as a function of annealing time at 120 °C, which appears only after the induction period of crystallization 30 min [12]

two middle-underlined letters indicate the conformations around the C – C bond on both sides of the carbon atom bonded to a phenyl group. If the polymer chains extend even in the induction period, it may be expected that during this period the TTTT band increases in intensity and the other bands including G conformation decrease with time because the chain extension is caused by the G to T conversion. This expectation can qualitatively be confirmed from Fig. 16; the 538 cm^{-1} band actually increases in

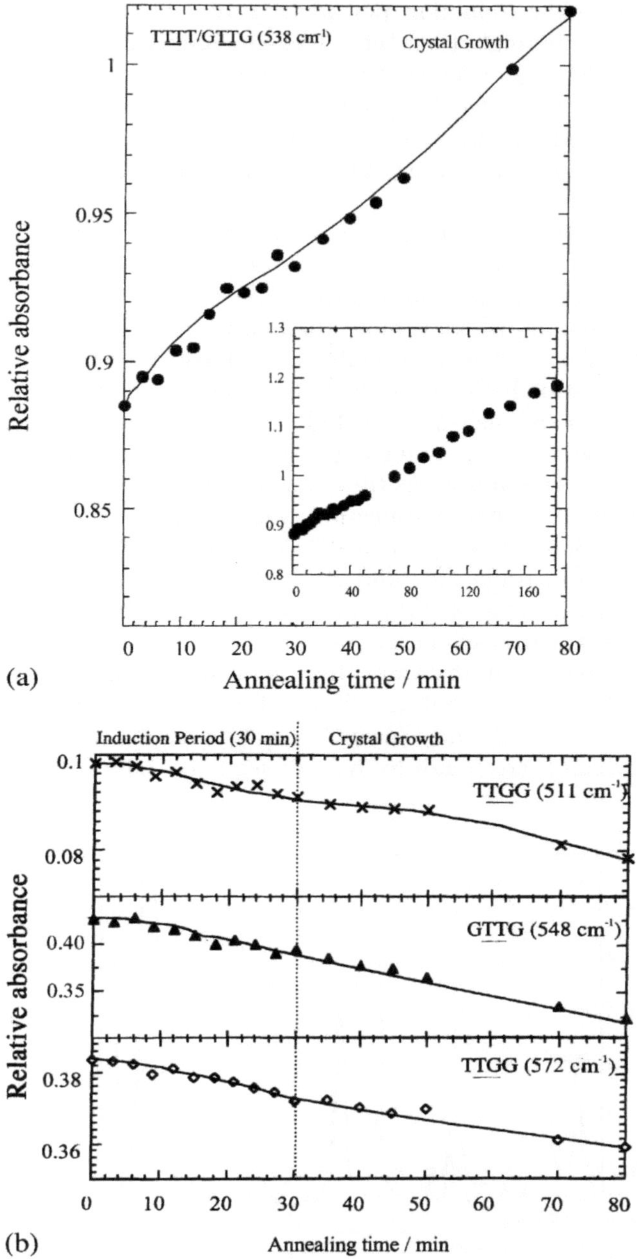

Fig. 16 537 cm^{-1} band of sPS assigned to TTTT/GTTG (**a**), and the 511, 548, 572 cm^{-1} bands assigned to TTGG, GTTG, TTGG, respectively (**b**) as a function of annealing time [12]. The intensity is normalized to that immediately after the temperature jump to the annealing temperature at 120 °C

intensity during the induction period (Fig. 16a) while all the other bands decrease correspondingly (Fig. 16b). The time dependence of the integrated intensity of DPLS for orientation fluctuations was as measured. The results show that the intensity growth is exponential until about 17 min, which corresponds to the early stage of SD. After 17 min, its growth rate is slightly lowered until 30 min; this time range may correspond to the late stage of SD.

In the case of iPS such a phenomenon is much clearer [13]. The crystallization temperature was 135 °C, 35 K above the glass transition temperature $T_g = 100$ °C, where the induction period determined from the crystallization isotherm was 70 min, which is more than two times as long as the induction period of sPS. This may be because the latter needs more time to form the 3/1 helical conformation before crystallization. Figure 17 is the comparison of the FT-IR spectra for the melt-quenched glassy sample and the sample annealed at 135 °C for 400 min. The conformational sensitive bands are seen in a range $500 \sim 600\ \mathrm{cm}^{-1}$. The annealing time dependence of this range is shown in Fig. 18, where the crystal modification was α form [57]. Four bands at 548, 562, 567, and 586 cm^{-1}, assigned to G<u>TT</u>G, T<u>TG</u>G/G<u>TG</u>G, G<u>TG</u>T, and G<u>TG</u>T, respectively, are distinguished and indicated with arrows. For the quantitative analysis, we decomposed the spectra into four components by assuming a Lorentzian shape for each band.

The time evolution of the absorption intensity of each component is plotted in Fig. 19. Since the conformation in the iPS crystal is 3/1 helix $(TG)_3$, two G<u>TG</u>T bands at 567 and 586 cm^{-1} may be assigned to a 3/1 helical conformation, which corresponds to the stiff segments. Here, it should be noted

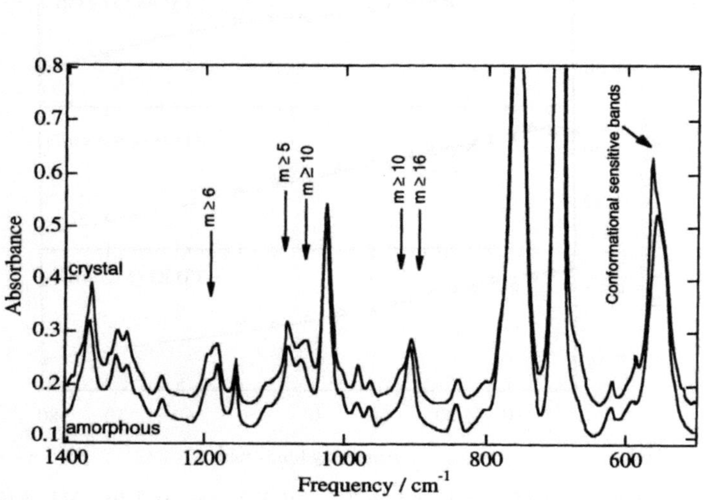

Fig. 17 FT-IR spectra of crystalline (*upper*) and amorphous (*lower*) iPS in the region 500–1400 cm^{-1} [13]: *m* indicates the minimum number of monomer units included in the 3/1 helix parts of the polymer chains

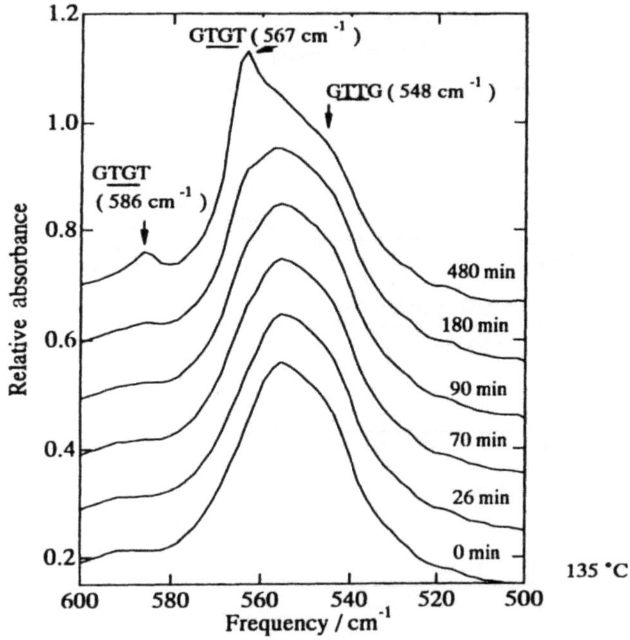

Fig. 18 FT-IR spectra of iPS in the range 500–600 cm^{-1} as a function of annealing time at 135 °C, 35 K above T_g [13]. The bands at 548, 562, 567, and 586 cm^{-1} are assigned to GTTG, TTGG/GGTG, GTGT, and GTGT, respectively

that the band at 567 cm^{-1} does not change until crystallization starts while the 586 cm^{-1} band already increases in intensity during the induction period. Hence, the former band can be assigned to the 3/1 helix in the crystalline state and the latter to the 3/1 helix independent of whether the state is noncrystalline or crystalline. Thereafter, these bands are called a crystalline band and an inherent band of the 3/1 helix, respectively. The latter also provides evidence that the noncrystalline helices or stiff segments actually begin to be formed during the induction period. The band relating to the random coil at 548 cm^{-1} decreases monotonously with annealing time, compensating for the monotonous increase of the inherent helix band. The 562 cm^{-1} band shows a complicated behavior making it rather difficult to assign. One possible assignment , however, may be proposed; this band is probably related to the 3/1 helix in the liquid crystalline state appearing as a precursory process to the crystal formation because it initially increases in the induction period and decreases with the beginning of crystallization; this may therefore be called a liquid crystalline band.

Furthermore, from the bands in the range 900 $\sim$ 1200 cm^{-1} we can roughly estimate the lengths of helical segments. All of these bands indicated by arrows are assigned to the 3/1 helix as well, but their frequencies depend on

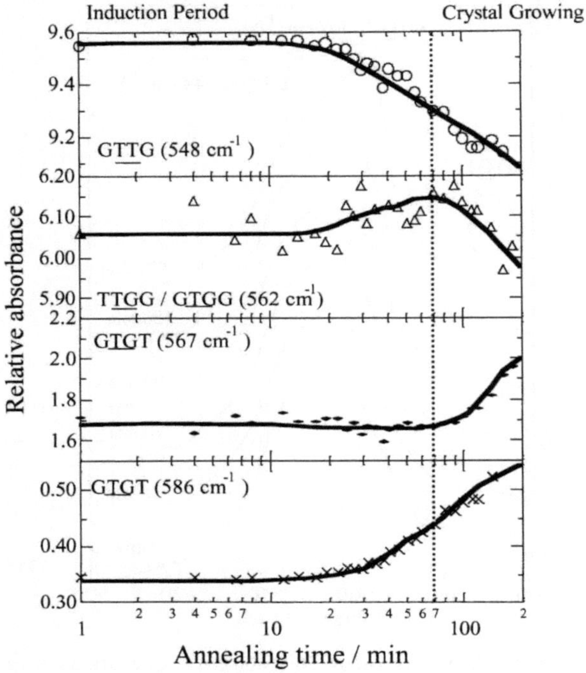

Fig. 19 Annealing time dependence of the IR intensities of 548 cm^{-1} (GTTG), 562 (TTGG/GTGG), and 586 cm^{-1} (GTGT) bands of iPS [13]. The intensity is normalized to the value immediately after the temperature jump to 135 °C

the sequence length of the helix part, i.e., the number m of monomer units in the helix part (Kobayashi et al. [59]). For example, the band with $m \geq 5$ (1081 cm^{-1}) does not appear until the number of monomer units in the helix part exceeds a critical figure of not less than 5. Hence, using these bands we can roughly determine the lengths of the helical sequences. Figure 20 shows the annealing time dependence of these bands. The bands for $m \geq 5$ to 10 increase in intensity abruptly by 3.0 min and then gradually in the induction period while the band for $m \geq 16$ does not change initially though it slightly increases later. These facts suggest that an average length of the initially formed helices is between $m = 10$ and 15. A tentative calculation of the critical concentration of rigid segments ν^* in Eq. 6 with a cross-sectional diameter (1.40 nm) of the 3/1 helix chain [60] gives a critical length $m_c = 10$, above which the segment orientation begins to occur. Since the band for $m \geq 10$ starts to increase after 3 min, the orientation fluctuations caused by the parallel ordering of the 3/1 helix segments will be expected after this time. Such fluctuations can actually be observed in the annealing time dependence of DPLS of Fig. 21, which corresponds to the orientation fluctuations. In the very early stage up to 3 min the invariant hardly changes, and after that it abruptly

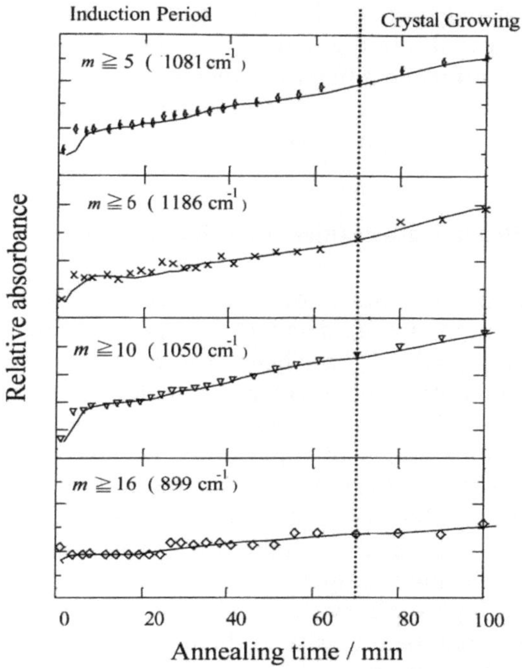

Fig. 20 Annealing time dependence of the IR bands for iPS 3/1 helix containing m or more monomer units: $m \geq 5$ (1081 cm^{-1}), $m \geq 6$ (1186 cm^{-1}), $m \geq 10$ (1050 cm^{-1}), $m \geq 16$ (899 cm^{-1}) [13]. The intensity is normalized to the value immediately after the temperature jump to 135 °C.

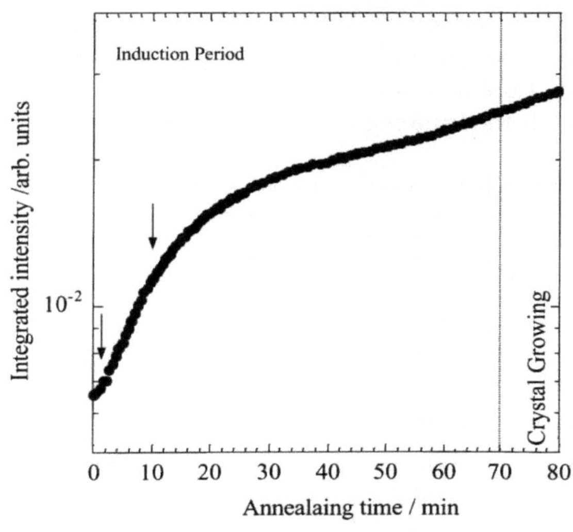

Fig. 21 Annealing time dependence of the integrated DPLS intensity for iPS in the induction period of crystallization at 135 °C [15]

increases, showing an exponential growth up to 10 min, which is characteristic of the early stage of SD. In the time period after 10 min the intensity shows a tendency to level-off while it begins to increase more steeply when the crystallization starts after 70 min. It is natural to consider that the steeper increase is caused by the growth of spherulites.

3.4 The Temperature Dependence of Orientation Fluctuations

In order to clarify the mechanism of SD occurring in the induction period of the glass crystallization in more detail, the temperature dependences of depolarized light scattering (DPLS) have been examined for several polymers such as poly(ethylene terephthalate) (PET), poly(ethylene naphthalate) (PEN), isotactic poly(styrene) (iPS), and syndiotactic poly(styrene) (sPS) [15]. The glass transition temperatures of these polymers determined by DSC were 75 °C for PET, 110 °C for PEN, 100 °C for iPS and sPS. The crystallization isotherms for these polymers as a function of annealing temperature are shown in Fig. 22, from which the induction periods were determined as in Table 1. The crystallization temperature dependence of the integrated DPLS intensity vs. annealing time curve for PET is shown in Fig. 23a as an example. These curves show that the intensity or the degree of segment orientation increases more rapidly with increasing temperature. In order to analyze such data quantitatively the curves were separated into several time regions; such an analytical method is indicated in Fig. 23b. As seen from this figure, the induction period may generally be separated into four regions. When the crossover times for these regions are designated t_1, t_2, and t_3, the individual time ranges are defined as follows. The first region ($0 \le t \le t_1$) is the time region where the integrated intensity I_{orient} hardly changes, the second ($t_1 \le t \le t_2$) is the time region where it increases exponentially with time, the third ($t_2 \le t \le t_3$) is the time region where it increases following a power law,

Table 1 Lengths of induction period determined by DSC for PET, PEN, sPS, and iPS [15]

Polymer	T/°C	t_{ind}/min	Polymer	T/°C	t_{ind}/min
PET	95	150	PEN	145	150
	100	60		150	70
	105	20		155	25
	110	10		160	10
sPS	115	110	iPS	130	150
	118	65		135	70
	120	30		140	25
	123	10			

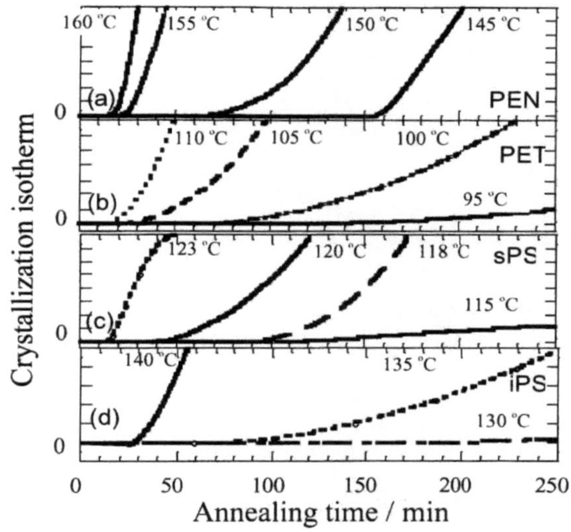

Fig. 22 Crystallization isotherm $\phi(t)$ as a function of annealing temperature for PEN (**a**), PET (**b**), sPS (**c**), and iPS (**d**), crystallized from the glassy state [15]

and the last ($t_3 \leq t \leq t_{\text{ind}}$) is the exponential increase region again where t_{ind} is the induction period. However, when the annealing temperature is slightly raised, the time region of the power law becomes shorter to eventually disappear at high annealing temperatures. These regions may be considered as follows.

Region I: here the molecular chains partly assume helical (stiff) conformation from random coils where the persistence length gradually increases.

Region II: the early stage of SD, i.e., the process where the helical (stiff) segments begin to orientate involving SD.

Region III, IV: the late stage of SD, i.e., the process where the oriented domains grow with self-similarity. The reason for the time dependences of the integrated intensity is unclear.

From the annealing temperature dependences of DPLS we estimated the activation energies of the individual ranges. The growth rate R_1 of the DPLS intensity for Region I was assumed to be proportional to a reciprocal time length t_1^{-1} of this range, i.e., $R_1 \propto t_1^{-1}$ because in this range orientation fluctuations hardly occur. Then, the activation energy ΔE_1 can be calculated by

$$R_1 = R_{10} \exp(- \Delta E_1 / k_B T) , \tag{11}$$

where R_{10} is a coefficient and k_B is the Boltzmann's factor. The activation energies for Regions II and IV are given by Arrhenius-type plots as

$$R_2 = R_{20} \exp(- \Delta E_2 / k_B T) , \tag{12}$$

$$R_4 = R_{40} \exp(- \Delta E_4 / k_B T) . \tag{13}$$

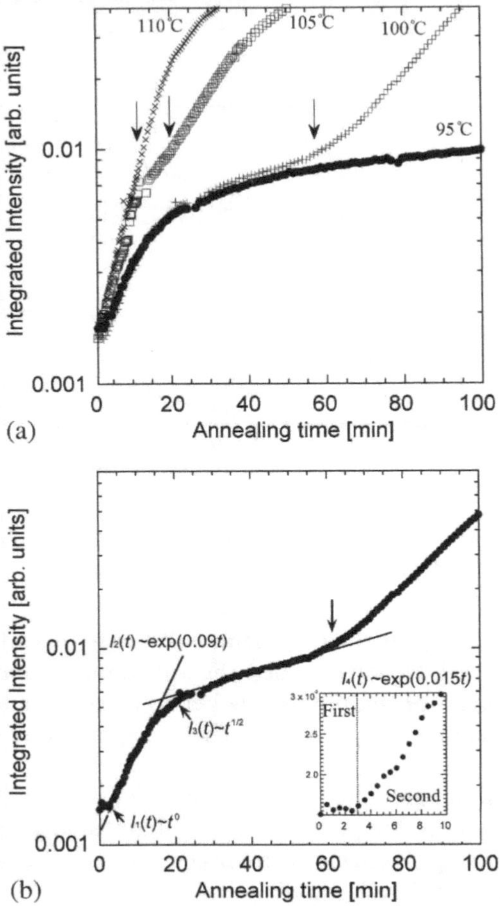

Fig. 23 Integrated DPLS intensity vs. annealing time curves for PET as a function of crystallization temperature [15] (**a**), and the analytical method for determination of the stages and their activation energies which is indicated for the crystallization of PET at 100 °C as an example (**b**)

The activation energy for Region III was not estimated because this region is very short or disappears completely especially at higher temperatures though the temperature dependence of this range should be given by the coefficient $R_3 = R_{30}(T)t^{1/2}$.

The resulting activation energies for all the polymers are summarized in Table 2. The apparent activation energies for the chain segments to assume helical structures characteristic of the polymers (Region I) are in the range of 35 to 40 kJ/mol or 8 to 10 kcal/mol. These values are three to four times as large as the potential barrier of a single C – C bond rotation, suggesting that three to four C – C bonds need to rotate simultaneously in order to form the

Table 2 Apparent activation energies for each stage of the induction period of glass crystallization just above T_g for PET, PEN, sPS, and iPS [15]

Region	Activation energy /kJ mol^{-1}			
	PET	PEN	sPS	iPS
I	40.4	39.0	34.9	37.1
II	34.9	24.8	46.0	50.0
IV	296	406	222	177

helical structure. The values for Region II are in the range of 25 to 50 kJ/mol or 6 to 12 kcal/mol. Such parallel orientation is considered to represent the rotation around the normal to the segment axis, which couples with density fluctuations. The apparent values for sPS and iPS are higher than those for PET and PEN. This may be due to the large side groups of benzene rings of the formers which reduce the axial ratios of their stiff segments; the smaller the axial ratio, the slower the rate of orientation. The apparent activation energies for Region IV depend on the polymer species; those for polystyrenes (sPS and iPS) are about 200 kJ/mol while those for PET and PEN are about 300 and 400 kJ/mol, respectively. Nevertheless, all of these values are by one order of magnitude larger than those in the other two ranges. Such larger activation energies may be understood in the framework of the growth mechanism by Binder [48]; the dense or oriented domains (clusters) grow by cluster reactions and cluster diffusion due to the stochastic exchange processes of atomic groups (see Fig. 3 in [48]). The cluster diffusion also occurs as a result of cluster reaction because the center of mass of the resulting larger cluster shifts. Such exchanging atomic groups probably correspond to the so-called "cooperatively rearranging region (CRR)" which was introduced to explain the glass transition of fragile liquids since the apparent activation energies for the α process are about 500 kJ/mol near the glass transition temperature [61], which are of the same order as those for Region IV. The sizes of these atomic groups calculated in terms of a conformer model proposed by Matsuoka and Quan [62] are 7.7 and 29 conformers for iPS and PET, respectively, where the numbers of conformers in a monomer are respectively 2 and 5, and hence the average numbers of monomers in the atomic group or CRR for iPS and PET are 3.8 and 5.8, respectively.

3.5 The Mechanism of Structure Formation During the Induction Period of Crystallization from the Unstable State

On the basis of the above described studies, we propose a model for the structure formation mechanism during the induction period when a polymer is crystallized from the unstable state. Figure 24 shows a schematic diagram of

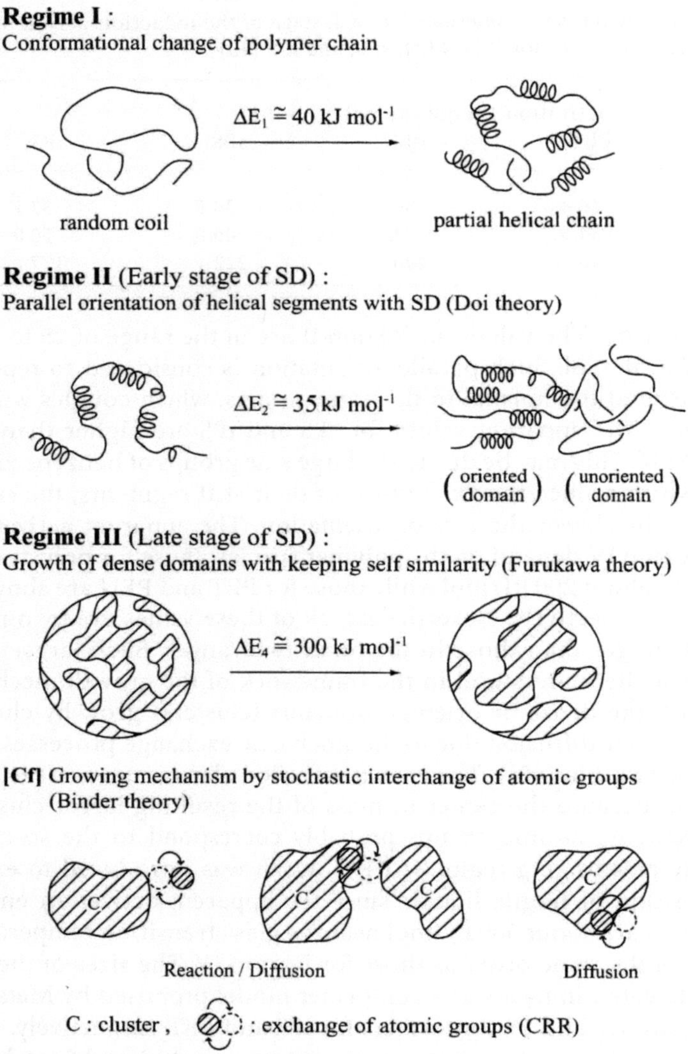

Fig. 24 Schematic diagram of a model for the structure formation mechanism in the induction period of crystallization from the unstable state

the model. In the very initial stage of Regime I molecular chains partly convert to stiff helical conformation from the amorphous random coils; about four C – C bonds internally rotate simultaneously to form the helix. In this stage the stiff segment orientation hardly proceeds. In the next step of Regime II (Doi's regime [24]) the stiff segments start to orient parallel to one another involving the SD-type microphase separation, which corresponds to the early stage of SD. Then, in Regime III (Binder's regime [48]) corresponding to Regions III and IV the segment-oriented dense domains produced by SD grow in size whilst maintaining self-similarity where the dense or oriented domains (clusters) grow by cluster reactions and cluster diffusion due to the stochastic exchange processes of the atomic groups (CRR).

Finally, we should note that the initial glassy sample had a structure with large density fluctuations with a correlation length around a few thousand Å, the scattering intensities of which were subtracted in the present analysis. Such a density fluctuation was originally found by Debye [50] and is now called the Fischer's cluster [45, 46]. Presently, however, nobody knows the true cause of these fluctuations. Generally, it is considered that they represent a frozen structure of thermal fluctuations in the supercooled liquid at higher temperatures below the melting temperature since they disappear above the melting temperature [63]. However, there is a possibility that these fluctuations might correspond to a spinodal structure formed by the primary phase separation during the quenching process from the melt. This is an important problem for the understanding not only of polymer crystallization but also the structure of the glass and is to be confirmed in the future.

Therefore, in the case of glass crystallization just above T_g we may possibly see a secondary phase separation of the spinodal decomposition (SD) type occurring inside the dense region caused by the first SD.

4 Crystallization at Higher Temperatures

We have so far described structural formation during the induction period of crystallization when the melt-quenched amorphous samples were annealed just above the glass transition temperature T_g. In this case the SD-type microphase separation due to the orientation fluctuations of stiff segments was observed in the induction period where the characteristic wavelengths were of the order of tens of nm. However, what happens when a sample is crystallized from the glassy or molten state at much higher temperatures than T_g? We have investigated this problem using PET.

This problem is very important, but it is extremely difficult to make SAXS measurements at higher temperatures because the induction period becomes too short to observe the time evolution of SAXS intensities. For example, as was seen in Sect. 2.2, the induction period was only 100 s when the PET glass was crystallized even at 115 °C, 40 K higher than T_g, where a detailed analysis of the SAXS data was impossible. Of course, as the crystallization temperature approaches the melting temperature, the induction period is expected to become longer. However, as will be shown below, no characteristic peaks of SD could be detected in SAXS curves either. This is probably because the crystallization temperature was not in the unstable state, or the characteristic wavelength was much larger compared with the lower resolution limit of the SAXS camera. If the latter case is true, we could expect to observe SD patterns directly by optical microscopy. We have therefore carried out such two experiments of SAXS and optical microscopic observations. Here, we would like to emphasize that the latter observations are technically very difficult, especially for the crystallization from melt, since extremely rapid quenching is essential in order to avoid possible crystallization at higher temperatures than a given crystallization temperature. For this purpose we developed a special rapid temperature-jump apparatus to make possible in situ optical microscopic observations.

4.1 SAXS Observations of Melt Crystallization Near the Melting Temperature

Figure 25 shows the time evolution of the difference SAXS intensity which was observed in situ when a PET sample was crystallized by cooling down from

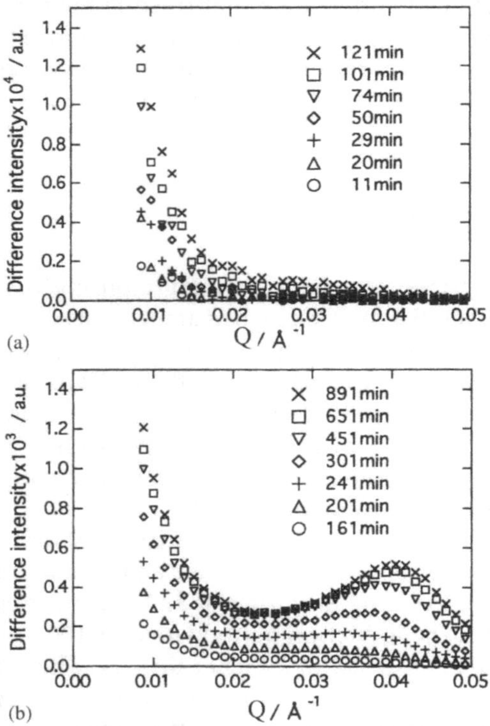

Fig. 25 Annealing time evolution of the difference SAXS intensity in the induction period (**a**) and the crystallization period (**b**) for the melt crystallization of PET at 244 °C [18]. This system corresponds to crystallization from the metastable state where a nucleation and growth type of primary phase separation first occurs followed by the spinodal decomposition type secondary phase separation

the molten state at 290 °C to 244 °C, 23 K lower than $T_m \sim 267$ °C, in a SAXS camera; the induction period was about 110 min in this case. As seen from Fig. 25, the intensities initially increase with time at low Qs below 0.025 Å, but no peak is observed in this Q range while a broad peak starts to appear above 0.025 Å near the crystallization time; the maximum position of this peak shifts from around 0.03 Å^{-1} to 0.04 Å^{-1} with increasing intensity with time. As will be discussed in Sect. 4.2 and shown in Fig. 27, the melt at this crystallization temperature is not in the unstable state but in the metastable state; the strong intensities at low Q's may come from the droplets caused by the primary phase separation of the N&G type though no detailed analysis was made. As described in the Introduction and as will be discussed in Sect. 5, the broad peak at around 0.03 Å^{-1} may be considered to be attributed to the secondary phase separation of SD type inside the droplets which were produced earlier by the primary phase separation of the N&G type.

Another interesting problem is that the broad peak at around 0.03 Å^{-1} shifts to a higher Q with time (see Fig. 25b). Gehrke et al. [64], who found this phenomenon for the first time, assigned this peak to the well-known crystalline long period and explained it in the following way. Initially formed crystalline lamellae have wavy surfaces, but as the crystallization proceeds, their surfaces become smooth resulting in a slight decrease of the long period. However, a more reasonable explanation for this would be as follows. The initial broad peak is due to the adjacent spacing of the small particles formed by the secondary phase separation of SD type. Thereafter, these particles fuse with the adjacent particles to form crystalline lamellae with a long period which is shorter than the average interparticle spacing.

4.2 Optical Microscopic Observations of the Melt Crystallization

In the previous subsection, we referred to a possibility of observing SD patterns by optical microscopy. After considerable efforts, we have at last succeeded in observing such a SD structure.

Before discussing our results let us describe some details of the experimental measurements. A pure PET sample, warranted to be additive-free by the supplier, Toray Co. Ltd. (Lot HT91342), was used. The melting temperature T_m and the glass transition temperature T_g of this sample were determined to be 255 °C and 77 °C, respectively, using DSC, Perkin Elmer DSC7. The temperature jump experiments were attained by a successful combination of in situ optical observation and rapid deep quench; the temperature jump was driven simply by rapidly transferring the sample of a melt or a glass onto a heat capacitor block being held at a given temperature. A typical initial cooling rate, actually measured inside the sample, was of the order of 10 000 °C/min when quenched from $T = 280$ °C to 130 °C. Apparatus assembled on the basis of our prototype is commercially available from

Japan Hightech and Linkam (LK-300). The time evolution of optical micrographs was recorded using a Nikon Optiphot2-Pol with a CCD camera and a video recorder. The usual quenching experiments with a steady cooling rate (20 °C/min) were carried out using a Mettler FP82HT hot stage.

Figure 26 demonstrates that a rapid quenching is crucially important for unveiling the SD pattern clearly. When a PET melt is quenched moderately at a rate of –20 °C/min in the usual hot stage of the microscope, the temperature of the melt gradually decreases from 280 °C to a given temperature 130 °C as shown with the dotted line in Fig. 26a. In this case no pure structure inherent to a given crystallization temperature is obtained, but an overlapping structure of randomly distributed crystal entities produced at higher temperatures is observed as in Fig. 26b. On the other hand, when the melt is rapidly quenched within a few seconds in the above-mentioned special temperature-jump apparatus, a completely different pattern appears; quasi-periodic density fluctuations with a period of ∼ 2 μm emerge uniformly in the whole visual field in a short time (see Fig. 26c). Of course, these fluctuations can be considered to be due to the orientation fluctuation of the stiff polymer segments as described earlier. The fluctuation contrast is enhanced with time to attain a maximum after around 30 s, and then it strangely begins

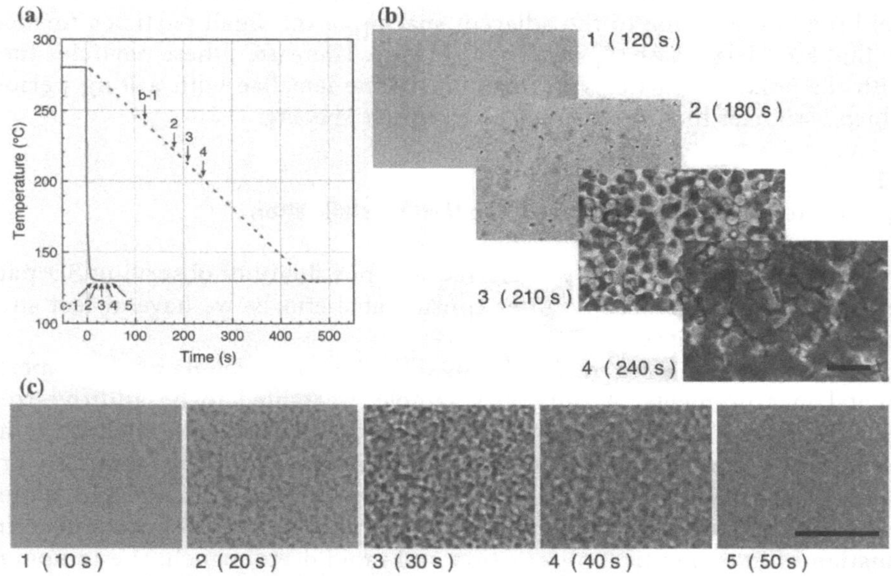

Fig. 26 Effect of quenching speed on the morphology that develops [16]. **a** Time-temperature chart recorded in the sample spot. **b** Sequence showing morphological development when a PET melt is quenched moderately (see *dotted line* in **a**). *Scale bar* represents 50 μm. **c** Sequence showing morphological development after a PET melt is quenched rapidly to $T_x = 130\,°C$ (see *solid line* in **a**). *Scale bar* represents 20 μm

to decrease. Thus, the fluctuations are only transiently observable. The reason for this is not very clear, but it seems that when the crystallization starts, even the less dense unoriented (or isotropic) regions produced by phase separation also begin to transform to the denser oriented (or nematic) phase by the driving force of further extension of stiff segments, resulting in the decrease of the contrast between the dense and less dense regions. This may be a characteristic feature appearing concomitantly with the phase separation due to orientation fluctuations in one component system.

The above-mentioned quasi-periodicity can be considered as corresponding to the characteristic wavelength of SD. Next, we will present evidence for this. First, the periodicity hardly changed until the contrast attained the maximum (Fig. 26c). Second, the integrated DPLS intensity increases exponentially with time until 20 s. These facts agree with the theoretical predictions for the early stage of SD. However, the characteristics for the late stage of SD are not obvious because the rapid start of crystallization at high temperatures hinders the further evolution of SD. In fact, the glitter of microcrystals was observed on the pattern of the quasi-periodic fluctuations under a polarizing microscope.

Further evidences supporting SD come from the crystallization temperature dependence of the optical micrographs of PET. Figure 27 shows the

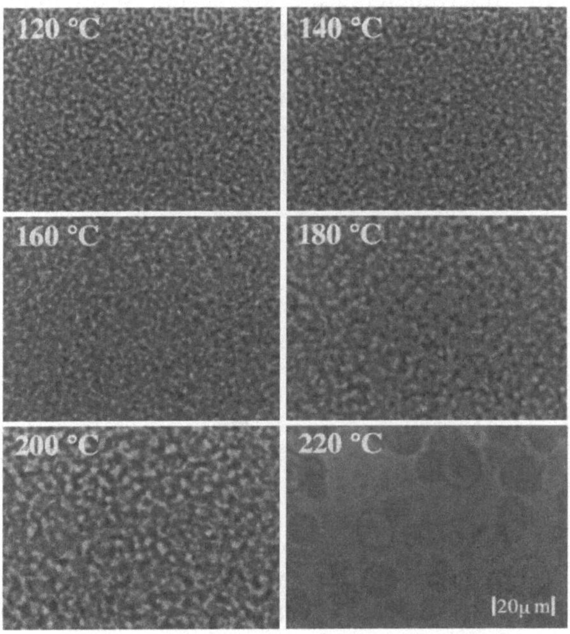

Fig. 27 Morphological change of PET observed by optical microscopy as a function of quenching temperature T_x [16, 19]. *Scale bar* represents 20 μm

micrographs with maximum contrast when the molten PET was quenched at various temperatures of 120 to 220 °C. Below 200 °C, all the micrographs indicate spinodal-like patterns with self-similarity; they increase in periodicity when increasing the quenching temperature. Of course, the time evolution of their patterns is essentially the same as that observed in Fig. 26c. If these spinodal-like patterns are actually due to the SD-type microphase separation, the temperature dependence of the characteristic wavelength Λ should fit in with the van Aartsen equation [22, 23] derived for the early stage of SD according to the mean-field treatments. This equation is given as follows:

$$\Lambda = \left(\frac{2\pi}{\sqrt{3}}\right) l_{\mathrm{rmi}}/[1 - T_{\mathrm{x}}/T_{\mathrm{s}}]^{1/2}, \tag{14}$$

where l_{rmi} is a range of molecular interactions, T_{x} the quenching temperature (or crystallization temperature), and T_{s} the spinodal temperature below which the system critically enters the unstable state. The observed Λ values for the melt crystallization of PET are plotted with solid circles in Fig. 28 where a dotted line is a fitting curve of Eq. 14 based on the SD theory [22]; the fitting is very good and the fitting parameters are $\Lambda_0 \equiv (2\pi/\sqrt{3})$, $l_{\mathrm{rmi}} = 0.89$ µm or $l_{\mathrm{rmi}} = 0.245$ µm and $T_{\mathrm{s}} = 213 \pm 5$ °C. This is a strong support for SD, and therefore we believe that the quasi-periodic fluctuations at the micrometer scale are certainly attributed to SD. One more interesting phenomenon is also observed in Fig. 27; the SD pattern is suddenly transformed to spherulitic structure above $T_{\mathrm{s}} = 213$ °C between 200 and 220 °C. Such a sudden change strongly supports the phase dia-

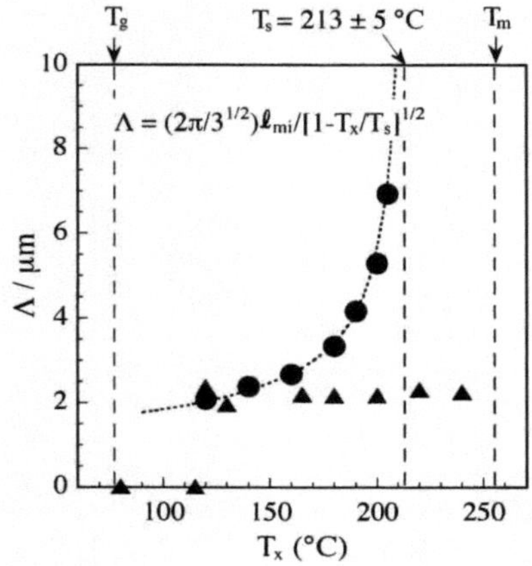

Fig. 28 Characteristic wavelength Λ as a function of quenching or jumping-up temperature T_{x} [16]. (•) quenched from the molten state, (▲) jumped up from the glassy state. The *dotted line* represents a fitting curve with Eq. 14 for the melt quenching

gram proposed by Olmsted et al. [4] as well, since it predicts the spinodal line.

4.3
Optical Microscopic Observations of Glass Crystallization

We measured the characteristic wavelengths Λ for the glass crystallization as well, i.e., PET was crystallized by jumping up to given temperatures from the glassy state. The results are plotted with solid triangles also in Fig. 28. In this case the spinodal structure was strangely held even above T_s until the melting temperature. More surprisingly, the characteristic wavelengths are almost independent of the crystallization temperature and do not follow van Aartsen's equation, Eq. 14. This may mean that the spinodal pattern formed somewhat above T_g is held to T_m above which it will be destroyed. The cause for this peculiar phenomenon is not confirmed at the moment, but this may be because crystallites fixing the texture formed somewhat above T_g on heating do not melt until T_m whereas in the case of melt crystallization crystallites are never formed to a given crystallization temperature because of much undischarged heat remaining in the sample due to the sufficiently rapid quench. Actually, it was observed under the optical microscope that the melt is violently stirred by thermal agitation until the temperature drops to a given T_x. In other words the structure formed by melt crystallization undoubtedly corresponds to that formed at T_x. Anyway, it should be emphasized that glass crystallization gives a very useful method to industrially produce high performance materials with a homogeneous ideal microstructure.

The above observations would give one possible explanation for the great difference between melt and glass crystallization in the number densities of spherulites, which was pointed out by van Krevelen [21]. In the melt crystallization from the liquid-crystal coexistence region at high temperatures large spherulites at least more than tens of μm in diameter are produced while in the glass crystallization from the unstable region at lower temperatures crystal entities smaller than one half of Λ, i.e., less than one μm in diameter are produced. The size difference is one or two orders of magnitude between these two cases, leading to a number density difference of three to six orders of magnitude because of its three dimension, which roughly agrees with the Krevelen's note.

Furthermore, it should be noted for glass crystallization that there might exist a boundary crystallization temperature T_{bx} between 115 and 120 °C because below 115 °C Λ is of the order of tens of nm while above 120 °C it becomes a few μm; they are two orders of magnitude different. The cause of this is not clear at the moment, but one possibility is due to the range l_{rmi} of molecular interaction involved in Eq. 14. In order to understand this the detailed mechanism of the glass transition would be necessary. Although the problem of the glass transition is not completely solved, the concept has greatly progressed in recent years (for examples, see [65, 66]). According to the mode-coupling theory (MTC) for the glass transition [67], it is predicted that there exists a critical temperature, called the mode-coupling critical temperature T_c, around which the so-called "dynamic cages" are formed and moving particles are often trapped by them. This delays the molecular motion greatly, resulting in interaction distances that are extremely limited within very small regions. This critical temperature is slightly higher than the calorimetric glass transition temperature T_g, i.e., $T_c = 1.1 \sim 1.3T_g$ [68], and the above critical crystallization temperature T_{bx} is in this range: $T_{bx} \cong 1.1T_g$.

However, it should be again emphasized here that the initial glassy sample had a structure with large density fluctuations with a correlation length around a few thousand Å, the scattering intensities of which were subtracted for the present analysis. These fluctuations might correspond to a spinodal structure formed by the primary phase separation during the quenching process from the melt. Hence, the analysis we made here might be for the secondary phase separation of the SD type occurring inside the dense region caused by the primary SD-type phase separation. In this case a characteristic wavelength of SD corresponding to μm might be observed, which should be confirmed in the future.

5
Crystallization from the Metastable Melt

As described in the previous Sect. 4.3, the SD patterns were observed below the spinodal temperature $T_s = 213\,°C$, but above this temperature the morphology suddenly changed from the SD pattern to the spherulite-like one as seen from Fig. 27. The phase diagram for a polymer melt in Fig. 30, which was proposed by Olmsted et al. [4], shows that there is the metastable region above the spinodal line T_s and below the binodal line T_b. It may therefore be considered that this sudden morphological change corresponds to the transition from the unstable state to the metastable state of the polymer melt. Thus, spherulitic objects of a few tens of μm in diameter may correspond to the droplets with ordered regions of a nematic-like structure caused by the primary phase separation of N&G; they are produced in crystallization from the metastable state of the polymer melt. If so, it is expected that crystal nucleation first occurs inside these spherulitic objects where we could observe the crystal nucleation processes. In order to confirm this idea we investigated the inner structure of these spherulitic objects. Figure 29 shows the scanning electron micrograph (SEM) of a fracture surface of the spherulitic object produced at 220 °C by quenching a PET melt at 290 °C [36]. This quenching temperature is somewhat higher than the spinodal temperature $T_s = 213\,°C$ and corresponds to the crystallization temperature at which the spherulitic objects were observed by optical microscopy in Fig. 27. As seen from this

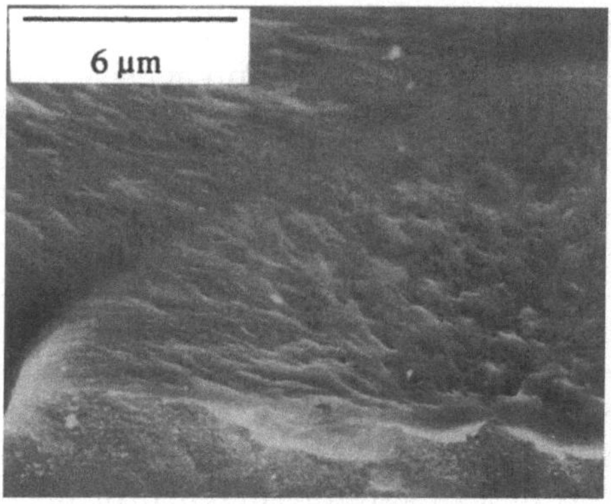

Fig. 29 Fractured morphology of spherulitic objects in a thin film of PET crystallized at 220 °C [36]. On the fractured surface many small particles with a diameter of 0.2 ~ 0.3 μm are seen while on the spherulite surface there is a fibril structure 0.2 ~ 05 μm thick

SEM, many small particles with a diameter of 200 ~ 300 nm are produced inside the spherulitic objects while on the surface a fibril structure has already developed, which may be formed by fusion of the small particles. These particles are not an artifact from when the sample was prepared for SEM observation because immediately after quenching no such particles were observed and the fracture surface was completely smooth though the details will be reported elsewhere. What are these small particles? It may be assumed that when the droplet size becomes larger than a critical value these small particles are formed probably at the final stage of the secondary SD-type phase separation inside the droplets because the small particle size is rather homogeneous. Furthermore, they may correspond to the so-called nodular structure observed by several authors. For examples, Kanig [69] reported a nodular structure in a PE melt, which was observed when cooled down from 150 °C to 120 °C (see Fig. 1 of [69]). Wang et al. [28] observed a cluster-like structure with a size of about 10 nm in a mesomorphic iPP sample which was melt-extruded and quenched at 8 °C. Though the sizes are of an order of tens of nm in the case of flexible polymers, which are one order of magnitude smaller than that of the PET particles, the correspondence would be reasonable because the characteristic wavelength of SD determining the average particle size strongly depends on the quenching temperature as well as on the polymer species. The inner structure of such a small particle is not clear at the moment but it may be presumed that initially it has a smectic structure because the secondary phase separation corresponds to the transition from the nematic structure to the smectic structure of liquid crystal with excluding polymer chain entanglements included in the nematic structure; the exclusion of entanglements would cause such a phase separation.

Some evidences for the formation of the smectic phase prior to crystallization were given by several authors. A comprehensive review of this problem was given by Geil [43]. The original idea for this was proposed by Bonart in 1966 [70], who studied structural formation in the crystallization processes of uniaxially stretched noncrystalline PET by WAXD, and found the appearance of a meridional 001′ reflection with a spacing of $d_{\mathrm{Bragg}} = 1.07$ nm as an evidence of smectic structure. Asano et al. [71] also concluded from WAXD and SAXS observations of cold-drawn noncrystalline PET as a function of annealing temperature that the nematic phase transferred into a smectic phase before the well-known crystalline triclinic phase was formed. Furthermore, Mahendrasingam et al. [72] investigated the structural change during stretching at 90 °C by means of a time-resolved synchrotron radiation technique, finding that the 001′ reflection, with a spacing of 1.02 nm, appeared prior to crystallization and increased in intensity with time, but it began to reduce after the start of crystallization. Recently Fukao [73] has carried out a similar study, where a noncrystalline PET film cold-drawn by four times was annealed at various temperatures between 63.6 and 76.3 °C. He showed that when annealed at 71.2 °C, a meridional 001′ reflection with a scattering vector near $Q = 6.08\ \mathrm{nm}^{-1}$ ($d_{\mathrm{Bragg}} = 1.03$ nm) appeared initially, increasing in intensity until a SAXS four-point pattern emerged, and then decreasing in intensity. He also concluded that crystallization proceeds through three steps of phase transition: amorphous to nematic, nematic to smectic, and smectic to crystalline triclinic. All of these studies support the nematic-to-smectic phase transition prior to crystallization. After crystallization this period, of course, grows to the well-known long period which can be observed by SAXS. In addition, we should point out that this mechanism may essentially correspond to the crystallization mechanism of pre-formation of mesomorphic structure proposed by Strobl [26].

Next, we will discuss the point that the secondary phase separation is of the SD type. First, some authors [27–31] have shown that in the initial stage of crystallization of iPP a SAXS peak appears. As we described in the Introduction, this is the case for crystallization from the metastable state, the primary phase separation is the N&G type. Thus, as the first step the droplets with a nematic-like structure are formed sporadically in the isotropic matrix and grow in size with time. These droplets, however, do not cause a SAXS peak because their distribution in the matrix is not homogeneous and their size of the order of tens of μm is out of the low angle resolution of SAXS cameras; they show only strong central scattering intensities due to their shape factors. After that, the secondary phase separation occurs inside each droplet, resulting in the small particles. This phase separation gives a SAXS peak because the small particles are distributed homogeneously in the droplet and the size of the order of tens of nm is in the range of SAXS resolutions; the SAXS peak is due to the interference between these densely packed small particles. However, these small particles are not produced directly from the nematic phase but through the secondary phase separation of SD type inside the droplet; they are formed at the final stage of the secondary SD owing to the surface tension. This SD also gives a SAXS peak near the same position because of its characteristic wavelength. A strong evidence for the secondary phase separation of SD type seems to have been given in terms of a Cahn–Hilliard (C–H) plot of $R(Q)/Q^2$ versus Q^2 by Ryan et al. [29–31] (also see Introduction). They showed that this plot provides a maximum at a certain Q value, designated as q_{ic} by the authors [31], above which the straight line with a negative slope was obtained. This maximum can be considered to be due to the overlap of the primary N&G- and secondary SD-type phase separations; the part above this critical Q value may correspond to the secondary phase separation of SD type occurring inside the droplet and the part below it to the N&G process. The reason for this is as follows. The critical Q values are about $0.02 \sim 0.014$ Å^{-1} ($300 \sim 450$ Å in spacing) at $138 \sim 145$ °C depending on the crystallization temperature, which are roughly equal to the inter-cluster distance of about 200 Å obtained for the sample quenched at 8 °C by Wang et al. [28]. Before the emergence of such small particles or clusters, SD begins inside the droplet giving a characteristic wavelength nearly equal to the inter-cluster distance, so that the dynamic SAXS observations mean that we are seeing the growing process of the characteristic peak of SD. Conversely, the fact that the linearity with a negative slope is valid above the critical Q value means that the SD occurs inside the droplet. If this is the case, the line with a negative slope will represent strong evidence for the secondary phase separation of SD type. Furthermore, we would like to point out that a plot of $R(Q)$ versus Q for the iPP (the inset of Fig. 7a of [29]) shows a maximum above q_{ic}, indicating that the system is conserved. This again supports the secondary SD phase separation.

Finally, it should be noted that the spinodal temperature obtained from a plot of D_{eff} versus $1/T$ by Ryan et al. [29] corresponds to the secondary phase separation. This secondary spinodal temperature T_{s2} for PET was reported as 226 °C by them which is 13 K higher than the primary spinodal temperature of 213 °C obtained by the present authors [16]. The former temperature may possibly correspond to the binodal temperature above which the system enters the liquid-crystal co-existence region where single crystals might be formed. Fortunately, we can show one evidence for this in the case of iPP. According to Ryan et al. [29], T_{s2} for iPP is 142 °C above which it would be expected that single crystals are produced if T_{s2} is a binodal temperature on the co-existence curve of the phase diagram by Olmsted [4]. Nishida et al. [17] have recently succeeded in preparing such single crystal-like objects by crystallizing a mesomorphic iPP at 162 °C above T_{s2}; a molten iPP sample was first quenched at about 0 °C so as to yield the mesomorphic phase and reheated to 162 °C, keeping the temperature there for 10 min, and then fixed by further crystallization at 130 °C for 2 min. The obtained crystals showed a "bamboo leaf" (BL)-like network structure (see Fig. 1 of [17]) and the size of one leaf was a few tens of μm and about 5 μm in length and width;

the length corresponds to the diameter of a spherulite. These BL crystals with a high WAXD crystallinity seem to be single crystals though this should be confirmed using an electron diffraction method in the future. Of course, a direct crystallization from the melt at 130 °C, which is below $T_{s2} = 142$ °C, gave the ordinary spherulites with a diameter of a few tens of μm (see Fig. 2 of [17]). Therefore, it is expected that the melt crystallization of PE above $T_{s2} = 135$ °C [29] also provides single crystals though it may take a long time.

6 The General Concept of Polymer Crystallization Based on a Phase Diagram

Much experimental data shows that below the equilibrium melting temperature liquids and glasses can be maintained out of equilibrium without crystallization, which corresponds to the induction period. In this state the Gibbs free energy of the liquid is higher than that of the crystal phases, and hence such undercooled liquids should undergo a phase transformation because they are thermodynamically less stable. It is well known that in the case of binary systems such as solutions and blends, the route leading to crystals depends on the degree of undercooling [74]. When the degree of undercooling is not so great, direct transformation to the crystal phase occurs in the co-existence region, but as it increases, indirect transformation to the crystal phase occurs through the phase separations by nucleation and growth (N&G) in the meta stable region or by spinodal decomposition (SD) in the unstable region. Such indirect transformations can occur even in the case of one-component polymer systems, which was predicted theoretically by Olmsted et al. [4]. Below we will consider this theory and then consider a general model for the crystallization mechanisms of bulk polymers.

6.1 The Phase Diagram of the Polymer Melt by Olmsted

In 1998 Olmsted et al. [4] proposed a generic temperature-density diagram for a polymer melt as reproduced in Fig. 30. Here, it should be noted that the value on the horizontal axis indicates the normalized density ρw, which is defined as the average mass density ρ of the melt multiplied by the specific volume w of the monomer core, in other words the average density divided by the maximum density w^{-1} for the state of the closest packing of monomer cores, and so it does not exceed unity for any state including the crystalline one. The key concept of this theory is to take into account that the free energy of the change in chain conformation from coil to helix, that latter being almost the same as crystalline conformation. This phase diagram is similar to that for a binary component system, but the contents of the order parameter of density are different. In the case of the binary component system the

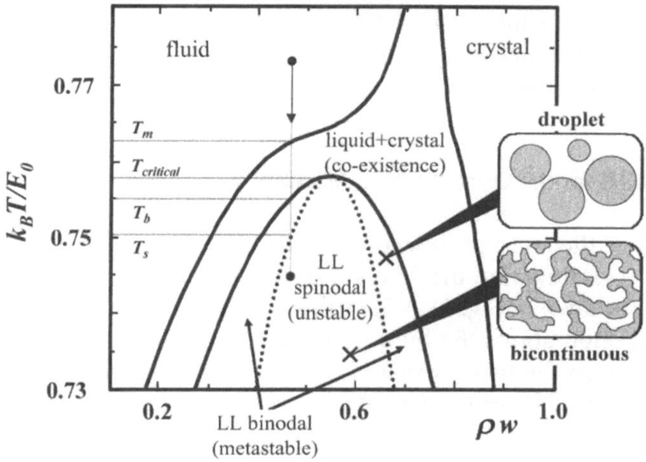

Fig. 30 Generic temperature-density diagram for a polymer melt proposed by Olmsted et al. [4]. The *horizontal axis* indicates the normalized density ρw, which is defined as the average mass density of the melt multiplied by the specific volume w of the monomer core

order parameter can be given by the ratio of two components (or the concentration of each component) while in the case of the polymer melt consisting of one component it is mainly governed by the chain conformation, or roughly speaking the length and the degree of orientation of stiff segments [24, 25] because the density couples with the chain conformation [4]. As described above, the crystallization routes are different as well in this case, depending on the degree of undercooling. When a polymer melt is quenched to a temperature in the co-existence region between T_m and the binodal temperature T_b, the crystal nucleation occurs directly from the melt, which is well known as the usual homogeneous crystal nucleation mechanism. However, when quenched to a temperature below T_b, the state of the melt first changes prior to crystal nucleation. Thus, the quench into the metastable region between T_b and the spinodal temperature T_s causes the phase separation into the unoriented (isotropic) phase and the oriented (or nematic) phase by a mechanism of nucleation and growth (N&G), while the quench to a temperature within the unstable region below T_b causes phase separation by a mechanism of spinodal decomposition (SD). After these phase separations, crystal nucleation initiates in the oriented phase and probably the isotropic phase will later transform to the oriented phase where crystal nucleation occurs later as well.

Though it is impossible to apply this phase diagram quantitatively to real polymers, we can understand the easiness of phase separation qualitatively. For example, it may be presumed that SD in PE is probably more difficult than that in PET; the normalized density of PE is $\rho w \cong 0.685$ [4] which is quite far from the critical point of $\rho w \cong 0.53$ and so SD would occur at considerably lower temperatures while that of PET may be nearer to the critical point due to the longer persistence length or a lower ρw and hence SD can be expected to occur at higher temperatures.

6.2
A General Model for the Crystallization Mechanism

On the basis of the concept described above, we propose a model for the homogeneous crystallization mechanism of one component polymers, which is schematically shown in Fig. 31. When the crystallization temperature is in the coexistence region above the binodal temperature T_b, crystal nucleation occurs directly from the melt, which is the well-known mechanism of polymer crystal nucleation. However, the rate of crystallization from the coexistence region is considered to be extremely slow, resulting in single crystals in the melt matrix. Crystallization at a greater rate always involves phase separation; the quench below T_b causes phase separations. The most popular case

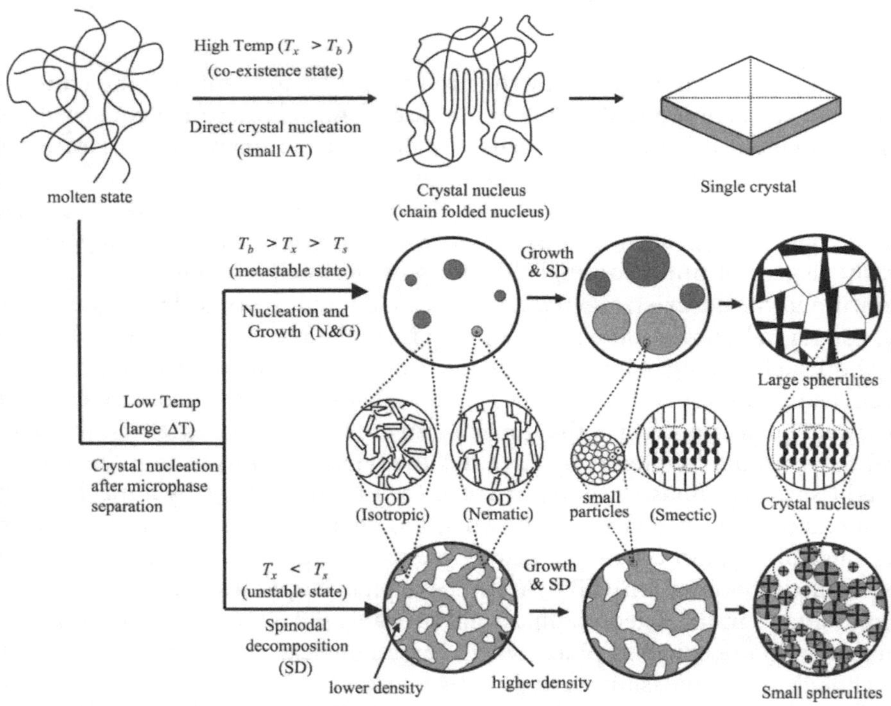

Fig. 31 Structural formation model for the initial stage of polymer crystallization [19]. N&G: nucleation and growth of oriented domains, SD: spinodal decomposition into oriented and unoriented domains, T_b, T_s, and T_x : bimodal, spinodal, and crystallization temperatures, respectively; I: isotropic, N: smectic, and C: crystalline

for flexible polymers such as PE and PP is the quench into the metastable region between the binodal temperature T_b and the spinodal temperature T_s where the droplets of the oriented (nematic) domains are first produced in the isotropic matrix by the N&G-type phase separation. On the other hand the quench into the unstable region below T_s causes an SD-type phase separation into the unoriented (isotropic) and oriented (nematic) phases. After these phase separations, the resulting nematic phase inside the droplet of N&G or the oriented domain of SD undergoes the secondary phase separation of SD type again prior to crystal nucleation; it separates into a smectic phase and an amorphous phase to provide small particles for the final stage. During this process the molecular entanglements contained in the nematic phase should be excluded from the smectic phase domains into the amorphous domains; in this case a sliding mechanism of molecular chains along the molecular axis seems to work. This secondary microphase separation should provide a long period just before crystal nucleation, which is the so-called "SAXS before WAXD". In the case where N&G is the primary phase separation the droplet develops into a large spherulite by fusing the small particles while in the case of SD the fusion of the particles is limited with the characteristic wavelength of the primary SD to produce small spherulites.

7 Conclusions

This review has been focused on explaining the structural formation processes in the induction period of one component homopolymer crystallization. The structures produced in this period, which affect the crystalline morphologies subsequently grown, greatly depend on crystallization temperature and the initial state of the sample, molten or glassy. In the case of crystallization from the melt, there are three different temperature regions into which the sample is quenched. For the higher temperature side they are the coexistence, metastable, and unstable regions. In the coexistence region above the binodal temperature crystal nuclei are produced directly from the melt, which is the commonly known mechanism of crystal nucleation. On the other hand, the quench into the metastable or unstable region below the binodal temperature causes binodal or spinodal phase separation, respectively, both into isotropic and nematic phases before crystallization. Although it is not recognized that crystallization from the metastable region is the most common case for flexible polymers, this concept is very important for understanding the large number of observations on crystallization obtained to date. In addition, for stiff or semi-stiff polymers such as PET and PEN another type of crystallization can be observed comparatively easily, i.e., crystallization from the unstable region showing the typical SD patterns in the optical micrographs whose characteristic wavelengths follow the theoretical equation by van Aartsen [22, 23] (the derivation of the equation is based on SD theory). Surprisingly, crystallization from the glass indicated no temperature dependence of the characteristic wavelength; it was constant until the melting temperature, meaning that even above the spinodal temperature the SD pattern appears. The reason for this was tentatively explained.

Finally, it should be emphasized that the spinodal-type of crystallization produces a bicontinuous structure consisting of regions with higher and lower degrees of crystallinity, which provides industrially important high-performance materials with an ideal homogeneous microstructure.

Acknowledgements This study was supported by the Grant-in-Aid for Scientific Research on Priority Area "Cooperative Phenomena in Complex Liquids" (1995–1997), Grant-in-Aid for Scientific Research on Fundamental Research (A-2) (1998–2001), and Grant-in-Aid for Scientific Research on Priority Area "Mechanism of Polymer Crystallization" (2000–2002) from the Ministry of Education, Science, Sports and Culture of Japan, and by the International Joint Research Grant for the Project "Fundamental Studies on Crystallization of Polymers" (1995–1998) and Industrial Technology Research Grant (ITRG) Program "Control of Higher Order Structure of Polymer Materials" (2001–2004) of the New Energy and Industrial Technology Development Organization (NEDO), Japan.

References

1. Bassett DC (1981) Principles of Polymer Morphology. Cambridge Univ Press, Cambridge
2. Lauritzen Jr JI, Hoffman JD (1973) J Appl Phys 44:4340–4352
3. Hoffman JD, Thomas Davis G, Lauritzen Jr JI (1976) In: Hannay NB (ed) Treatise on Solid State Chemistry, vol 3. Plenum Press, New York, p 497–614
4. Olmsted PD, Poon WCK, McLeish TCB, Terrill NJ, Ryan AJ (1998) Phys Rev Lett 81:373–376
5. ten Wolde PR, Frenkel D (1997) Science 277:1975–1978
6. Imai M, Mori K, Mizukami T, Kaji K, Kanaya T (1992) Polymer 33:4451–4456
7. Imai M, Mori K, Mizukami T, Kaji K, Kanaya T (1992) Polymer 33:4457–4462
8. Imai M, Kaji K, Kanaya T (1993) Phys Rev Lett 71:4162–4165
9. Imai M, Kaji K, Kanaya T (1994) Macromolecules 27:7103–7108
10. Imai M, Kaji K, Kanaya T, Sakai Y (1995) Physica B 213/214:718–720
11. Imai M, Kaji K, Kanaya T, Sakai Y (1995) Phys Rev B 52:12696–12704
12. Matsuba G, Kaji K, Nishida K, Kanaya T, Imai M (1999) Macromolecules 32:8932–8937
13. Matsuba G, Kaji K, Nishida K, Kanaya T, Imai M (1999) Polymer J 31:722–727
14. Matsuba G, Kanaya T, Saito M, Kaji K, Nishida K (2000) Phys Rev E 62:R1497–R1500
15. Matsuba G, Kaji K, Kanaya T, Nishida K (2002) Phys Rev E 65:061801-1∼7
16. Nishida K, Kaji K, Kanaya T, Matsuba G, Konishi T (2004) J Polym Sci B Polym Phys 42:1817–1822
17. Nishida K, Konishi T, Kanaya T, Kaji K (2004) Polymer 45:1417–1421
18. Kaji K, Imai M (1998) In: Yonezawa F, Tsuji K, Kaji K, Doi M, Fujiwara T (eds) The Physics of Complex Liquids. World Scientific, Singapore, p 258–273

19. Kaji K (2002) In: Fakirov S (ed) Handbook of Thermoplastic Polyesters, vol 1. Wiley, Weinheim, p 225–251
20. Kaji K, Nishida K, Matsuba G, Kanaya T, Imai M (2003) J Macromol Sci B42:709–715
21. van Krevelen DW (1990) Properties of Polymers. Elsevier, Amsterdam, p 585–623
22. van Aartsen JJ (1970) Eur Polym J 6:919–924
23. Smolders CA, van Aartsen JJ, Steenberger A (1971) Kolloid-Z u Z Polymere 243:14–20
24. Doi M, Edwards SF (1986) The Theory of Polymer Dynamics. Oxford University Press, Oxford, Chapters 9, 10, p 350–380
25. Shimada T, Doi M, Okano K (1988) J Chem Phys 88:7181–7186
26. Strobl G (2000) Eur Phys J E3:165–183
27. Wang Z-G, Hsiao BS, Sirota EB, Agarwal P, Srinivas S (2000) Macromolecules 33:978–989
28. Wang Z-G, Hsiao BS, Srinivas S, Brown GM, Tsou AH, Chen SZD, Stein RS (2001) Polymer 42:7561–7566
29. Ryan AJ, Fairclough JPA, Terrill NJ, Olmsted PD, Poon WCK (1999) Faraday Disccus 112:13–29
30. Heeley EL, Kit Poh C, Li W, Maidens A, Bras W, Dolbnya IP, Gleeson AJ, Terrill NJ, Fairclough JPA, Olmsted PD, Ristic RI, Hounslow MJ, Ryan AJ (2002) Faraday Discuss 122:343–361
31. Heeley EL, Maidens AV, Olmsted PD, Bras W, Dolbnya IP, Fairclough JPA, Terrill NJ, Ryan AJ (2003) Macromolecules 36:3656–3665
32. Muthukumar M, Welch P (2000) Polymer 41:8833–8837
33. Muthukumar M (2003) Phil Trans R Soc London A 361:539–556
34. Li L, de Jeu WH (2003) Macromolecules 36:4862–4867
35. Li L, de Jeu WH (2004) Phys Rev Lett 92:075506-1–075506-3
36. Imai M (1993) PhD Thesis, Kyoto University; Imai M, Kaji K, Polymer (in press)
37. Yeh GSY, Geil PH (1967) J Macromol Sci Phys B1:235–249
38. Yeh GSY (1972) J Macromol Sci Phys B6:465–478
39. Uhlmann DR (1979) Faraday Disc Chem Soc 68:87–95
40. Geil PH (1979) Faraday Disc Chem Soc 68:141–144
41. Flory PJ (1956) Proc Roy Soc A 234:60–73
42. See for example, Kaji K (2000) In: Gabrys BJ (ed) Applications of Neutron Scattering to Soft Condensed Matter. Gordon and Breach Science Publisher, Amsterdam, p 107–161
43. Geil PH (2002) In: Fakirov S (ed) Handbook of Thermoplastic Polyesters. Wiley, Weinheim p 105–224
44. Lee S, Miyaji H, Geil PH (1983) J Macromol Sci Phys B22:489–496
45. Fischer EW (1990) In: Colmenero J, Alegra A (eds) Basic Features of the Glassy State. World Scientific, Singapore p 172–191
46. Fischer EW (1993) Physica A 201:183–206
47. Cahn J, Hilliard JE (1958) J Chem Phys 28:258–267
48. Binder K (1974) Phys Rev B 15:4425–4447
49. Furukawa H (1984) Physica 123A:497–515
50. Debye P, Bueche AM (1949) J Appl Phys 20:518–525
51. Komura S, Osamura K, Fujii H, Takeda T (1985) Phys Rev B31:1278–1301
52. Strobl GR, Schneider M (1980) J Polym Sci Polym Phys Ed 18:1348–1359
53. Flory PJ (1969) Statistical Mechanics of Chain Molecules. Interscience, New York, Chapter, 1, p 1–29
54. Stein RS, Wilson PR (1962) J Appl Phys 33:1914–1922
55. Koberstein J, Russel TP, Stein RS (1979) J Polym Sci Polym Phys Ed 17:1719–1730

56. Kobayashi M, Nakaoki T, Ishihara N (1989) Macromolecules 22:4377–4382
57. Greis O, Xu Y, Asano T, Petermann J (1989) Polymer 30:590–594
58. Natta G, Corradini P, Bassi IW (1960) Nuovo Simmento, Suppl 15:68–70
59. Kobayashi M, Akita K, Tadokoro (1968) Makromol Chem 118:324
60. Einaga Y, Koyama H, Konishi T, Yamakawa H (1989) Macromolecules 22:3419–3424
61. Ediger MD, Angell CA, Nagel SR (1996) J Chem Phys 100:13200–13212
62. Matsuoka S and Quan X (1991) Macromolecules 24:2770–2779
63. Kanaya T, Patkowski A, Fischer EW, Seils J, Glaeser H, Kaji K (1994) Acta Polymer 45:137–142
64. Gehrke R, Riekel C, Zachmann HG (1989) Polymer 30:1582–1590
65. Donth E (2001) The Glass Transition: relaxation dynamics in liquids and disordered materials. Springer, Berlin, p 1–418
66. Kanaya T, Kaji K (2001) Adv Polym Sci 154:87–141
67. Goetze W (1991) In: Hansen JP, Levesque D, Zinn-Justin (eds) Liquids, Freezing and Glass Transition. Elsevier Science Publishers BV, Amsterdam, p 287–503
68. Perez J (1998) Physics and Mechanics of Amorphous Polymers. AA Balkema Publishers, Rotterdam, p 280–285
69. Kanig G (1983) Colloid Polym Sci 261:373–374
70. Bonart R (1966) Kolloid-Z Z Polym 213:1–11
71. Asano T, Balta-Calleja FJ, Flores A, Tanigami M, Mini MN, Sawatari C, Itagaki H, Takahashi H, Hatta I (1999) Polymer 40:6475–6484
72. Mahendrasingam A, Martin C, Fuller W, Blundell DJ, Oldham RJ, MacKerron DH, Harvie JL, Riekel C (2000) Polymer 41:1217–1221
73. Fukao K (2003) Macromol Sci-Phys 42:717–731
74. Kelton KF (1991) Crystal Nucleation in Liquids and Glasses. Solid State Physics, vol 45. Academic Press, New York

Adv Polym Sci (2005) 191: 241–274
DOI 10.1007/12_008

Published online: 28 October 2005

Modeling Polymer Crystallization

M. Muthukumar

Polymer Science and Engineering Department, University of Massachusetts,
Amherst, MA 01003, USA
muthu@polysci.umass.edu

Abstract We summarize the salient conclusions derived from Langevin dynamics simulations of many flexible polymer molecules undergoing crystallization from solutions. These simulations reveal molecular mechanisms of nucleation and growth, and the accompanying free energy barriers, during the very early stages of crystallization. The simulation results are also analyzed by statistical mechanics theories. Major conclusions on the growth of density fluctuations in the primordial stage, birth of baby nuclei, which then mature into lamellae through a stage of smectic pearls, and spontaneous selection of finite equilibrium lamellar thickness are addressed. Furthermore, selection of shapes is addressed using a novel Monte Carlo algorithm for polymer crystallization in solutions. In addition, details of free energy landscape just in front of the growth front are summarized, based on Langevin dynamics simulations. The mechanism of growth is seen to be an adsorption process, in contrast to previous beliefs. Finally, the role of externally imposed flow on polymer crystallization is addressed by considering the molecular mechanisms behind the formation of shish-kebab morphology in extensional flows. The major conclusions from the reviewed simulation results are qualitatively different from the established models of polymer crystallization.

1 Introduction

Recent advances [1–16] in experimental investigation of very early stages of polymer crystallization have led to a flurry of discussions involving the merits of existing paradigms and to new hypotheses. Complementing these experimental investigations to explore the early stages, molecular modeling has played a significant role. Careful molecular modeling offers an opportunity to monitor the manifestation of atomic forces and macromolecular details in morphological ordering. Generally speaking, there have been several approaches [17–39] to molecular modeling of polymer crystallization, using molecular dynamics, Brownian dynamics, Langevin dynamics, and Monte Carlo techniques. In these simulations, different issues such as crystallization from solutions and the nature of the crystalline-amorphous interface have been the foci. In this review, we focus only on the efforts of the author's research group to understand the molecular mechanisms of polymer crystallization from solutions. Most of the results reviewed here are based on Langevin dynamics simulations. However, these results are robust in terms of conceptual conclusions, because simulations by other research groups using different methodologies have yielded similar results.

The most popular paradigm of polymer crystallization is embodied in the celebrated Lauritzen–Hoffman (LH) theory [5, 6, 40–46]. The basic features of the model used in the LH theory are as follows. Let the starting template for further growth of a lamella be a frozen growth front of thickness L and width L_p. Polymer chains diffuse to this growth front and attach at the growth front after negotiating a free-energy barrier. This process is imagined to take place by a successive deposition of stems, with each stem having a length L. The deposition of the first stem of a chain is assumed to be equivalent to a secondary nucleation process involving a free-energy barrier, and is assumed to occur at a rate i. In writing the free energy for the deposition of the first stem, continuum thermodynamics is assumed to be valid at the monomeric level. The deposition of the second stem of the chain is assumed to have another free-energy barrier and occurs at a lateral spreading rate g. The flux of the stems is calculated in the steady state in terms of the parameters L, L_p, free energy gain in the formation of stems, lateral and fold free energies associated with formation of surfaces, and the height of imposed free energy barriers. From the steady state flux, the mean lamellar thickness and the distribution of lamellar thickness (both being determined kinetically), and the growth rates are calculated analytically. The key final results are that the lamellar thickness is determined kinetically (and allowing a divergence of lamellar thickness at a finite lower temperature), and that there are potentially three regimes of lamellar growth depending on the relative dominance of i with respect to g. There has been a notable exception [47] to the LH theory by Allegra who had investigated the thermodynamic stability of bundles of stems.

The natural questions that arise are what constitutes the stems, how the stems (if they exist) get attached at the growth front, what the free-energy barriers are, etc. The molecular modeling has provided vivid details for these questions, and has shown that the assumptions of the LH model are not valid in any universal way. The simulation results have clearly underscored the need for new theories of polymer crystallization by providing credence to the criticisms of the LH theory.

In addition, a suggestion [8–12] that polymer crystallization is initiated by a spinodal mode prior to growth of lamellae in one-component polymer melts has led to a more active exploration. Again, simulations have shed light on this debate by showing how the observed scattering data are consistent with the nucleation and growth process for topologically correlated systems such as polymers. Further, the molecular details of the birth of the "shish-kebab" morphology [48–60] in polymer crystallization under flow have also been unraveled by molecular modeling.

2 Simulation Methods

We have used two types of simulations to follow the polymer crystallization. The first type uses details at the monomeric level and the second type utilizes coarse-grained approaches where details at length scales below chain dimensions are integrated out. In the first, molecular dynamics and Langevin dynamics are typically used. All of these simulations provide a converging viewpoint regarding the nature of the initial stages of polymer crystallization. In the second type of simulation, Monte Carlo methods are employed to explore the growth kinetics. We review only the Langevin dynamics methodology for the first type. Using this method, we identify the generic features of how single chains fold and attach to the growth front. We then use these results as building blocks for the growth of large lamellae of macroscopic dimensions, by implementing a Monte Carlo method. If we were to continue with Langevin dynamics simulations, monitoring of the growth of very large lamellae would be prohibitively difficult.

2.1 Langevin Dynamics Methods

In our Langevin dynamics simulations, the polymer crystallization is modeled by following the competition between the attraction among non-bonded monomers and the torsional energies along the chain backbone. The simulation model attempts to incorporate just enough detail to observe chain-folding without impeding the efficiency of the simulation. As a result, the united-atom model for polyethylene is chosen for a polymer chain, in which each methylene unit is treated as a bead in a bead-spring model of N beads. The total potential energy of a chain consists of the potential energy of each bond arising from bond stretch U_r, bond angle U_θ, and bond torsion U_ϕ, and non-bonded bead-bead interaction which is taken to be the Lennard-Jones interaction $U_{L\text{-}J}$. The potential energy associated with bond stretch is taken to be

$$U_r = k(r - r_0)^2 , \tag{1}$$

where r is the bond length and r_0 is the equilibrium bond length. The spring constant k is taken to be 115 kcal/mol Å^2 and $r_0 = 1.53$ Å. The potential energies associated with bond angle θ and torsion angle ϕ are assumed to be of the form

$$U_\theta = k_\theta(\cos\theta - \cos\theta_0)^2 \tag{2}$$

and

$$U_\phi = k_1(1 - \cos\phi) + k_2(1 - \cos 2\phi) + k_3(1 - \cos 3\phi) , \tag{3}$$

where $\theta_0 = 109°$, $k_\theta = 60.0$ kcal/mol, $k_1 = 3.02$ kcal/mol, $k_2 = -0.560$ kcal/mol, and $k_3 = 2.58$ kcal/mol. These values of k_1, k_2, and k_3 were used in [29], whereas 0.8, − 0.43, and 1.62 kcal/mol were used respectively in [22]. The particular choice of these parameters determine the time needed to form chain-folded states and the lamellar thickness. Since we are generally interested in global features of nucleation in the early stages instead of trying to figure out how a specified polymer crystallizes, we are yet to systematically vary the values of these parameters and then to establish a relation between the characteristic ratio of a polymer and its lamellar dimensions. The Lennard-Jones potential $U_{L\text{-}J}$ is

$$U_{L\text{-}J} = \varepsilon_0[(\sigma_0/r)^{12} - 2(\sigma_0/r)^6] , \tag{4}$$

where the interaction strength ε_0 is set to 0.112 kcal/mol, and r is the bead-bead distance. The equilibrium distance σ_0 is 4.53 Å for beads further than five repeat units apart along the chain backbone. In order to enhance computational stability, beads that are closer than 5 repeat units along the chain interact with a σ value equal to 1.54 Å. In our simulations, solvent molecules are not explicitly treated. In the absence of excluded volume forces from solvent molecules, the chain can collapse into a globule at lower temperatures. The choice of two different values of σ_0 with a cut-off length of 5 repeat units along the backbone is partially to account for the size of solvent molecules. We account for solvent molecules only through their frictional forces on the beads. The motion of each bead is described by the Langevin equation (Eq. 5) which consists of inertial term, force field, frictional drag, and noise, respectively [61].

$$\ddot{r}_i = -\nabla U_i - \Gamma \dot{r}_i - W_i(t) \tag{5}$$

The Langevin dynamics method simulates the effect of individual solvent molecules through the noise W, which is assumed to be Gaussian. The friction coefficient Γ is related to the autocorrelation function of W through the fluctuation-dissipation theorem,

$$\langle W_i(t) \cdot W_j(t') \rangle = \delta_{ij}\delta(t - t')6k_B T\Gamma\,. \tag{6}$$

Furthermore, we set Γ to be 1, between the over-damped regime and the purely deterministic regime. We used the velocity Verlet finite-differencing scheme [61] for integration. All simulation results given below are in reduced units (united-atom mass m of 1, equilibrium bond length r_0 of 1, and Lennard-Jones ε_0 of 1). The reduced temperature, T^*, is equal to kT/ε_0 (with kT being the Boltzmann constant times the absolute temperature), the reduced free energy is F/ε_0 and the reduced time is $t\sqrt{\varepsilon_0/m\sigma_0{}^2}$. It must be remarked that the following simulation results are for particular sets of values of various parameters ($r_0, \varepsilon_0, \sigma_0$, etc.). Different choices give different quantitative details. The primary objective of our simulations is not to predict *ab initio* melting temperature, lamellar thickness, growth rate, etc. Instead, we hope to capture the underlying universal molecular mechanism behind polymer nucleation in the very early stages.

The protocol of a typical Langevin dynamics simulation is as follows. The first step is the determination of the melting temperature, T_m^*, for the model chains (for chosen values of N and force field parameters). An initially created random configuration is equilibrated at $T^* = 15.0$ ($> T_m^*$). The chain is then quenched to $T^* = 9.0$ and crystallization is allowed to take place. Once a single chain-folded structure is obtained, several runs are performed at heating rates ranging from 0.0001 to $0.002T^*$/time units. Discontinuities are observed in the slopes of both the total potential energy and global orientational order parameter at the onset and end of melting. The equilibrium melting temperature is estimated by an extrapolation of the observed melting temperatures to the zero heating rate. This temperature is approximately $T^* = 11.0 \pm 0.2$. After determining T_m^*, a collection of chains (or an individual chain) is quenched to a prescribed T^* below T_m^* and the chain configurations are followed.

Throughout the simulations, data are collected at periodic intervals. These include the radii of gyration, the lamellar thickness, the kinetic and potential energies, the single chain form factor $S(q)$, the orientational order parameter, and free energy. $S(q)$ is calculated as

$$S(q) = \frac{1}{N^2}\sum_{i=1}^{N}\sum_{j=1}^{N}\frac{\sin(qr_{ij})}{qr_{ij}}\,, \tag{7}$$

where r_{ij} is the distance between the i and j beads, and q is the magnitude of the scattering wave vector.

In order to monitor how stems are oriented with respect to each other in a growing lamella, we identify an orientational vector at each bead i,

$(\vec{r}_{i+1} - \vec{r}_{i-1})$, connecting the positions of neighbors of i. For a stem with only *trans* conformations, the linearity of the stem is better captured by this choice of orientational vector than the bond vectors $(\vec{r}_{i+1} - \vec{r}_i)$. We then construct an order parameter s by calculating the angles Φ between the various orientation vectors and performing the average over all possible pairs of orientation vectors,

$$s = \frac{\langle 3\cos^2(\Phi) - 1\rangle}{2} . \tag{8}$$

We have monitored the value of s for the whole chain (and the aggregate for the case of many chains) and we call this the global order parameter. In addition, we have also computed the local order parameter where the volume element extends only over $2\sigma_0$ but is averaged over throughout the chains. Further, we have calculated the free-energy landscape as a function of a measure, L, of lamellar thickness of single chains at a given quench depth and utilizing a histogram technique [62]. L is the radius of gyration along the axis parallel to the chain backbone within the crystal. The free energy $F(L)$ is estimated as $F(L) = -kT\ln\left(\frac{n(L)}{\mathcal{N}}\right)$, where $n(L)$ is the number of times the system visited states between L and $L + \Delta L$, and $\mathcal{N}$ is the total number of states visited. Typically $\Delta L/r_0$ is 2 and the estimate is constructed by performing nine different simulations for 60 000 reduced time units. Samples were recorded every 20 time units.

Attempts were made to include all hydrogen atoms explicitly in the simulations. This computationally demanding explicit-atom model shows (Fig. 1) that the crystal symmetry is orthorhombic, in agreement with the well-known experimental result for polyethylene single crystals, instead of the hexagonal symmetry seen in united-atom model simulations.

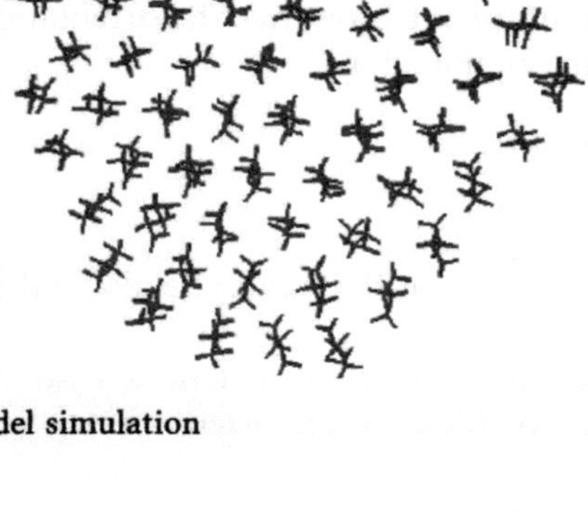

Fig. 1 Explicit-atom model simulation

However, the essential mechanisms of lamellar formation and growth are found to be the same in both the united-atom and explicit-atom models. Only the united-atom model simulation results are therefore discussed in Sect. 3.

2.2
Monte Carlo Simulations

As discussed later in Sect. 3, the Langevin dynamics simulations of the united-atom model show that the growth kinetics are dominated by adsorption of new chains at the growth front. In view of this observation, we have developed [34] a coarse-grained model of lamellar growth and implemented the Monte Carlo algorithm. The algorithm is as follows. We start with a cubic box of volume Ω containing one initial square lamella of prescribed lateral dimension R_0 (in units of the lamellar thickness, taken as the grid size) situated at the origin of the coordinate system, and n folded chains randomly distributed in Ω. The chain orientation of the lamella is taken to be along the z-axis. The orientations of the n chains are randomly distributed among the three axes. The folded chains are allowed to undergo diffusion and if any of the chains would approach the growth front with the correct orientation then the chain will absorb with a probability of unity. Once adsorbed, the chain can freely slide at the growth front if the neighboring sites are free. An adsorbed chain can desorb with a probability of $\exp(-\varepsilon_2/T)$, where ε_2 is a parameter, if the neighboring sites on the new growth layer are free. If an adsorbed chain has a neighboring site occupied by another chain on the new growth layer, then it can desorb with a probability of $\exp(-2\varepsilon_2/T)$. Using these Monte Carlo rules, the evolution of the growth fronts of the nucleus is followed for the anisotropic adsorption model described above.

3
Results

3.1
Nucleation in the Very Early Stage

We first summarize the salient features of the Langevin dynamics simulation results followed by a theoretical analysis.

3.1.1
Simulations

As reported in [22] and [29], Fig. 2 shows a typical sequence of images depicting nucleation of a lamella by a single chain of $N = 700$ beads as obtained from Langevin dynamics simulations. The chain is quenched to $T^* = 9.0$

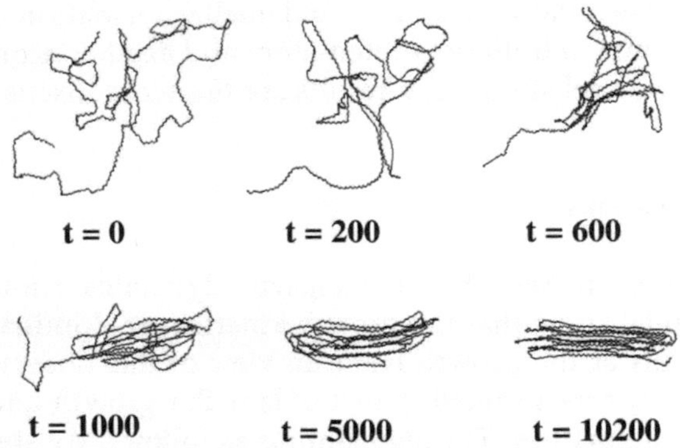

Fig. 2 Snapshots of nucleation by a single chain ($N = 700$) [22]

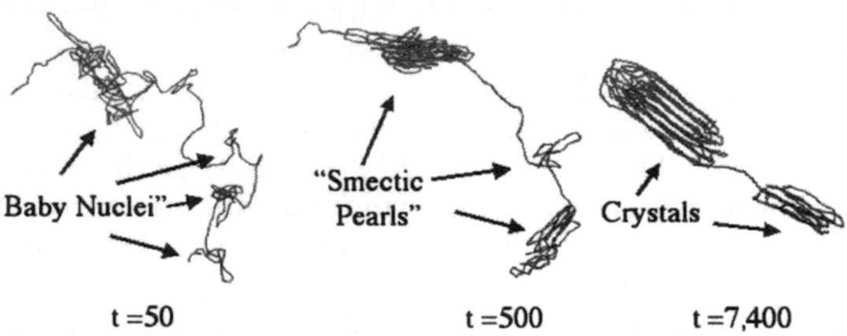

Fig. 3 Growth of smectic pearls by reeling in the connector ($N = 2000$) [29]

after equilibration at $T^* = 12.0$. Another example of $N = 2000$ (quenched to $T^* = 9.0$ from 20.0) is given in Fig. 3. The time steps shown in the sequence are selected from representative configurations during the course of crystallization.

As seen in these figures, several "baby nuclei" are formed, connected by the same single chain. The strands connecting these baby nuclei are flexible with considerable configurational entropy. As time progresses, the monomers in the flexible strands are reeled into the baby nuclei while the orientational order in each nucleus increases making them "smectic pearls". Simultaneously, the competition between nuclei for further growth dissolves some nuclei. Eventually, a folded-chain structure emerges. Thus, the description is essentially the same as nucleation and growth encountered in small molecular systems, except that the polymer now is long enough to participate in several nuclei. Immediately after the quench ($t < 500$ in Fig. 3), we observe that the average distance between baby nuclei does not change with time.

But the number of monomers in the connectors is reduced, accompanied by an increase in segmental orientation inside the nuclei as time increases.

To quantify this result, the structure factor $S(\boldsymbol{q}, t)$ at time t and the initial structure factor $S(\boldsymbol{q}, 0)$ were computed. As seen in experiments, a scattering peak at $q_{\max}$ was observed. In these simulations, $q_{\max}$ was found to correspond to the spacing between baby nuclei and the peak position is essentially independent of time in the very early stages.

Figure 4 contains a plot of Ω_q/q^2 vs. q^2, where Ω_q is the rate of growth of monomer density fluctuations with wave vector $\boldsymbol{q}$. According to the linearized theory of spinodal decomposition for mixtures [6], $S(\boldsymbol{q},t) \propto \exp(2\Omega_q t)$, where $\Omega_q \propto q^2(1-\kappa q^2)$, where κ is a positive constant. Therefore, a plot of Ω_q/q^2 vs. q^2 must be linear with a negative slope if spinodal decomposition is present. Some experimentalists have used this criterion to claim that spinodal decomposition is the mechanism of polymer crystallization at the early stage. As in experiments, we also observe that Ω_q/q^2 vs. q^2 is linear with a negative slope. However, this is not an evidence for spinodal decomposition because this behavior is observed for only intermediate values of $\boldsymbol{q}$. Our results show that for small $\boldsymbol{q}$, $\Omega_q \propto q^4$, in agreement with experiments but in disagreement with the predictions of spinodal decomposition.

To get more insight into the further growth of smectic pearls, typical configurations at various times are presented in Fig. 5 ($t = 500$, 1550, 7400, 10 300, 12 850, 13 350).

For the sake of clarity, we have used two shades for the polymer although the chain is a homopolymer. As pointed out already, monomers in the connectors are transferred into the growing nuclei in the very early stage.

This process continues until the connector is essentially stretched out while keeping the average inter-nuclei distance the same. Then, the connector is pulled into the nuclei to varying degrees until the nuclei impinge against each other. This is followed by a cooperative reorganization by which nuclei merge to form a single lamella. The mechanism of the merger is not by sequentially placing stems, but by a highly cooperative process involving all stems of the lamella.

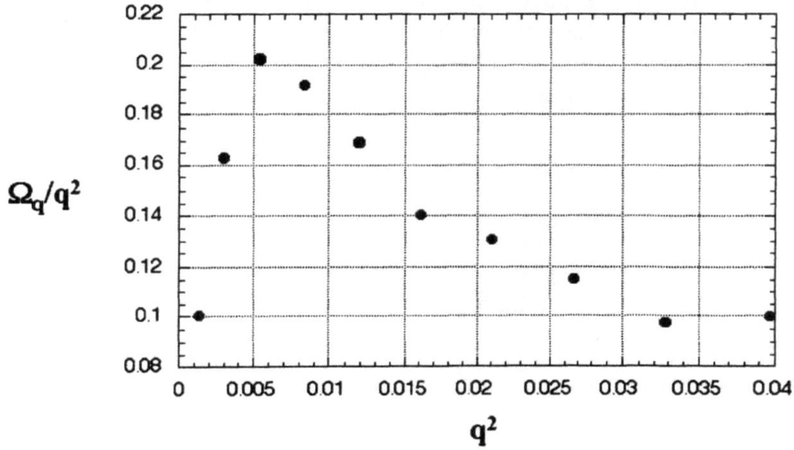

Fig. 4 Dependence of growth rate of density fluctuations on the square of wave vector [29]

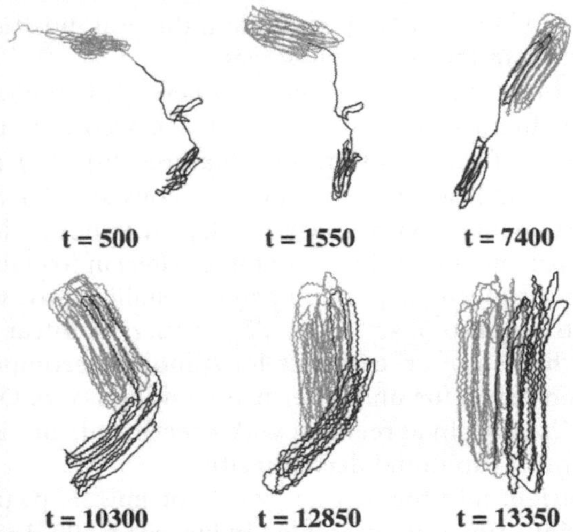

Fig. 5 Merger of smectic pearls [29]

3.1.2
Theory

We now present the simplest analytical model [32] for the origin of q_{max}, the mechanism of growth of smectic pearls, and the growth of density fluctuations in the very early stages of nucleation of lamella.

Origin of q_{max}

To address why a certain average distance is maintained between two smectic pearls at a very early stage, let us consider a model chain of N beads with only two smectic pearls (containing N_1 and N_2 beads) connected by a strand of m $(= N - N_1 - N_2)$ beads (Fig. 6).

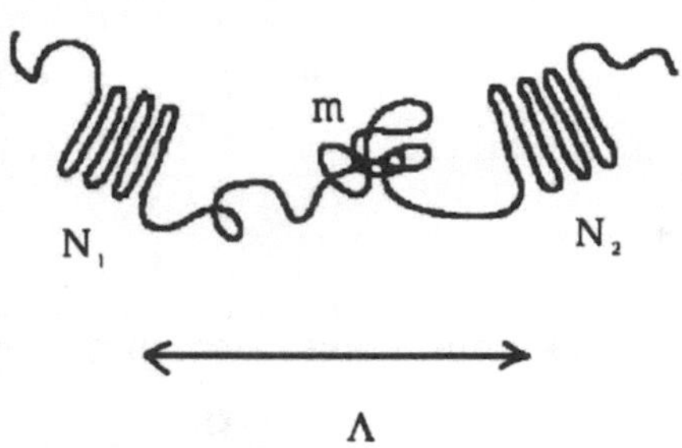

Fig. 6 Model to consider connector entropy

Let the end-to-end distance of the strand be Λ, which is comparable to the average distance between the smectic pearls. Let the average energy of a bead in either of the smectic pearls be ε.

The free energy F_0 of this configuration is given by

$$\frac{F_0}{k_B T} = -(N - m)\varepsilon + \frac{3}{2}\frac{\Lambda^2}{ml^2}, \tag{9}$$

where the second term on the right hand side is the entropic part from the strand (assuming Gaussian chain statistics and l being the Kuhn length, a multiple of r_0). Minimization of F_0 with respect to m gives the optimum value of m ($= m^*$) for the configuration of Fig. 6,

$$m^* = \sqrt{\frac{3}{2\varepsilon}}\left(\frac{\Lambda}{l}\right). \tag{10}$$

Since Λ is roughly proportional to $\sqrt{m^*}$ according to the Gaussian statistics valid approximately before the quench, we expect

$$\frac{\Lambda}{l} \sim \sqrt{m^*} \sim \frac{1}{\sqrt{\varepsilon}}. \tag{11}$$

Thus, the initial selection of average distance between the smectic pearls is determined by ε (and consequently quench depth, proportional to ε).

Kinetics of Growth of Smectic Pearls

Although arguments based on equilibrium are used above to estimate q_{max}, the conformation discussed above is not in equilibrium and it evolves further by reeling in the connector. To address how this process takes place, let us consider the time-dependent probability $W_m(t)$ of finding m beads in the connector at time t. Let k_1 be the rate constant for one bead to detach from either of the smectic pearls, and k_1' be the rate constant for one bead to attach to either of the smectic pearls. Using a detailed balance to express k_1' in terms of k_1 and letting m be a continuous variable, a mapping [32] with the standard arguments of the classical nucleation theory gives the Fokker–Planck equation,

$$\frac{\partial W_m(t)}{\partial t} = k_1\left[\frac{\partial}{\partial m}\frac{\partial(F_0/kT)}{\partial m} + \frac{\partial^2}{\partial m^2}\right] W_m(t) \tag{12}$$

where F_0 is given in Eq. 9. The prediction of Eq. 12 (solid curve) is compared with the simulation data in Fig. 7.

In the comparison, Λ is taken as an input from the simulations, ε is a parameter and $k_1 t$ is the reduced time. The agreement is good, providing qualitative support to the present theoretical model, in the initial stages. For reduced times greater than 4000, the mechanism is not reeling-in, and consequently, simulation data deviate from the solid curve.

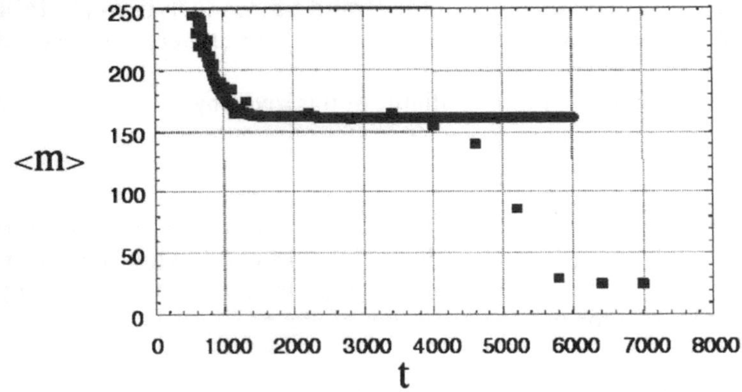

Fig. 7 Time-dependence of average number of monomers $\langle m \rangle$ in the connector between smectic pearls. The simulation data (*filled square*) are from 8 simulations corresponding to the conditions of Figs. 3 and 6; the *solid line* is calculated from Eq. 12

Growth of Density Fluctuations

We now generalize the model of Fig. 6 to account for the wave-vector dependence. There are three contributions to the free energy, F: (a) density difference ψ between the "baby nuclei" and the amorphous background giving a free-energy contribution that is proportional to $-\Delta T \psi^2 (\Delta T \equiv T_m^0 - T)$; (b) interfacial free energy given by the square gradient of ψ, proportional to $q^2 \psi_q^2$ (where q is the scattering wave vector); and (c) monomer–monomer correlation arising from the chain connectivity of the connector participating in multiple nuclei, leading to a free-energy contribution that is proportional to $q^{-2} \psi_q^2$ (as in the Debye structure factor for length scales shorter than R_g). Therefore, the free energy of a system with "baby nuclei" connected by strands is

$$F \sim \sum_q (-\Delta T + q^2 + \frac{1}{q^2}) \psi_q^2 , \tag{13}$$

where all the prefactors are left out. At this juncture of the early stage of nucleation and growth, ψ evolves with time, in accordance with the relaxation of the chemical potential gradient

$$\frac{\partial \psi(r,t)}{\partial t} \simeq -\nabla \cdot (-\nabla \frac{\partial F}{\partial \psi}) \tag{14}$$

so that

$$\frac{\partial \psi_q(t)}{\partial t} = -q^2(-\Delta T + q^2 + \frac{1}{q^2}) \psi_q(t) . \tag{15}$$

Therefore, we expect the scattered intensity, $I(q,t)$, proportional to $\langle\psi_q^2(t)\rangle$ to be exponential in time, $I(q,t) \sim \exp(2\Omega_q t)$, with the rate $\Omega_q = q^2(\Delta T - q^2 - \frac{1}{q^2})$ with both lower and upper cut-offs in q. If the above arguments are valid, Ω_q/q^2 should rise sharply with q^2, reach a maximum and then decrease at higher q values. These predictions are fully consistent with the experimental [8] and simulation [30] observations on $I(q,t)$ and Ω_q (Fig. 4). If the mechanism is simply a spinodal decomposition into two liquid phases, then Ω_q/q^2 should show a monotonic linear decrease from a finite positive value at $q \to 0$ with a slope independent of quench depth, which is not experimentally observed during polymer crystallization. Thus, the mechanism of polymer crystallization, even in the very early stage, is nucleation and growth with an additional contribution arising from chain connectivity. When the original "baby nuclei" have grown into lamellae comparable to or larger in size than R_g, their further growth is dictated essentially by the nature of the growth front.

3.2
Spontaneous Selection of Lamellar Thickness

3.2.1
Simulations

Many simulations of n chains, each with N beads, such that $nN \leq 15\,000$ and the volume fraction of the polymer, $\phi \leq 0.5$, have been performed at different quench depths. The key observations are summarized below.

Quantization of Lamellar Thickness

The initial lamella formed as described in Sect. 2, is typically thin and small. However, over a period of time, it thickens.

The lamellar thickening proceeds through many metastable states, each metastable state corresponding to a particular number of folds per chain, as illustrated in Fig. 8. In the original simulations of [22], R_g was monitored. R_g is actually very close to the lamellar thickness due to the asymmetric shape of the lamella. The number of folds indicated in Fig. 8 were identified by inspection of the coordinates of the united atoms. This quantization of the number of folds has been observed in experiments [50], as already mentioned. The process by which a state with p folds changes into a state with $p-1$ folds is highly cooperative. The precursor "lives" in a quiescent state for a substantial time and "suddenly" it converts into the next state. By a succession of such processes, crystals thicken. If the simulation is run for a reasonably long time, the lamella settles down to the "equilibrium" number of folds per chain.

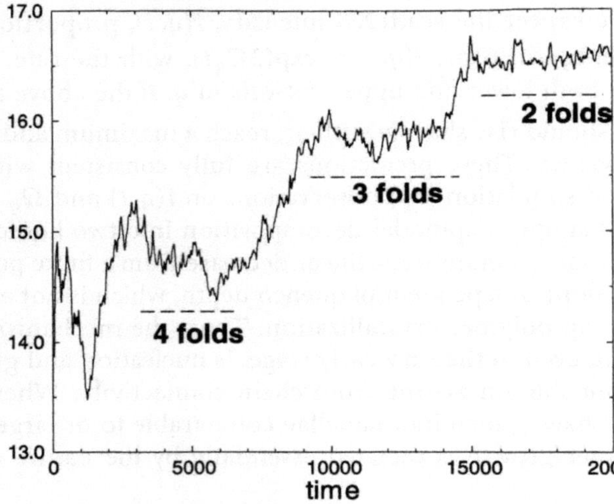

Fig. 8 Quantized lamellar thickening. R_g is the radius of gyration of the lamella [22]

Lamellar Thickness and Quench Depth

The lamellar thickness L, after the thickening is apparently complete, is found in the simulations to obey the law

$$L = \frac{C_1}{\Delta T^*} + C_2 , \tag{16}$$

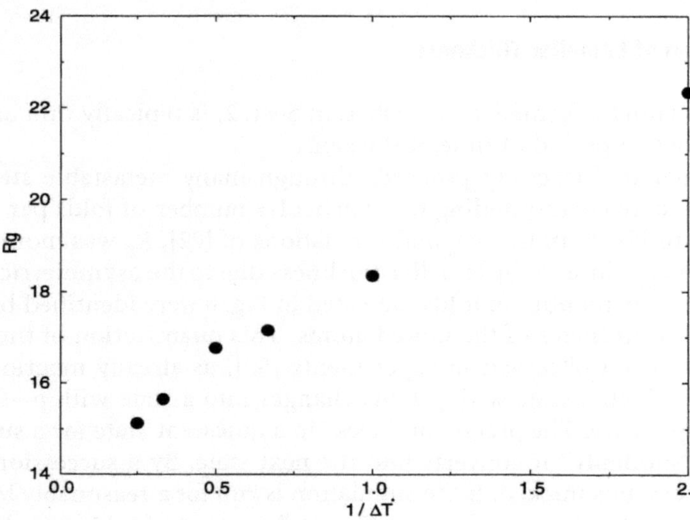

Fig. 9 Lamellar thickness (expressed as R_g) versus inverse undercooling for chains of $L =$ 500. Each data point is the average of four chains. Curve follows the $1/\Delta T^*$ relationship

where C_1 and C_2 are constants. Figure 9 is a plot of the radius of gyration averaged over four isolated chains at each undercooling against $1/\Delta T^*$, where $\Delta T^* = 11.0 - T^*$. The plot is approximately linear.

This relation is consistent with previous results observed experimentally [49]. Although the kinetic theory of Lauritzen and Hoffman predicts the same law as Eq. 16, it predicts a divergence in L at lower undercoolings. The simulations do not show any evidence for such a catastrophe.

Free-Energy Landscape

In an effort to quantify the free energies of different quantized states and free-energy barriers separating these states, the free-energy landscape has been calculated as a function of a measure, L, of lamellar thickness of single chains at a given quench depth and utilizing a histogram technique [30]. For example, the estimated free energy $F(L)$ for $N = 200$ at a quench depth of approximately 2.0 is given in Fig. 10a exhibiting several wells. Each well corresponds to a different number of stems in the lamella. For example, six, five, and four stem structures are observed for chains composed of 200 united atoms. Increasing the number of united atoms results in the addition of more wells. For example, the free-energy profile (Fig. 10b) for $N = 300$ displays additional wells. As N increases, the chains increase the number of stems in the crystal to accommodate the optimum crystal thickness.

The minimum in $F(L)$ is observed to be near $L/r_0 \simeq 10$ for all chain lengths examined in our simulations. Although it is seen in Fig. 10 that the free-energy minimum occurs at a higher L for the larger N, we are yet to establish the quantitative relation between the thickness corresponding to the free-energy minimum and chain length. It is to be noted that this free-energy minimum is the global minimum and the barrier between this state and other thicker lamellae increases prohibitively as the thickness increases. These simulations strongly suggest that a lamellar thickness that is much smaller than the extended chain thickness is actually an equilibrium result.

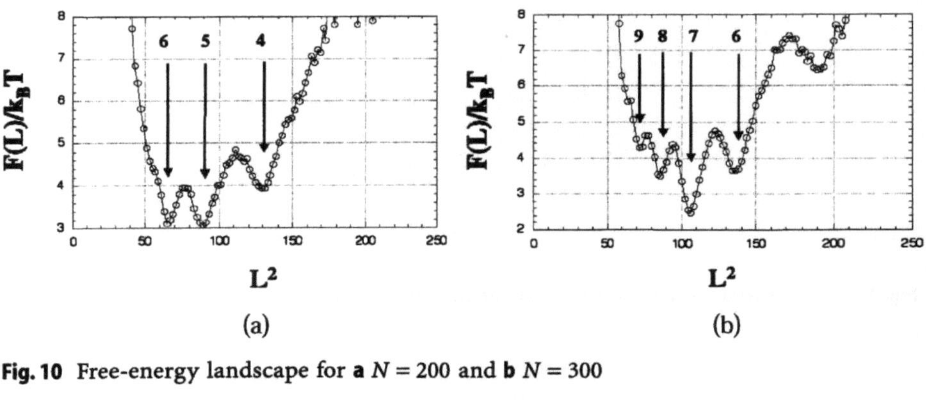

Fig. 10 Free-energy landscape for **a** $N = 200$ and **b** $N = 300$

3.2.2 Theory

Motivated by our simulation results, we now consider a theoretical model [32] which allows an exact calculation of the equilibrium lamellar thickness. Consider a nucleus sketched in Fig. 11, of thickness $L = ml$ and radius R. For each of the n chains, let there be μ stems (and $\mu - 1$ loops and two chain tails), each of length L.

Let $\varepsilon > 0$ be the energy gain per segment in the nucleus (in units of kT) and σ' be the lateral surface free energy per unit area. The free energy $F_{m,\mu}$ per chain in the nucleus of Fig. 11 is given by

$$\frac{F_{m,\mu}}{kT} = -\mu m\varepsilon + \sigma\sqrt{\mu}m - \ln Z_{m,\mu}\,, \tag{17}$$

where $\sigma = 2\sqrt{\pi}\sigma' l^2$. The third term on the right hand side is due to the entropy associated with different ways of realizing loops and tails on the two fold surfaces. The partition sum of a loop of p monomers in semi-infinite space with ends at R_{11} and X_{11} (both located on the fold surface) is given by

$$g_{\text{loop}}(p) = 2\left(\frac{3}{2\pi p l^2}\right)^{\frac{3}{2}} \mathrm{e}^{-\frac{3(R_{11}-X_{11})^2}{2pl^2}} \left[1 - \sqrt{\pi}\Gamma \mathrm{e}^{\Gamma^2}\operatorname{erfc}(\Gamma)\right] \tag{18}$$

with $\Gamma = c\sqrt{6p}$ and c is the strength of the interaction pseudopotential at the fold surface. For a tail of p segments in semi-infinite space, the partition sum is

$$g_{\text{tail}}(p) = \mathrm{e}^{\Gamma^2}\operatorname{erfc}(\Gamma)\,. \tag{19}$$

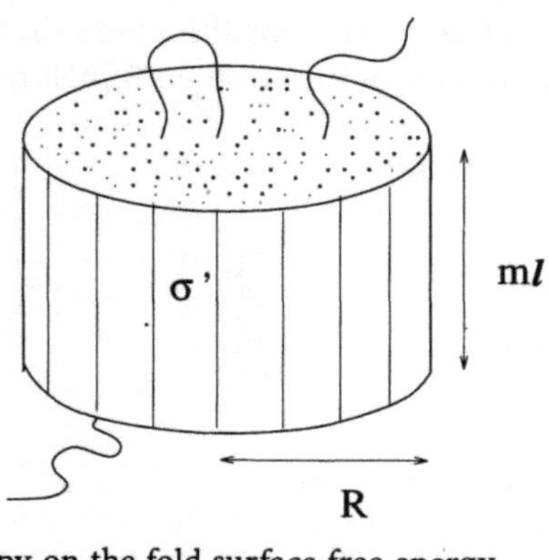

Fig. 11 Role of loop entropy on the fold surface free energy

Since the contour lengths of loops and tails are quite short ($c\sqrt{p}$ is small), as evident in the simulations, g_{loop} and g_{tail} approach the limits

$$g_{\mathrm{loop}}(p) \to 2\left(\frac{3}{2\pi p l^2}\right)^{\frac{3}{2}} \mathrm{e}^{-\frac{3(R_{11}-X_{11})^2}{2pl^2}}$$

and

$$g_{\mathrm{tail}}(p) \to 1\,. \tag{20}$$

By following the field-theoretic technology of [32] and choosing a cut-off of l_c for $(R_{11} - X_{11})$, the partition sum $Z_{m,\mu}$ for distributing $(N - m\mu)$ segments among $(\mu - 1)$ loops and 2 tails without breaking the chain connectivity is given by

$$Z_{m,\mu} = 4\left(\frac{1}{\nu b_0}\right)^{\mu-1}(N - \mu m)\left[\left(\frac{z^2}{2} + \frac{1}{4}\right)\mathrm{erfc}(z) - \frac{z}{2\sqrt{\pi}}\mathrm{e}^{-z^2}\right], \tag{21}$$

where

$$z = \frac{(\mu - 1)b_0}{2\sqrt{N - \mu m}}, \tag{22}$$

with $b_0 = \sqrt{6}l_c/l$ and $\nu = \pi l^3/3\sqrt{6}$.

Substitution of Eq. 22 into Eq. 17 gives the free-energy landscape in terms of the lamellar thickness ($\sim m$) and width ($\sim \mu$) per chain for a given choice of ε, σ, and l_c. The remarkable consequence of the entropic part of $F_{m,\mu}$ is that $F_{m,\mu}$ has a global minimum for a finite value of m.

This is illustrated in Fig. 12, where $F_{m,\mu}/k_{\mathrm{B}}T$ is plotted against m and μ for a representative set of $\varepsilon = 1$, $\sigma = 5$, $N = 1000$, $l_c/l = \sqrt{32/3}$, $\nu/l^3 = \pi/3\sqrt{6}$. For the case of Fig. 12, the global minimum (the ground state) is at $m^* = 14.26$ and $\mu^* = 45.3$.

This result is to be contrasted with the standard model [53] of Fig. 13, where the fold surfaces are simply treated as planar interfaces with fold surface free energy σ_f per unit area. In the latter case, the free energy of the nucleus is given by

$$\frac{\Delta F}{k_{\mathrm{B}}T} = -\,\mu m\varepsilon + \sigma\sqrt{\mu}m + 2\mu\sigma_f\,. \tag{23}$$

In terms of the critical nucleus ($m_c = \frac{4\sigma_f}{\varepsilon}$, $\mu_c = (\sigma/\varepsilon)^2$, $\Delta F_c/k_{\mathrm{B}}T = \varepsilon\mu_c m_c/2$), ΔF becomes

$$\overline{\Delta F} = -\,2\overline{\mu}\overline{m} + 2\overline{m}\sqrt{\overline{\mu}} + \sqrt{\overline{\mu}}\,, \tag{24}$$

where $\overline{m} = m/m_c$, $\overline{\mu} = \mu/\mu_c$ and $\overline{\Delta F} = \Delta F/\Delta F_c$. The free-energy landscape of Eq. 25 is given in Fig. 14 as a contour plot of $\overline{\Delta F}$ against $\overline{m}$ and $\overline{\mu}$, and the lamella grows into infinitely large dimensions in all directions. In contrast, the exactly solved model of Fig. 11 and Eqs. 9–23 show that finite lamellar

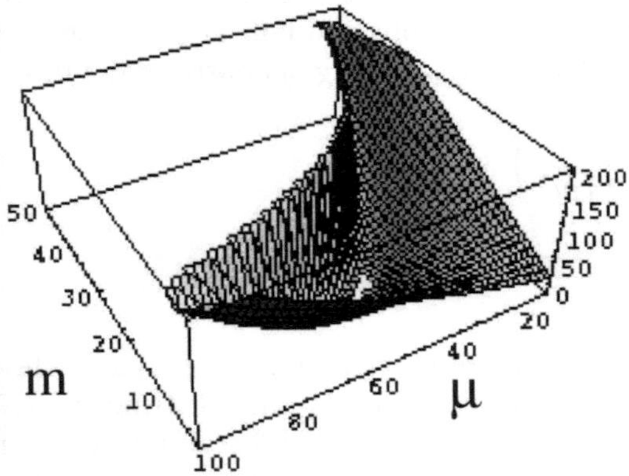

Fig. 12 Chain entropy leads to thermodynamic stabilization of finite lamellar thickness

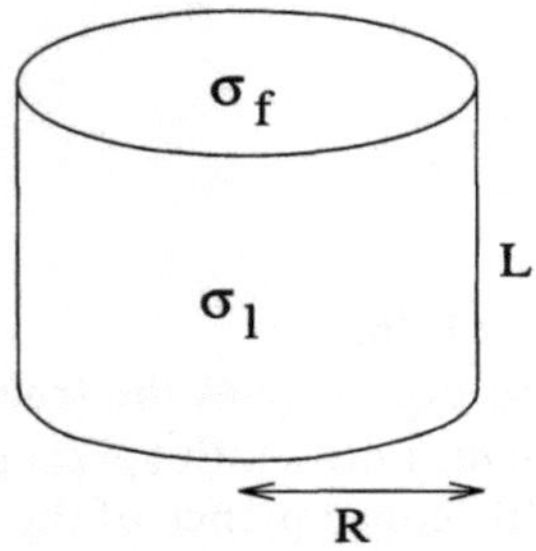

Fig. 13 Cylindrical nucleus

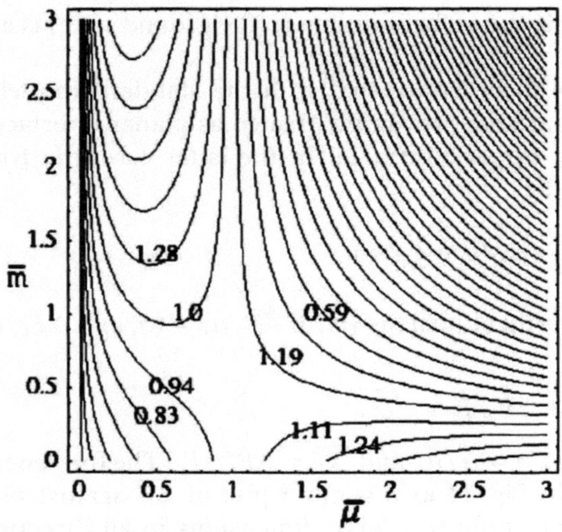

Fig. 14 Free-energy landscape for a cylindrical nucleus

thickness (much smaller than the extended chain value) is actually the equilibrium description. According to the results of Fig. 12, the equilibrium values of m^* and μ^* show that roughly 30% of monomers are in the loops for the chosen values of the parameters of the model. Different choices of values of parameters and an additional inclusion of fold surface energy lead to different values of m^* and μ^*, the details of which are to be presented in a future publication. However, the key result is that the extended chain dimension is not the equilibrium lamellar thickness at non-zero temperatures and in the absence of external pressure fields.

3.3 Addition of a Chain to the Growth Front

To simulate the regime of secondary nucleation, we isolate the model to where the growth is taking place, namely, the growth front at the edges of lamellae. The growth front is modeled by a two-layer-thick wall consisting of 20 extended-chains of 50 beads each. All beads on the growth front are fixed in space so that a stable growth front can be simulated with a minimum of beads. While this growth front does not have folds and irregularities that actual lamellae would have, it is well-defined and suited for this study.

In addition to visualizing the crystallization process, we also seek to determine the effect of varying the chain length relative to the lamellar thickness. Since the inclusion of chain ends inside the crystal is energetically unfavorable (akin to the inclusion of impurities), we can choose a case where the chain length is not an integer multiple of the lamellar thickness and see how the chain is accommodated on the growth front.

For the first case, we have chosen a chain of length 100 at $T^* = 8$, which should fold once on a growth front of length 50. Figure 15 shows the sequence of one such event and the values of time are indicated in the frames.

The chain at first appears to be loosely captured by the growth front but the bulk of the chain soon aligns with the orientation of the substrate and adopts a once-folded configuration. The last frame (f) shows the final configuration where the chain has a single hairpin fold and is in perfect registration with the substrate. Frame (e) is particularly interesting because the kink near the bottom shows one mechanism by which the chain stem migrates from one site on the substrate to another.

For the second case, we have chosen a chain of length 125, which is not an integer multiple of 50. As seen in Fig. 16, the chain explores many possible energy-minimizing conformations. Frame (c) shows two hairpin folds, similar to the previous case except with half a stem folded on the crystal. Because the stability of the crystallized chain is dependent on the competition between the energy gained from attached stems versus the energy lost from creating the hairpin turns, the energy of the half-stem is not enough to offset the energy of the additional fold. This causes the chain to explore other

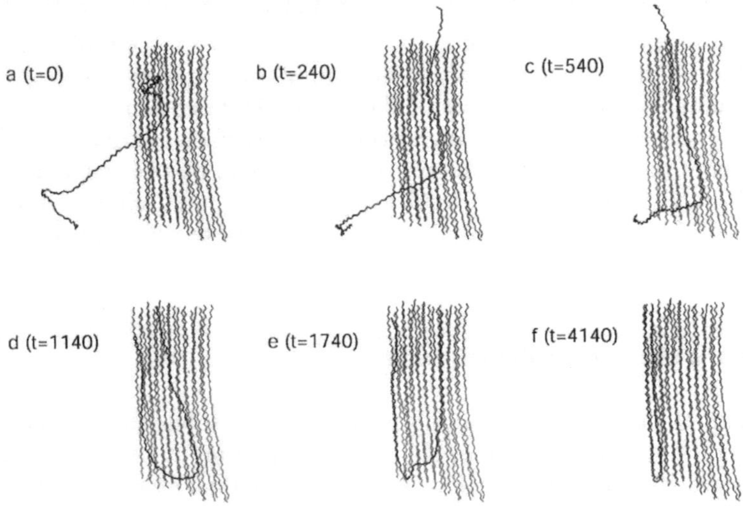

Fig. 15 Attachment of $L = 100$ chain onto $L = 50$ growth front model. Growth front chains are immobilized. The chain exhibits significant mobility (**a–e**) before it establishes perfect registration with the surface (**f**). The values of time t are indicated in each frame

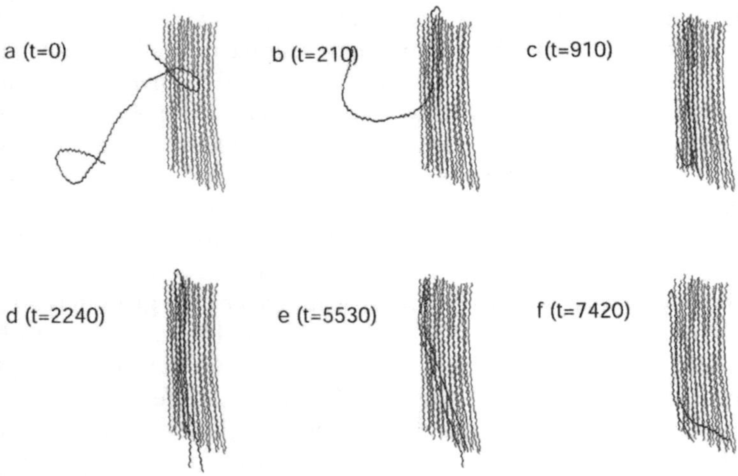

Fig. 16 Attachment of a $L = 125$ chain onto $L = 50$ growth front model. Growth front chains are immobilized. The chain adopts only metastable configurations. The values of time t are indicated in each frame

conformations, such as in frame (d) where the chain ends dangle off the substrate, and frames (e) and (f) where the chain incurs an energy penalty in losing its registration with the surface.

The stability of the chains can also be observed in the time plots of the local order parameter (Fig. 17). For the chain of $L = 100$, the local order pa-

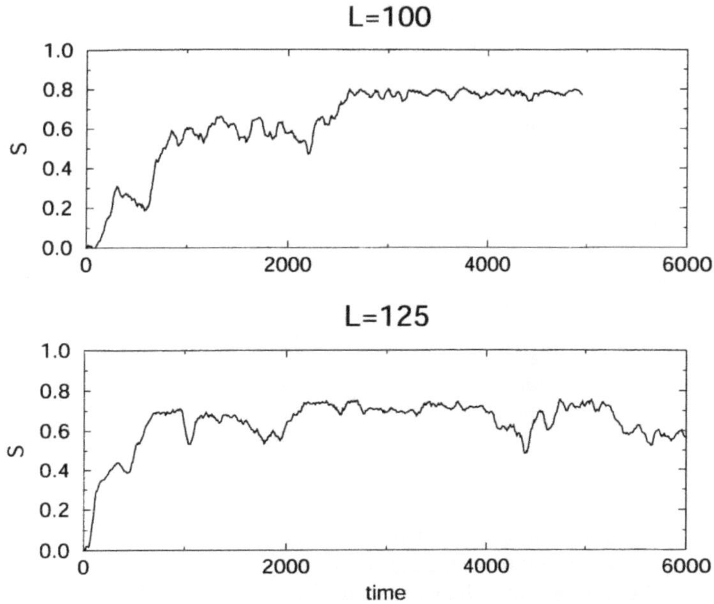

Fig. 17 Comparison of local order parameter for $L = 100$ and $L = 125$ chains from Figs.15 and 16. The stability differences between chains of integer multiple lengths versus mismatched chains is evident

rameter reaches a constant maximum at about $t = 2500$. The chain of $L = 125$, however, exhibits periods of metastability followed by large fluctuations in the local order parameter. Moreover, the maximum of its order parameter never equals or exceeds the maximum of the $L = 100$ case.

In secondary nucleation, one major difference in the assumptions of the current models is whether the simultaneous adsorption of many chains on the growth front has a disruptive effect on the growth rate or the perfection of the crystal. In an effort to address this issue, we have considered the simultaneous crystallization of 20 chains of length 140 onto a template. As an example, the template is prepared in the following manner. First, 8 chains of length 250 are allowed to crystallize together. In the resulting crystal, each chain has folded into 4 stems whose length is about 60 united atoms. The crystal is then pressed flat using hard boundary conditions into a two-layer thick crystal. This flat crystal is then fixed in space and placed near one surface within a cubic box 100 bond lengths on each side. Into this box we randomly place 20 disordered chains of length 140 and commence the simulation at $T^* = 10$. The results are shown in Fig. 18.

From the initial state, chains which are close to the growth front are pulled in and rapidly crystallized, adding one layer to it. Chains which are far from the attractive influence of the growth front undergo homogeneous nucleation.

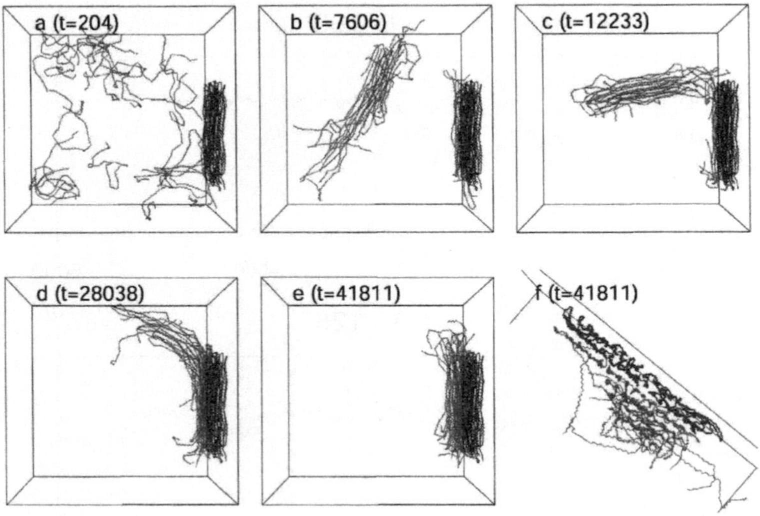

Fig. 18 Simulation of 8 chains of $L = 250$ near a fixed crystalline surface. **a–e** show simultaneous homogeneous and heterogeneous nucleation and subsequent interaction of the two nuclei. **f** is the end-view of **e** showing arrangement of stems on the surface. The values of time t are indicated in each frame

Through random forces, the free nucleus eventually drifts near enough to the growth front to interact with it. The chain ends act as "feelers" which align the free nucleus with the growth front. Interestingly, the free nucleus merges with the growth front by sliding itself in from the side. It is not known how common this kind of event is. Close observation of individual chains on the growth front reveals substantial stem mobility on the surface but insignificant activity inside. Figure 18f is an end-view showing the arrangement of stems on the crystal. At lower temperatures there would be insufficient mobility even at the surface, arresting the growth process. It was also observed that during the merging of the two nuclei, the motion of individual chains of the free nucleus is akin to that of a worm, with surges of forward motion extending the chain followed by relaxation periods in which the tail pulls up. This worm-like collective behavior is unexpected and merits further study.

3.4
Kinetics at the Growth Front

Very long simulations have been carried out with as many as 15 000 united atoms with the following protocol. First, a single chain crystal is placed at the origin. Next, a self-avoiding random chain is placed at a random location on a sphere whose radius is 1.5 times the radius of gyration of the crystal. The new system is equilibrated with the Langevin dynamics algorithm for 5000 time units. If the chain fails to add any segments to the crystal by the end of the addition period, the run is rejected and the crystal's coordinates are reset to their values at the beginning of the period. A new attempt to add a chain is then made. If the chain adds to the crystal, the process is repeated by moving the crystal to the origin and adding a new self-avoiding random chain to the simulation. Figure 19 illustrates the addition of the 40th chain to a 39 chain crystal for $k_{\mathrm{B}}T/\varepsilon_0 = 9.0$.

The crystal reels in the chain one segment at a time, and then crystallographically attaches each to the growth face. This process continues until the entire chain is incorporated into the crystal. Once adsorbed, the chain continues to rearrange until its fold length is commensurate with that of the growth face. The rate limiting step for the addition of the chain to the crystal is the diffusive contact with the surface. Once a few segments have come into contact with the crystal, the chain rapidly adds to the growth front. The numerical estimate of the free energy $F[s]$ as a function of the number of segments added to the crystal is given in Fig. 20.

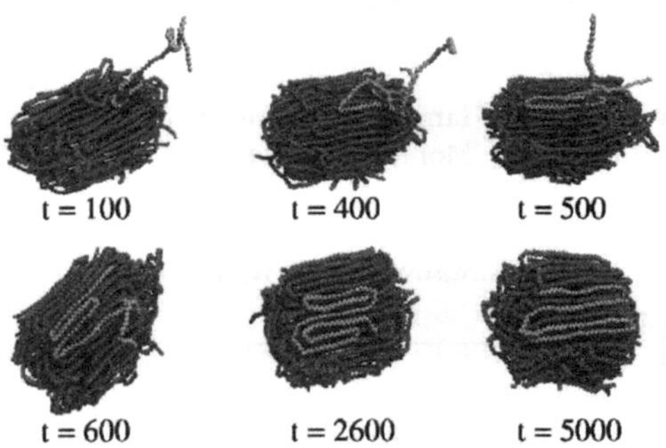

Fig. 19 Adsorption of a new chain at the growth front [30]

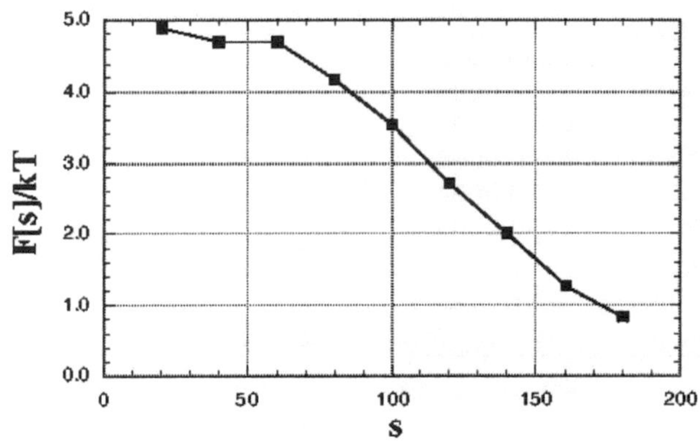

Fig. 20 Absence of free-energy barrier for attachment of a new chain at the growth front [30]

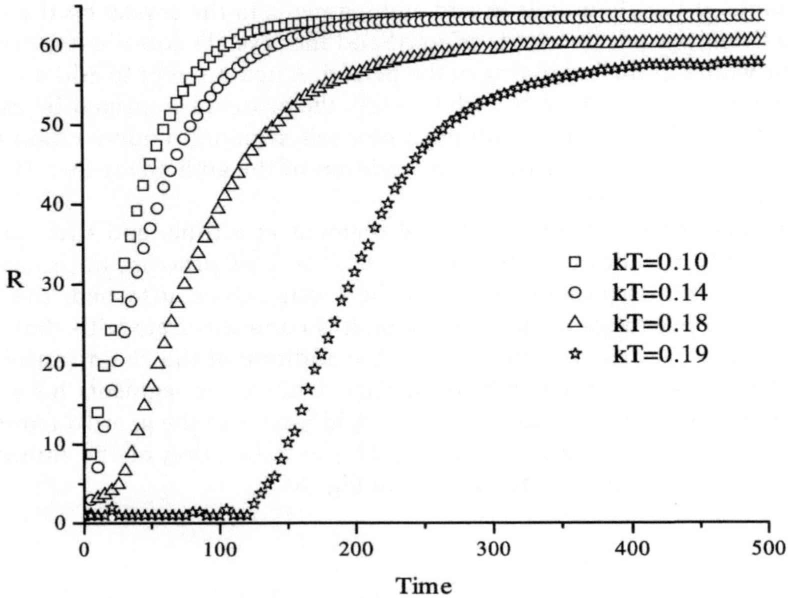

Fig. 21 Time evolution of lateral lamellar dimension as a function of temperature at $C = 0.0005$. Time is in units of 10^4 Monte Carlo steps

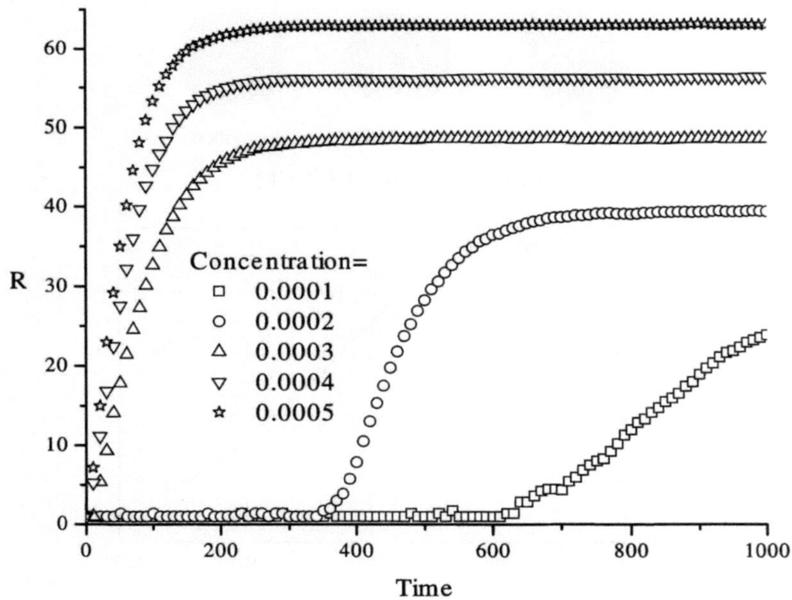

Fig. 22 Time evolution of lateral lamellar dimension as a function of initial concentration at $kT = 0.15$. Time is in units of 10^4 Monte Carlo steps

The addition of a new chain at the growth front is not hindered by a barrier, in contradiction with the underlying assumptions of the LH theory. Simultaneous to the addition of new chains at the growth front, chains inside the lamella move cooperatively. The center of mass of the lamella diffuses in space while the lamella thickens by a process of internal rearrangements; for details, see [30]. The mean squared displacement of a labeled monomer varies with the elapsed time, t, with an effective power law of $t^{0.74}$ by shuffling back and forth between the lamellar and amorphous regions.

The lamellae grown in these Langevin dynamics simulations are very small in comparison with experimentally investigated lamellae. In view of this, we have developed the coarse-grained anisotropic adsorption model described

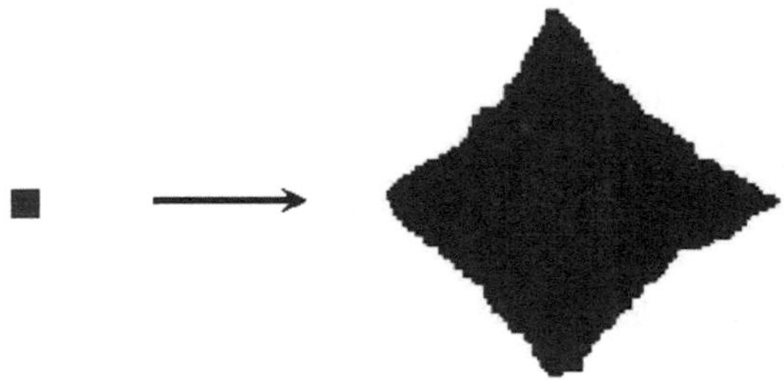

Fig. 23 Initial lamellar seed of $R_0 = 7$ grows into a large lamella of $R = 60$ in $t = 150 \times 10^4$ Monte Carlo steps for $kT = 0.09$ and $C = 0.0005$

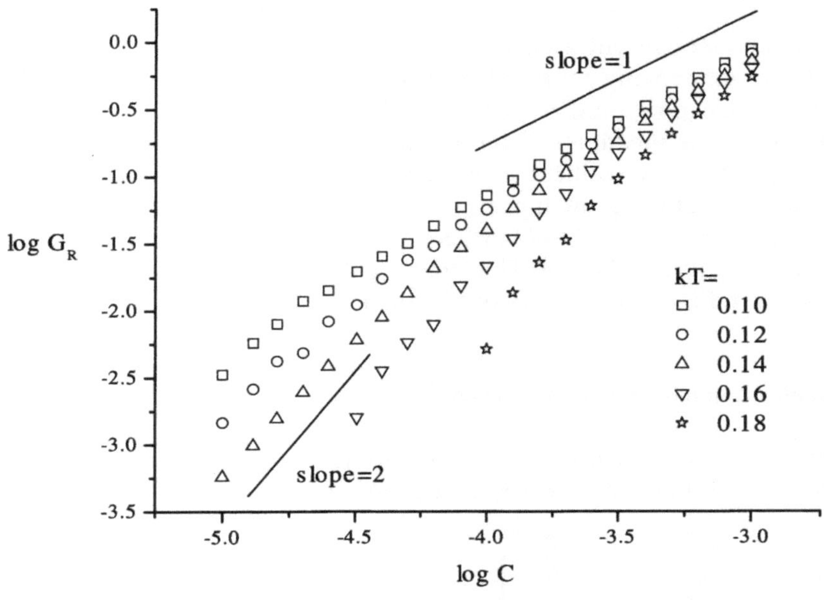

Fig. 24 Concentration dependence of the radial growth rate

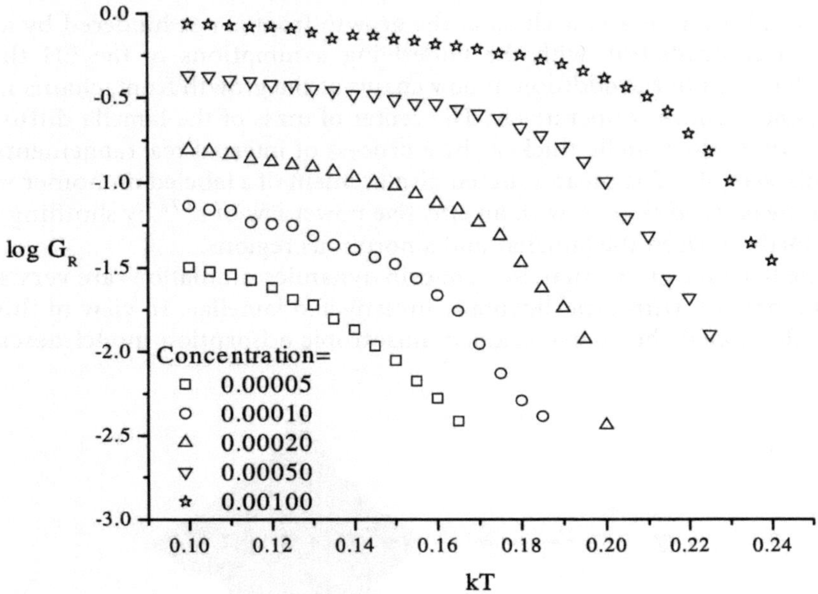

Fig. 25 Temperature dependence of the radial growth rate

in Sect. 2. The typical results from the Monte Carlo simulations of this model are as follows. The time-evolution of the lateral dimension of the lamella from R_0 at $t = 0$ to R at t is given in Figs. 21 and 22 at different temperatures and initial polymer concentrations.

The accompanying morphological evolution is illustrated in Fig. 23, for polymer concentration $C = 0.0005$ and $kT = 0.09$.

The results of the Monte Carlo simulations, as shown in Figs. 21, 22, and 23, are in qualitative agreement with many experimental observations. Preliminary analysis of the growth curves in Figs. 21 and 22 shows that the growth rate (G_R) is proportional to C^{α}, where α varies between 1 and 2 depending on T and C, as given in Fig. 24. The temperature dependence of the growth rate as shown in Fig. 25, does not exhibit any marked regimes.

3.5
Crystallization in an Elongational Flow

In the Langevin dynamics simulations [33], there is an additional force $\widehat{S}\boldsymbol{r}_i$ acting on the ith bead, where $\boldsymbol{r}_i$ is its position and

$$\widehat{S} = \dot{\varepsilon} \begin{pmatrix} -0.5 & 0.0 & 0.0 \\ 0.0 & -0.5 & 0.0 \\ 0.0 & 0.0 & 1.0 \end{pmatrix} \qquad (25)$$

with the parameter $\dot{\varepsilon}$ setting the flow rate. As expected, the chains undergo coil-stretch transition in the presence of flow, and the melting temperature is elevated. For example, at the reduced temperature of 11.0 (which is the extrapolated melting temperature in reduced units for the united atom model described above in the quiescent state) folded chains are readily formed in the presence of flow.

Several simulation runs were performed for a wide range of flow rates for chains of $N = 180$ beads. To ensure that the system is in the state of lowest free energy and to avoid the chain being in a metastable state, two initial conformations of the chain were chosen. One initial conformation is a random chain equilibrated at the given temperature without any flow. The second initial conformation is a fully extended chain obtained by equilibrating it at an extremely high flow rate ($\dot{\varepsilon} = 4.0$). Data were collected after the two chains with different initial conformation are in the same state, either coil or stretched polymer. The stagnation point at $\boldsymbol{r} = 0$ is unstable. In order to avoid the chain drifting away from the coordinate origin, the center of the mass of the polymer is fixed at $\boldsymbol{r} = 0$.

Figure 26 shows the square of the radius of gyration of the chain as a function of the flow rate $\dot{\varepsilon}$ at relatively low temperature ($T = 9.0$).

A discontinuous coil to stretch transition is evident at $\dot{\varepsilon}_c = 0.000725$. The transition point $\dot{\varepsilon}_c$ was found by using two different initial conformations as described above. For values lower than $\dot{\varepsilon}_c$, the random chain will eventually coil, form a folded chain crystalline structure and stay in that conformation until the end of the run for relatively long run times. On the other hand, a pre-stretched chain would fluctuate and eventually form a crystallized folded chain that is stable. Similarly, for flow rates higher than $\dot{\varepsilon}_c$, a pre-stretched chain will never coil and a random chain will eventually stretch.

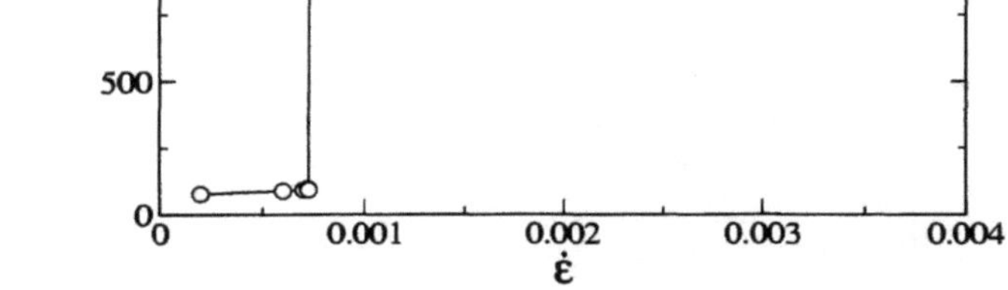

Fig. 26 Radius of gyration squared as a function of flow rate for a chain of $N = 180$ and $T = 9.0$

At a higher temperature $T = 11.0$, for flow rates near the transition rate $\dot{\varepsilon}_c$, the free-energy barrier between the coiled and stretched conformation is much lower than that for $T = 9.0$. The chain can therefore explore the phase space and jump back and forth from the coiled to the stretched state. Similar behavior has already been observed in [59] and [60]. Figure 27 illustrates this feature.

From Fig. 27, the free energy can be calculated according to:

$$F(R_g) = -kT \ln\left(\frac{\tau(R_g)}{\tau_{tot}}\right), \tag{26}$$

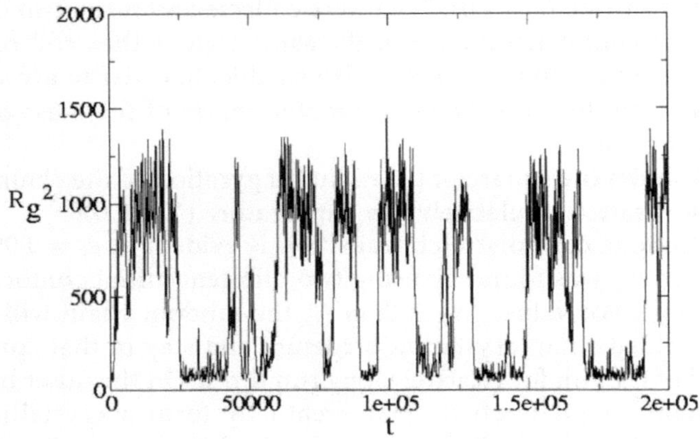

Fig. 27 Radius of gyration squared as a function of time for $N = 180$, $T = 11.0$, and $\dot{\varepsilon} = 0.00075$

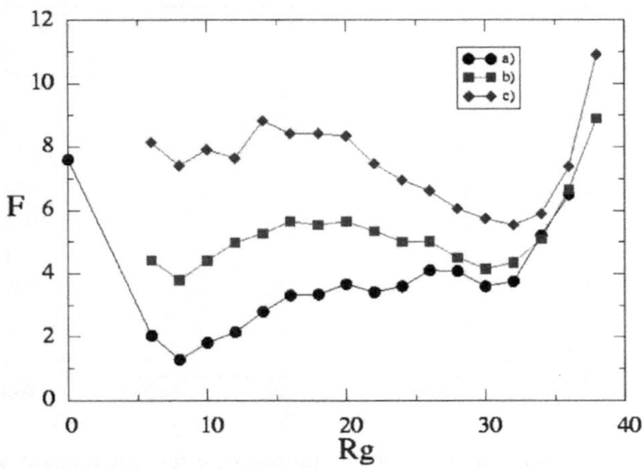

Fig. 28 The free energy F vs. $R_g (N = 180)$. **a** $\dot{\varepsilon} = 0.0006$, **b** $\dot{\varepsilon} = 0.00075$, and **c** $\dot{\varepsilon} = 0.0009$

where $\tau(R_g)$ is the time the system spends in states with a radius of gyration between R_g and $R_g + \Delta R_g$, and τ_{tot} is the total time. ΔR_g is chosen to be $2r_0$. Figure 28 shows the free energy of the chain at $T = 11.0$ for flow rates: (a) below, (b) very close to, and (c) above the transition. It is clear that at the transition the stretched and the folded states coexist.

Making the flow rate higher or lower will change from stable to metastable the folded or the stretched state, respectively. The effects of hysteresis associated with this first-order discontinuous transition play an important role in the formation of composite crystalline structures.

When simulations were performed with many chains of uniform length, some chains were stretched out and aggregated among themselves to form the shish, whereas other chains formed folded structures which in turn attached to the shish, initiating the formation of kebabs. This is attributed to the coexisting populations of stretched and coiled states. This feature is much more pronounced if the chains have different lengths. For a given flow rate, the longer chain is predominantly in the stretched state and the shorter chain is predominantly in the coiled state. Now the shish is formed by the crystallization of stretched chains and the kebabs are mostly from the shorter chains.

To follow the crystallization of kebabs around a shish, the dynamics of 10 short chains ($N = 180$) near a preformed shish (from 7 chains of length $N = 500$) were followed at $T = 9.0$, by fixing the center of mass of the shish. The initial position of the short free chains was chosen at random in a cylinder around the shish, with radius $30r_0$ and a height of $60r_0$. Each run started with different initial conditions. Figure 29 shows one such initial state.

The flow rate is then maintained at $\dot{\varepsilon} = 0.0001$ (lower than $\dot{\varepsilon}_c$ corresponding to $N = 180$) and the short chains are allowed to assemble on the shish.

Figure 30 illustrates nine examples of the structures obtained in these simulations. It is clear that the chains group into crystallized kebabs on the shish surface. There are very few areas where the chains are partially or completely stretched under the influence of the shish template. The dominant mode of crystal nucleation on the shish is the growth of folded chains grouped into lamellar nuclei. Also, some of the chains do not join the central structure but drift away from it leaving a large gap on the shish between them. These simulations show clearly that the presence of the ordered template (the shish) influences the nucleation of lamellae and the formation of kebabs.

Fig. 29 An example of the initial position of chains for the kebab formation simulation

Fig. 30 Nine examples of freely formed kebabs

Some of the chains, when in contact with the shish, will stretch almost completely. However, these highly stretched chains are not dominant compared to the highly folded bundles that form crystalline kebabs around the shish. Also, none of the highly stretched chains formed a structure with part of it stretched and attached to the shish and part of it in folded crystalline lamella. The formation of kebabs in these simulations is clearly growth of lamellae, nucleated on the shish.

Next, in order to study the stability of the kebabs, the flow rate was set at $\dot{\varepsilon} = 0.001$ and four, initially equilibrated (i.e. in a pre-crystallized conformation), short chains were added per $t = 2000$ with an initial position of $20r_0$ length units away from the stagnation point in the x and y direction. They form a kebab around the shish which was preformed. This procedure was repeated up to 44 short chains in the kebab as shown in Fig. 31.

The kebab is stable even though $\dot{\varepsilon}$ is larger than $\dot{\varepsilon}_c$ for a single short chain. The kebab has a uniform thickness and does not seem to resemble the flow contour. It must be stressed that the thickness of the kebabs formed in this way is determined independent of the presence of the shish. The short chains are pre-crystallized before they are incorporated in the kebab. The kebab formed in this way is only slightly influenced by the shish, except for the fact that it was nucleated on it. The presence of already-formed kebabs clearly modifies the flow, a feature that is not present in these simulations. The flow,

Fig. 31 A stable shish (7 chains) and kebab (44 chains)

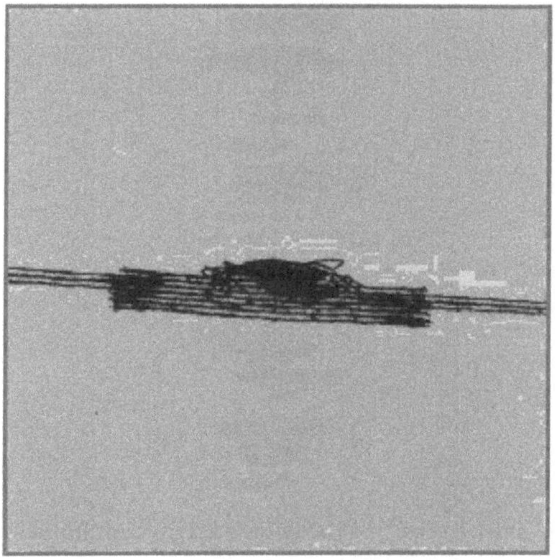

Fig. 32 An example of a completely stretched kebab grown with a very slow rate

however, is modified in a way that it must be zero in the already-formed shish and kebabs. This will result in a greater stability of the structures that are observed and therefore only emphasize the results in this section. Finally, when the rate of addition of the chains was lowered to one per $t = 5000$, most of the short chains stretched completely as shown on Fig. 32.

Similar to the discussion in Sect. 3.4, we have extended the anisotropic adsorption model for the formation of large "shish-kebab" structures from solutions under flow, using the Monte Carlo method. A typical result of the growth of a "shish-kebab" structure is shown in Fig. 33. As the diffusing molecular crystals attach and detach at the shish and kebabs, the size of the kebabs and the spacing between the kebabs change with time. For example, the time-dependence of the average spacing between the kebabs is shown in Fig. 34 at different temperatures. This result is in qualitative agreement with experimental observations.

4 Conclusions

Molecular modeling is an excellent tool for exploring the very early stages of polymer crystallization from solutions, during a time-duration inaccessible to the current experimental methods. Since large-scale simulations (corresponding to long time periods for the simulations, but short time periods for experiments) lead to results in qualitative agreement with experiments, the general mechanism for the birth of the initial nuclei presented in this

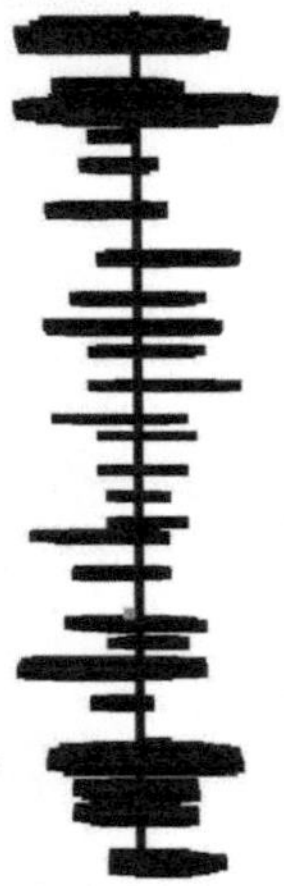

Fig. 33 Snapshot of a "shish-kebab" in the Monte Carlo simulation

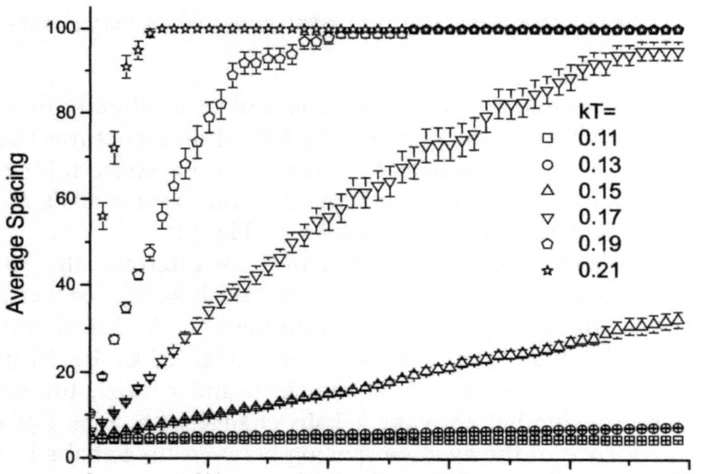

Fig. 34 Time-evolution of average spacing between kebabs at different temperatures

review is probably universally valid. This mechanism is qualitatively different from the LH model and a shift in paradigm is emerging for polymer crystallization.

Acknowledgements The author is grateful to all of his collaborators on the subject of the review, who are in chronological order, Chester Liu, Paul Welch, Ilija Dukovski, Jianing Zhang, and Arindam Kundagrami. Acknowledgement is made to NSF Grant DMR-0209256 for financial support.

References

1. Wunderlich B (1973) Macromolecular Physics, vols I and II. Academic Press, New York
2. Yoon DY, Flory PJ (eds) (1979) Special issue on the organization of macromolecules in the condensed phase. Faraday Discuss. Chem Soc 68:7
3. Bassett DC (1981) Principles of polymer morphology. Cambridge University Press, Cambridge
4. Phillips PJ (1990) Rep Prog Phys 53:549
5. Armitstead K, Goldbeck-Wood G (1992) Adv Polym Sci 100:219
6. Muthukumar M (2004) Adv Chem Phys 128:1
7. Strobl G (1996) The physics of polymers. Springer, Berlin Heidelberg New York
8. Imai M, Kaji K, Kanaya T (1994) Macromolecules 27:7103
9. Ezquerra TA, Lopez-Cabarcos E, Hsiao BS, Balta-Calleja FJ (1996) Phys Rev E 54:989
10. Terrill NJ, Fairclough JPA, Towns-Andrews E, Komanschek BU, Young RJ, Ryan AJ (1998) Polymer 39:2381
11. Olmsted PD, Poon WCK, McLeish TCB, Terrill NJ, Ryan AJ (1998) Phys Rev Lett 81:373
12. Ryan AJ, Fairclough RPA, Terrill NJ, Olmsted PD, Poon WCK (1999) Faraday Discuss. Chem Soc 112:13
13. Akpalu Y, Kielhorn L, Hsiao BS, Stein RS, Russell TP, van Egmond J, Muthukumar M (1999) Macromolecules 32:765
14. Strobl G (2000) Eur Phys J E 3:165
15. Lotz B (2000) Eur Phys J E 3:185
16. Muthukumar M (2000) Eur Phys J E 3:199
17. Toxvaerd S (1990) J Chem Phys 93:4290
18. Kavassalis TA, Sundararajan PR (1993) Macromolecules 26:4144
19. Boyd RH, Gee RH, Han J, Jin Y (1994) J Chem Phys 101:788
20. Sundararajan PR, Kavassalis TA (1995) J Chem Soc Faraday Trans 91:2541
21. Toma L, Toma S, Subirana JA (1998) Macromolecules 31:2328
22. Liu C, Muthukumar M (1998) J Chem Phys 109:2536
23. Yamamoto T (1998) J Chem Phys 109:4638
24. Takeuchi H (1998) J Chem Phys 109:5614
25. Balijepalli S, Rutledge GC (1998) J Chem Phys 109:6523
26. Doye JPK, Frenkel D (1999) J Chem Phys 110:2692
27. Fujiwara S, Sato T (1999) J Chem Phys 110:9757
28. Sommer J-U, Reiter G (2000) J Chem Phys 112:4384
29. Muthukumar M, Welch P (2000) Polymer 41:8833
30. Welch P, Muthukumar M (2001) Phys Rev Lett 87:218302
31. Yamamoto T (2001) J Chem Phys 115:8675
32. Muthukumar M (2003) Phil Trans R Soc Lond A 361:539
33. Dukovski I, Muthukumar M (2003) J Chem Phys 118:6648
34. Zhang J, Muthukumar M (manuscript in preparation)
35. Yamamoto T (2003) J Macromol Sci B42:629
36. Yamamoto T (2004) Polymer 45:1357
37. Lavine MS, Waheed N, Rutledge GC (2003) Polymer 44:1771
38. Hu W, Frenkel D, Mathot VBF (2003) Macromolecules 36:8178
39. Hu W, Frenkel D (2004) Macromolecules 37:4336
40. Hoffman JD, Davis GT, Lauritzen JI (1976) In: Hannay NB (ed) Treatise on solid state chemistry, vol 3, Chapt 7. Plenum Press, New York, p 497–614
41. Lauritzen JI, Hoffman JD (1960) J Res Nat Bur Std 64A:73

42. DiMarzio EA (1967) J Chem Phys 47:3451
43. Passaglio E, DiMarzio EA (1987) J Chem Phys 87:4901,4908
44. Point JJ, Dosière M (1989) Macromolecules 22:3501
45. Sadler DM, Gilmer GH (1984) Polymer 25:1446
46. Hoffman JD, Miller RL (1997) Polymer 38:3151
47. Allegra G (1977) J Chem Phys 66:5453
48. Keller A, Kolnaar HWH (1997) In: Meijer HEH (ed) Processing of Polymers, vol 18. VCH, New York
49. Organ SJ, Keller A (1985) J Mater Sci 20:1571
50. Ungar G, Stejny J, Keller A, Bidd I, Whiting MC (1985) Science 229:386
51. Ergoz E, Fatou JG, Mandelkern L (1972) Macromolecules 5:147
52. Ross GS, Frolen LJ (1980) Methods of Experimental Physics, vol 16B. Academic Press, New York, 363 pp
53. Mandelkern L (1964) Crystallization of Polymers. McGraw-Hill, New York
54. Pennings AJ, Kiel AM (1965) Kolloid Z Z Polym 205:160
55. Pennings AJ (1977) J Polym Sci: Polym Symp 59:55
56. Liu T, Lieberwirth I, Petermann J (2001) Macromol Chem Phys 202:2921
57. Hobbs JK, Miles MJ (2001) Macromolecules 34:353
58. Hobbs JK, Humphris ADL, Miles MJ (2001) Macromolecules 34:5508
59. Perkins TT, Smith DE, Chu S (1997) Science 276:2016
60. Pennings AJ, van der Mark JMAA, Booij HC (1970) Kolloid Z Z Polym 236:99
61. Allen MP, Tildesley DJ (1987) Computer Simulation of Liquids. Clarendon, Oxford
62. Kumar S, Bouzida D, Swendsen RH, Kollman PA, Rosenberg JM (1992) J Comput Chem 13:1011

Author Index Volumes 101–191

Author Index Volumes 1–100 see Volume 100

Subject Index